CALCULUS

CALCULUS

Kuldeep Chaudhary

Assistant Professor
Department of Mathematics
Amity Institute of Applied Sciences
Amity University, Noida (UP)

Shashank Goel

Assistant Professor
Department of Mathematics
Amity Institute of Applied Sciences
Amity University, Noida (UP)

I.K. International Publishing House Pvt. Ltd.

NEW DELHI

Published by

I.K. International Publishing House Pvt. Ltd.
4435-36/7, Ansari Road, Daryaganj,
New Delhi–110 002 (India)
E-mail: info@ikinternational.com
Website: www.ikbooks.com

ISBN: 978-93-85909-95-5

Published by Krishan Makhijani for I.K. International Publishing House Pvt. Ltd., 4435-36/7, Ansari Road, Daryaganj, New Delhi–110 002 and Printed by Rekha Printers Pvt. Ltd., Okhla Industrial Area, Phase II, New Delhi–110 020.

Preface

Calculus is a fundamental topic of mathematics and rich in applications. This book covers all the topics that are generally prescribed for undergraduate programmes (B.Sc. (H) Mathematics/B.Sc.).

Salient features:

- Each chapter is self-contained and written in a comprehensive style.
- Every topic is profusely illustrated with explanatory and easy-to-reproduce diagrams.
- The derivations are clear and complete, and are presented in a clear, easy and understandable way.
- Each chapter contains complete theory with a large number of solved examples along with graphs.
- Sketching of parametric curve, polar curve and conics are illustrated with graphs.
- Calculus is popular due to its applications in different areas so, a specific attention has been taken throughout to illustrate how and where the results can be applied.

The authors do not claim any originality. The authors have frequently consulted various journals and books on the related topics for the preparation of this book and would like to acknowledge all.

The authors are thankful to Dr. Ashok K. Chauhan (Founder President, Amity University), Dr. Atul Chauhan (Chancellor, Amity University), Prof. Balvinder Shukla (Vice Chancellor, Amity University) and Prof. Sunita Rattan (Addl. Director, AIAS, Amity University, Noida) for their encouragement and support.

The authors are also thankful to Dr. Prakriti Rai, HoD, Department of Mathematics, AIAS and all faculty members of Department of Mathematics, Amity University Noida for their valuable suggestions.

We are also indebted to Dr. Shiv Kumar Kaushik, Associate Professor, Department of Mathematics, Kirori Mal College, Delhi University for providing us the necessary counselling and support during writing this book.

We would like to mention special thanks to Mrs. Tripti Mittal Goel, Dr. Sudipa Chauhan and Ms. Divya Jain for their effort and support.

Last but not least, we would like to thank our parents for their cheerful support and blessings.

We hope that this book will cater to the needs of readers. In spite of careful effort some errors might have crept in. We shall be grateful for constructive criticism and helpful suggestions for the improvement of the book.

Kuldeep Chaudhary

Shashank Goel

Contents

1 CHAPTER

Functions

In Calculus, we find the relation between physical or mathematical quantities. To describe such relations, we construct graphs, formulas, numerical data and functions. Functions arise in every area of pure and applied mathematics, including mathematics applied to different areas of engineering, life sciences, economics, etc., whenever one quantity depends on another. Here, we present a general discussion on functions.

1.1 FUNCTIONS

A function f is a rule or correspondence or mapping that assigns to each element x in a set A exactly one element $f(x)$ in set B. Symbolically,

$$f : A \to B$$

is used to denote that f is a function from set A to B. The element x in set A is said to an independent variable, since it may have any value in the set A and $f(x)$ is said to be a dependent variable. The set A is called the **domain** and set B is known as **co-domain** of the function. The set of all possible values of $f(x)$ as x varies throughout the domain A is called **range**.

To visualize a function, we use graph of the function. The graph of the function f is a set of ordered pairs (see Figure 1.1)

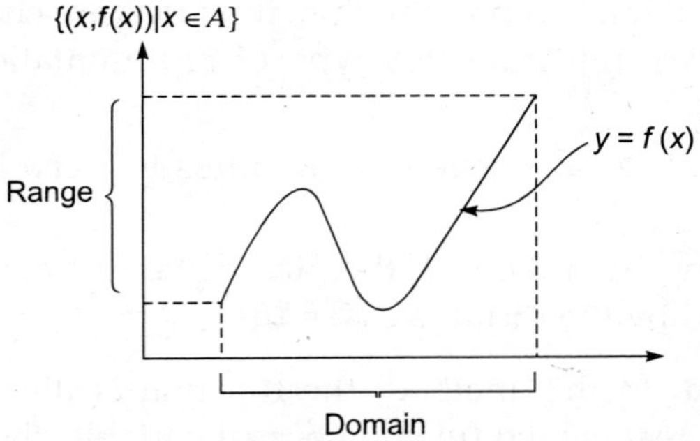

Figure 1.1. Graph of a function

EXAMPLE 1 What is the domain and range of $f(x)=\begin{cases} x, & x\geq 1 \\ 1, & x<1 \end{cases}$?

Solution:

Domain: Here, function is defined in stages, for $x\geq 1$: the formula $y=x$ is defined for all values of x greater than or equal to unity. The domain for this part is $[1,\infty)$.

For $x<1$: $y=1$ is defined for all values less than unity. Then domain of this part is $(-\infty,1)$.

Hence, domain of given function is $(-\infty,1)\cup[1,\infty)=(-\infty,\infty)$.

Range: For the part $y=x$, the range of this part is $[1,\infty)$ and second part $y=1$ is $\{1\}$. Therefore, the range of the complete function is $[1,\infty)$.

EXAMPLE 2 Find the domain and range of $f(x)=\dfrac{1}{|x|}$.

Solution: Here, $|x|=\begin{cases} x \text{ if } x\geq 0 \\ -x \text{ if } x<0 \end{cases}$

The given function is defined on all values except $|x|=0\Rightarrow x=0$. Then the domain of $f(x)=R-\{0\}$. The image of the points $x\in domain(f)$ is a set of positive real values, i.e., range = $(0,\infty)$.

1.2 REPRESENTATION OF FUNCTIONS

We know that if a function is well defined then there exist corresponding values in co-domain for all the given values in the domain. Generally, we can represent a function as follows:

(a) Analytically
(b) Graphically
(c) Numerically

(a) Analytical Method: In this representation, functions are characterized by specifying it through formulas. We illustrate this type of representation with the help of the following example:

(i) The area of a circle as a function of its radius is given by the algebraic formula $A(r)=\pi r^2$.

(ii) If a body is falling from the rest, then the distance s covered by it, as a function of time is defined by the formula $s(t)=16t^2$.

(b) Graphical Method: In this method, the function is illustrated as a curve in co-ordinate plane. If *f* is a real valued function of real variable, the graph of function is the collection of all points in the plane. The graph of a function can be defined as:

$$\textit{Graph of } f = \{(x, y) \;\; y = f(x) \;\&\; x \in Domain(f)\}.$$

EXAMPLE 3 The graph of function $f(x) = |x|, x \in R$ is shown in Figure 1.2.

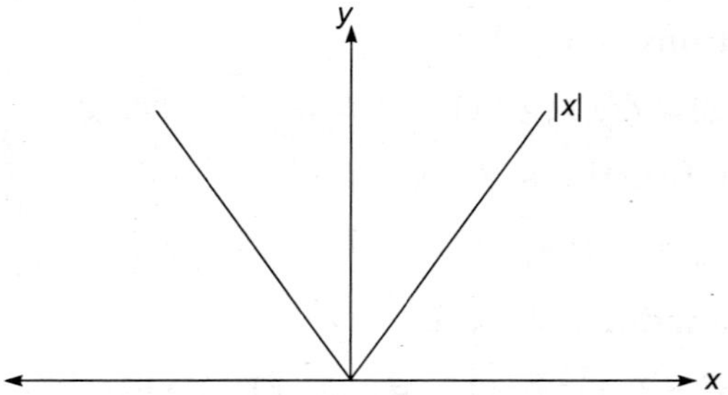

Figure 1.2

(c) Numerical Form (or Tabular Form): In this type of representation, function is defined by table values. This type of function is known as tabular function.

EXAMPLE 4 Let $c(x)$ be the total cost of producing x units of a product. The manager of the firm can have a systematic study of the cost structure by using the following (Table 1.1).

Table 1.1 Cost of production in terms of units

x(no units)	**0**	**2**	**4**	**6**	**8**	**10**	**12**
c(x) (in rupees)	20	70	120	170	220	270	320

Before we close this section, we would like to discuss that graph of a function is a curve in the *xy* plane. But which curve in *xy* plane is graph of the function. The answer to this question can be given by the following test.

"A curve in the xy plane is the graph of function iff no vertical line intersects the curve more than once."

This test is known as *vertical line test*. In Figures 1.3 and 1.4, graphs represent example of a function and not a function, respectively.

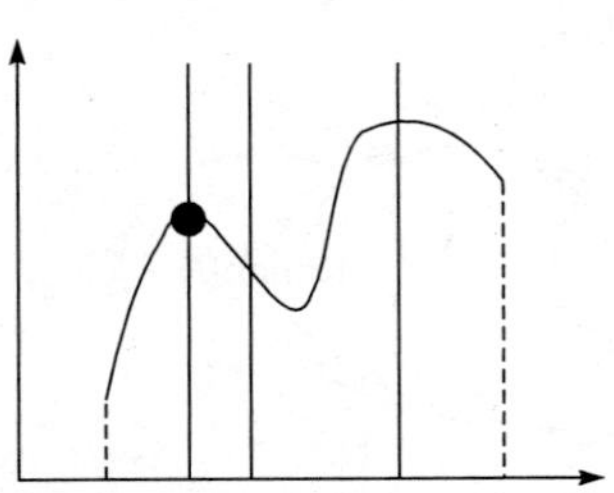

Figure 1.3 Graph is a function

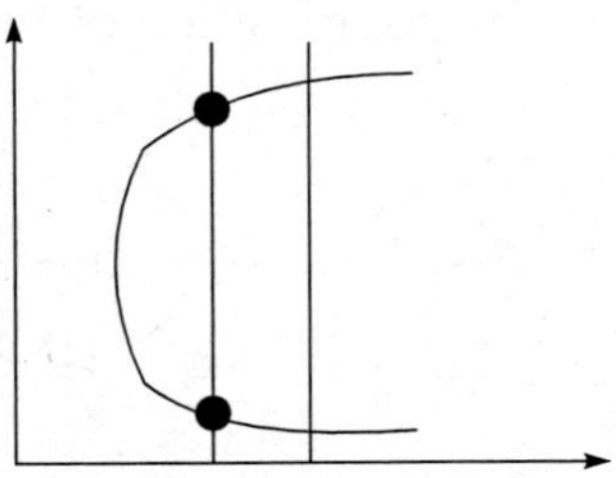

Figure 1.4 Graph is not a function

1.3 ALGEBRA OF FUNCTIONS

Let f and g be the two real valued functions defined on $dom(f)$ and $dom(g)$ respectively. Then, we define

(1) Sum of the functions $f+g$ by

$$(f+g)(x)=f(x)+g(x),\ x\in dom(f)\cap dom(g).$$

(2) Difference of the functions $f-g$ by

$$(f-g)(x)=f(x)-g(x),\ x\in dom(f)\cap dom(g).$$

(3) Product of the functions $f*g$ by

$$(f*g)(x)=f(x)*g(x),\ x\in dom(f)\cap dom(g).$$

(4) Quotient function f/g defined as

$$(f/g)(x)=\frac{f(x)}{g(x)},\ x\in\{dom(f)\cap dom(g)\}-\{x \mid g(x)=0\}.$$

1.4 EQUALITY OF FUNCTIONS

Let f and g be the two real valued functions defined on $dom(f)$ and $dom(g)$ respectively. The functions f & g are said to be equal *iff*:

(a) $dom(f)=dom(g)$, and

(b) $f(x)=g(x), \mathit{for\ each}\ x\in dom(f)$.

1.5 COMPOSITION OF FUNCTIONS

Let f and g be the two real valued functions defined on $dom(f)$ and $dom(g)$ respectively. Then the composite function fog is defined as

$$(fog)(x)=f(g(x)).$$

And domain of composite function is given by

$$dom(fog)=\{x \mid x\in dom(f)\ \&\ x\in dom(g)\}.$$

Remark: Composite function fog is also known as composition of f and g. Figure 1.5 shows the composition of two functions, fog.

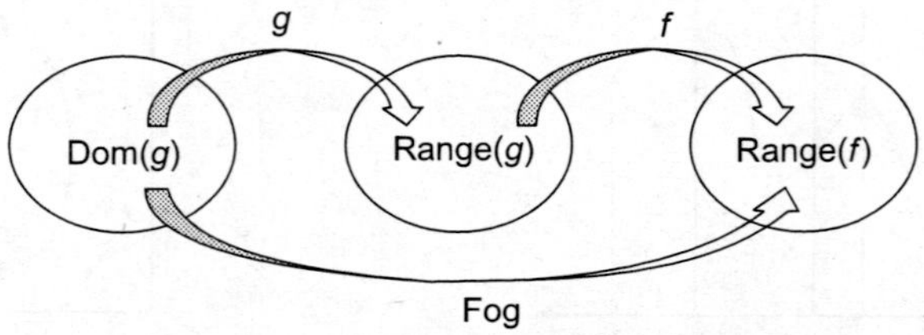

Figure 1.5 Composition of functions

EXAMPLE 5 If $f(x)=x^2$ and $g(x)=x-2$, find the composite functions *fog* and *gof*.

Solution: Using definition of composition of two function, we have

$$(fog)(x)=f(g(x))=f(x-2)=(x-2)^2$$
$$(gof)(x)=g(f(x))=g(x^2)=x^2-2.$$

Remark: From the above example we can see that in general, $fog = gof$.

In the following section, we describe some very important types of functions like one-one, onto and one-one onto functions.

1.6 INJECTIVE, SURJECTIVE AND BIJECTIVE FUNCTIONS

Definition: Let A and B be the two subsets of $\mathbb{R}$. If $f: A \to B$ is a function from A to B, then

(a) The function f is said to be injective (or one-one) whenever $x_1 \neq x_2$, then $f(x_1) \neq f(x_2)$.
(b) The function f is said to be surjective (or map A onto B) if $f(A)=B$; that is, if the range $R(f)=B$
(c) If f is both injective and surjective, then f is said to be bijective.

Remarks:

1. In order to show that a function f is injective, we must prove that:

   for all x_1, x_2 *in* A, if $f(x_1)=f(x_2)$, then $x_1 = x_2$.

   To do this, first we assume that $f(x_1)=f(x_2)$ and then we show that $x_1 = x_2$.
2. In order to show that a function f is bijective, we must prove that:

   for any $y \in B$, there exists at least one $x \in A$ such that $f(x)=y$.

EXAMPLE 6 Let $A=\{x \in \mathbb{R} : x \neq 1\}$ and $f: A \to B$ is defined as

$$f(x)=\frac{3x}{x-2} \textit{ for all } x \in A.$$

Show that f is an injective and bijective function.

Solution: To show that f is injective, we take x_1 and x_2 in A and assume that $f(x_1) = f(x_2)$. Thus, we obtain,

$$\frac{3x_1}{x_1-2}=\frac{3x_2}{x_2-2}$$

$$\Rightarrow x_1(x_2-2)=x_2(x_1-1)$$

$$\Rightarrow \quad x_1 = x_2.$$

Therefore, f is an injective function.

To show that f is bijective, we determine the range of the function f. Then, we find the solution of equation $y=\dfrac{3x}{x-2}$ for x in terms of y. Thus, we obtain

$$x=\frac{2y}{y-3}.$$

The above function is defined for all y except $y\neq 3$. Thus, the range of the function f is set $B=\{y\in\mathbb{R}:y\neq 3\}$. Hence, f is bijective from set A to B.

1.7 INCREASING AND DECREASING FUNCTIONS

Definition: Let f be a real valued function defined on interval $I=[a,b]$. Then, the function f is called increasing on I, if

$$f(x_1)\leq f(x_2) \text{ whenever } x_1<x_2 \text{ in } I=[a,b]$$

and f is called decreasing on I, if

$$f(x_1)\geq f(x_2) \text{ whenever } x_1<x_2 \text{ in } I=[a,b].$$

For example, the function $f(x)=x^2$ is increasing function in interval $[0,\infty)$ and decreasing in interval $(-\infty,0]$ (Figure 1.6).

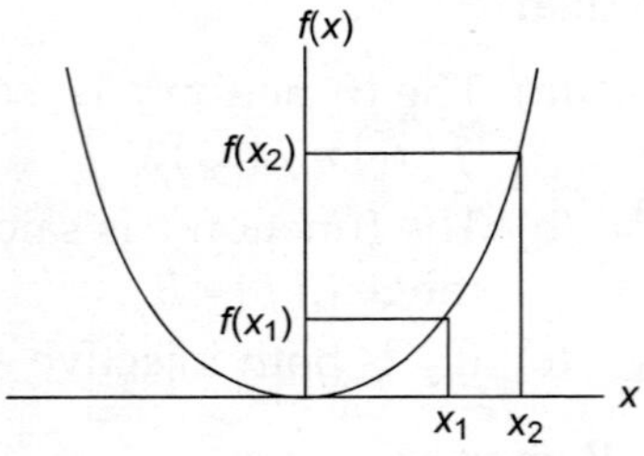

Figure 1.6

1.8 CLASSIFICATION OF FUNCTIONS

In this section, we will define some special types of functions which are frequently used in Calculus.

1.8.1 Polynomials

A function $P(x)$ is said to be a polynomial function of degree n, if

$$P(x)=a_nx^n+a_{n-1}x^{n-1}+\dots\dots\dots\dots\dots\dots+a_1x+a_0,\quad a_n\neq 0,$$

where n is a non-negative integer. The numbers $a_0,a_1,\dots\dots\dots,a_n$ are real constants and called coefficients of the polynomial function. The polynomials of degree $1,2,3,4$ and 5 are called **linear, quadratic, cubic, quartic** and **quantic** respectively. The natural domain of a polynomial in x is $(-\infty,\infty)$, since only addition and multiplication operation is involved. The range depends on particular polynomials.

For example, the function $P(x)=3x^2-x^4+3$ is a polynomial function of degree 4. We know that the graphs of polynomials of degree 0 and 1 are lines. Figure 1.7 shows the graphs of some polynomials of higher degree. From the graphs, we observe that number of peaks and valleys is less than the degree of the polynomials.

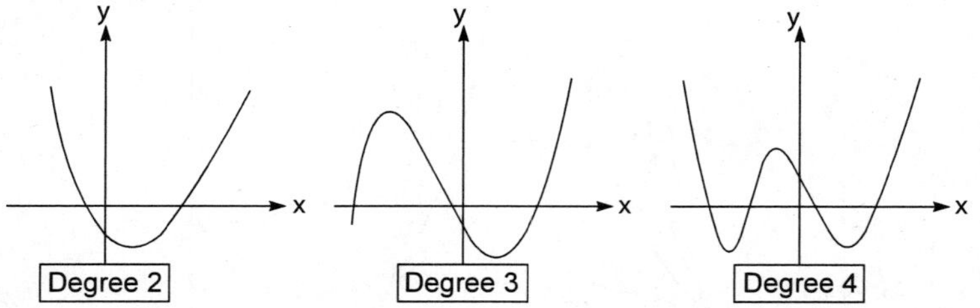

Figure 1.7 Graphs of some polynomials

1.8.2 Rational Function

Let $P(x)$ and $Q(x)$ be the two polynomial functions. A rational function $f(x)$ is a ratio of $P(x)$ and $Q(x)$:

$$f(x)=\frac{P(x)}{Q(x)}.$$

The domain of $f(x)$ contains all real values of x for which $Q(x)\neq 0$. A simple example of a rational function is $(x)=\dfrac{1}{1+x}$, whose domain is $\{x \mid x\neq -1\}$. The graph of two rational functions is shown in Figure 1.8.

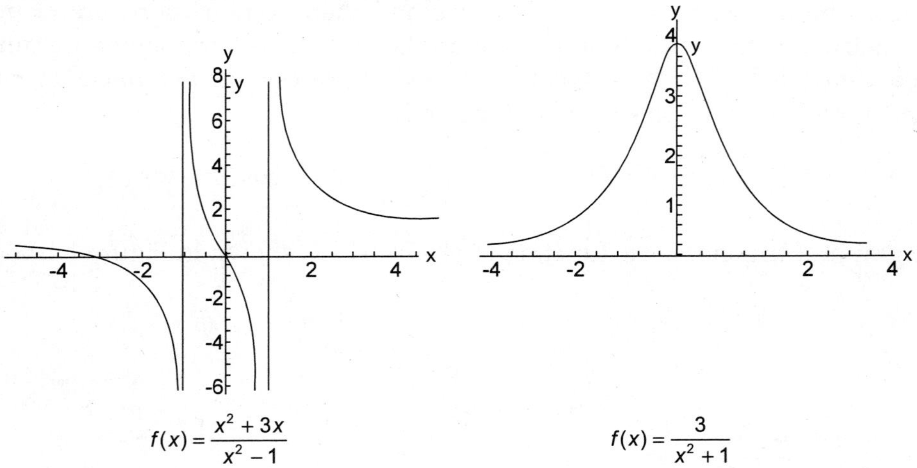

Figure 1.8 Graphs of rational functions

1.8.3 Power Functions

A function of the type $f(x)=x^b$, where b is a constant, is called a power function. The root function $f(x)=x^{\frac{1}{n}}$ is an example of power function, where n is a positive integer. The graph of some functions $f(x)=x^n$, when n is odd and even, is shown in Figure 1.9.

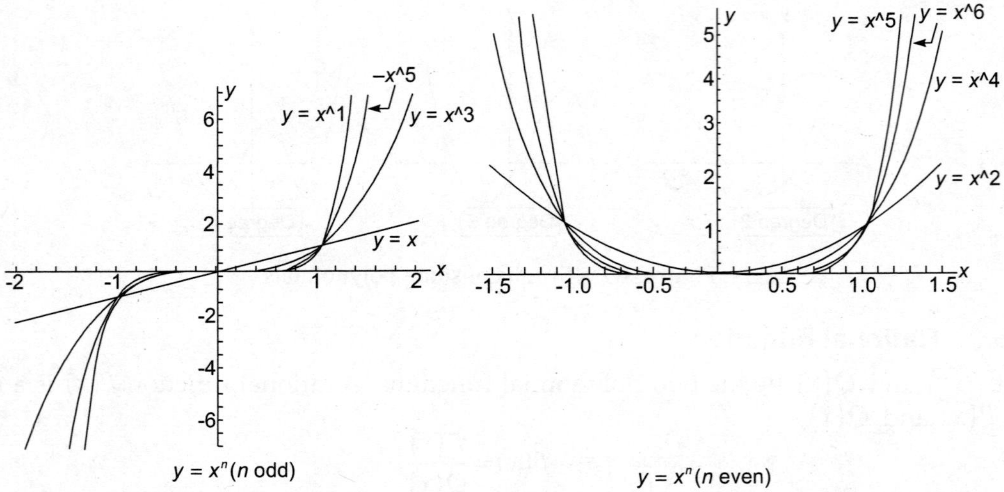

Figure 1.9

1.8.4 Trigonometric Functions

An angle is a measure of the amount of rotation. A function of angle is called trigonometric functions. The function such as *sine*, *cosine* and their combinations are examples of trigonometric functions. Many important applications of the trigonometric function of the type $y = A\sin(Bx - C) or y = A\cos(Bx - C)$ exist in physics and engineering problems. Some examples of trigonometric functions are:

$$f(x) = 3\sin 4x; f(x) = 2 + \cos\left(\frac{x}{4}\right); f(x) = \frac{1}{2}\cos(3x - \pi) \text{ etc.}$$

The graph of trigonometric function $f(x) = 2\sin 5x$ is shown in Figure 1.10.

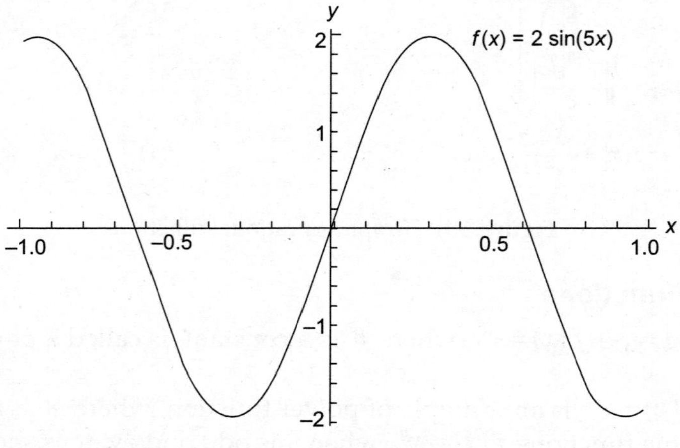

Figure 1.10

1.8.5 Exponential and Logarithmic Functions

A function of the form $f(x) = a^x$, where a is a positive constant, is called an exponential function. The function $f(x) = 2^x$ is example of an exponential function with domain $(-\infty, \infty)$ and range is $(0, \infty)$.

The logarithmic functions are the functions of the type $f(x) = \log_a x$, where the base a is a positive constant and is inverse function of the exponential function. The graph of some exponential and logarithmic functions is shown in Figure 1.11.

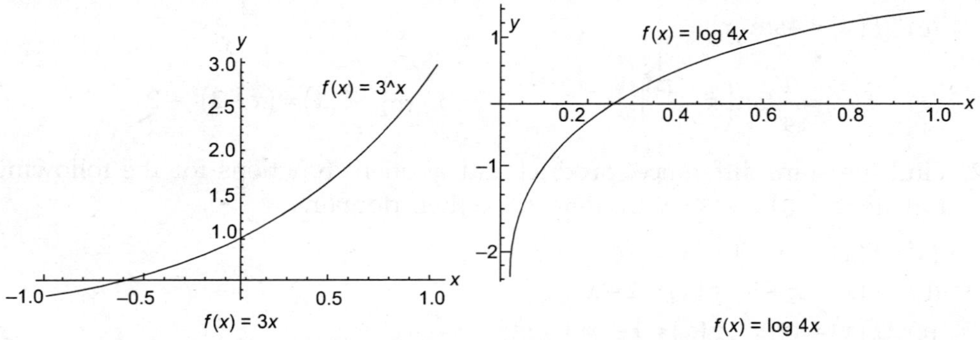

Figure 1.11

1.8.6 Greatest Integer Function

The function of the type $f(x) = [x]$, where $[x]$ is the greatest integer such that $n \le x < n+1$, is called the greatest integer function.

For example,

$$f(x) = [x], -2 \le x \le 2$$

$$\Rightarrow f(x) = \begin{cases} -2, & -2 \le x < -1 \\ -1, & -1 \le x < 0 \\ 0, & 0 \le x < 1 \\ 1, & 1 \le x < 2 \\ 2, & x = 2 \end{cases}$$

The graph of the function is shown in Figure 1.12.

The graph consists of line segments of unit length, all parallel to the x-axis.

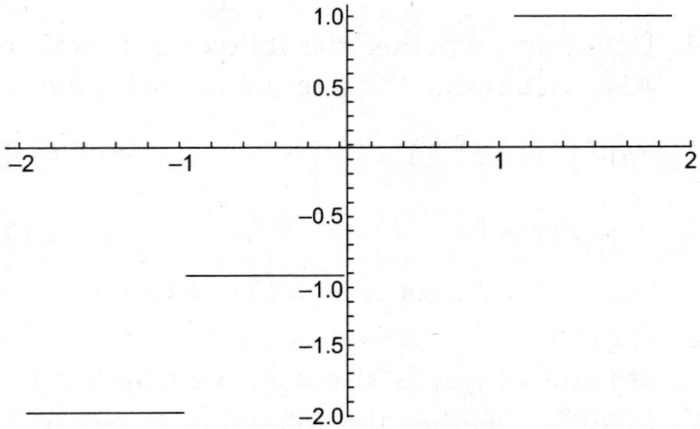

Figure 1.12

Exercise

1. Draw the graph of the following functions using software like Mathematica, Matlab, etc.

   (a) $f(x)=-x^3$ (b) $f(x)=1-x^2$

   (c) $f(x)=(x+1)^2$ (d) $f(x)=1+2\cos(x)$

   (e) $f(x)=4\sin(x)$ (f) $f(x)=\dfrac{2}{1+x}$

   (g) $f(x)=\dfrac{1}{4}\tan\left(x-\dfrac{\pi}{4}\right)$ (h) $f(x)=(x+2)^4+3.$

2. Find the sum, difference, product and quotient functions for the following two functions $f(x)$ & $g(x)$. Also determine their domains:

   (a) $f(x)=2x-4, g(x)=-4x$

   (b) $f(x)=x+3$, $g(x)=2-x$

   (c) $f(x)=1+x^2, g(x)=1-|x|$

3. Find the composite functions *fog* and *gof* and their domains for the following functions:

   (a) $f(x)=x^2-1$ and $g(x)=2x+1$ (b) $f(x)=\sin(x)$ and $g(x)=1-\sqrt{x}$

   (c) $f(x)=x-1$ and $g(x)=\dfrac{1}{x}$ (d) $f(x)=\sqrt{2x+3}$ and $g(x)=x^2+1$

4. Determine whether the following functions $f(x)$ are increasing or decreasing. Also, determine the interval in which the functions are increasing or decreasing.

   (a) $f(x)=x^2-1$ (b) $f(x)=\dfrac{1}{x}, x\in(0,1]$

   (c) $f(x)=|x|$ (w) $f(x)=x^3$

5. Show that the function $f(x)=3x+1$ is injective.
6. Show that if $f:A\to B$ is bijective and $g:B\to C$ is bijective, then the composition of *f* and *g* (gof) is also bijective map form *A* onto *C*.
7. Identify whether the following functions $f:R\to R$ are injective, surjective and bijective.

   (a) $f(x)=x^2$ (b) $f(x)=2x+3$

   (c) $f(x)=\dfrac{x^5}{5}+2$ (d) $f(x)=x^3$

   (e) $f(x)=|x|$ (f) $f(x)=\sqrt{x}$

8. Show that the function $f:X\to X$ such that $f(x)=\dfrac{x+1}{x-1}$, where X is a set of all real numbers except 1, is injective (one-one) and surjective (onto).

2 CHAPTER

Limit and Continuity

In this chapter, our main concerns are limits and continuity which are important aspects of analysis in mathematics and also, very important for application point of view. First, we start with a preliminary discussion of limits. Next, we take a closer look at the concept of limit and extend it in the direction of continuity.

2.1 AN INTUITIVE APPROACH OF LIMIT

To understand the basic meaning of limit of a function, we begin our discussion with some specific examples.

EXAMPLE 1 Suppose the function $f(x)=3x+5$. The value of function at $x=3$ is equal to $f(x)=14$. If we want to know that behaviour of function when x is near 3 but not equal to 3.

Table 2.1

x	$f(x)$	x	$f(x)$
2.5	12.5	3.5	15.5
2.6	12.8	3.4	15.2
2.7	13.1	3.3	14.9
2.8	13.4	3.2	14.6
2.9	13.7	3.1	14.3
2.99	13.97	3.01	14.03
2.995	13.985	3.001	14.003
2.999	13.997	3.0001	14.0003
2.9999	13.9997	3.00001	14.00003

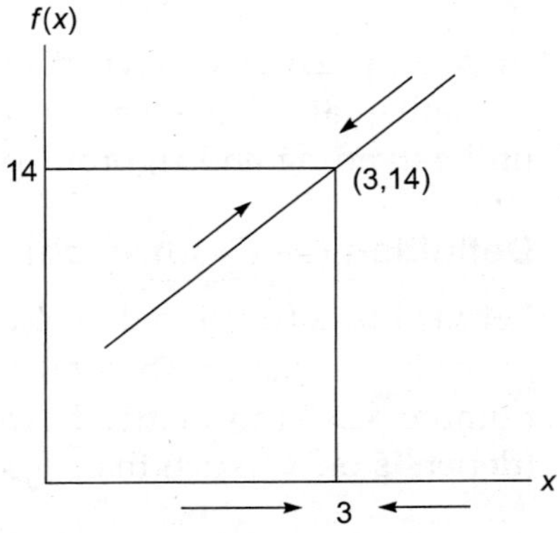

Figure 2.1

From the table and graph of the function shown in Figure 2.1, we see that when the value of x approaches 3 from the left or from the right, the value of function $f(x)$ approaches 14. It appears that we can make the values of $f(x)$ as close as we like to 14 by taking x sufficiently close to 3. Now, we express this by saying "the limit of the function $f(x)$ as x tends to 3 is equal to 14. To all we have discussed here, we give a notation for this

$$\lim_{x \to 3} 3x + 5 = 14.$$

In general, we use the following definition for intuitive approach.

Definition: If the function $f(x)$ gets arbitrarily close to some real value L as x gets sufficiently close to a (from either side of a) but not equal to a. We write it as

$$\lim_{x \to a} f(x) = L$$

and say that "limit of $f(x)$ is equal to L as x approaches a".

EXAMPLE 2 Evaluate the value of $\lim_{x \to 3} \frac{x^2 - 3}{x - 3}$.

Solution: Here, the function $f(x) = \frac{x^2 - 3}{x - 3}$ is not defined when $x = 3$, but it does not matter because as per the definition of limit we can take the values of x that are close to 3 but not equal to 3. So, we make guess that

$$\lim_{x \to 3} \frac{x^2 - 3}{x - 3} = \lim_{x \to 3} x + 3 = 6.$$

2.2 MATHEMATICAL APPROACH OF LIMIT

In section 2.1, we have discussed the definition of limit with intuitive approach. A mathematical description of limits must be based on analytical work for deeper understanding and rigorous proofs.

Definition ($\in - \delta$ Approach)

Let $f(x)$ be a function defined on some open interval containing a real number c, except possibly at c itself. Then $f(x)$ is said to tend to a limit L as x tends to c, if for every number $\in > 0$ (no matter however small it is) there is a corresponding number $\delta > 0$ (depends on $\in$) such that

$$|f(x) - L| < \in \quad \text{whenever} \quad 0 < |x - c| < \delta$$

The above definition is illustrated in Figure 2.2. If for a given number $\in > 0$, we can draw the horizontal lines $y = L + \in$ and $y = L - \in$ and the graph of the function $f(x)$. If limit exists, then we find a number $\delta > 0$ (say) such that we restrict x lie in the interval

$(c-\delta, c+\delta)$ and $x \neq c$, then the curve $y = f(x)$ lies between the lines $y = L + \in$ and $y = L - \in$. (see Figure 2.2).

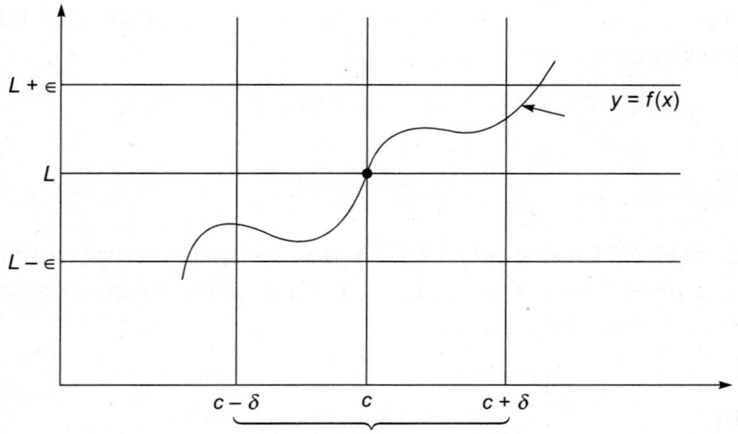

Figure 2.2

EXAMPLE 3 Prove that $\lim_{x \to 2} 3x - 5 = 1$.

Solution: Let $\in$ be a given positive number. According to $\in - \delta$ definition with $c = 2$ and $L = 1$, we can find a number $\delta > 0$ such that

$$|(3x-5)-1| < \in, \text{ whenever } 0 < |x-2| < \delta.$$

But $|3x-5-1| = |3x-6| = 3|x-2|$. Therefore, we have

$$3|x-2| < \in, \text{ whenever } 0 < |x-2| < \delta.$$

If we choose $\delta = \in/3$, then

$$|(3x-5)-1| < \in, \text{ whenever } 0 < |x-2| < \delta = \in/3$$

Therefore, by the definition of limit, we have

$$\lim_{x \to 2} 3x - 5 = 1.$$

Before we close this section, we give a very important result in limits that is called uniqueness theorem.

Uniqueness of Limits: If $\lim_{x \to c} f(x)$ exists, then it is unique.

Sometimes we see that some limits are best calculated by first finding the left and right-hand limits. In the next section, we define $\in - \delta$ definitions of left and right handed limits.

2.3 LEFT HAND AND RIGHT HAND LIMITS

Definition (Left Hand Limit): Let *f* be a function defined in an open interval containing *c*. Then $f(x)$ has a limit *L* from the left at *c*. If for a given positive number $\in > 0$, there exists a positive number $\delta > 0$ such that

$$|f(x) - L| < \in \quad \text{whenever } c - \delta < x < c.$$

Symbolically,

$$\lim_{x \to c^-} f(x) = L^-.$$

Definition (Right Hand Limit): Let *f* be a function defined in an open interval containing. Then $f(x)$ has a limit *L* from the right at *c*. If for a given positive number $\in > 0$, there exists a positive number $\delta > 0$ such that

$$|f(x) - L| < \in \quad \text{whenever } c < x < c + \delta.$$

Symbolically,

$$\lim_{x \to c^+} f(x) = L^+.$$

In definition of limit, we know that limit exists if and only if both the one sided limits exist and are equal. The following theorem represents the limit of function at a point.

Theorem: $\lim_{x \to c} f(x) = L$ if and only if $\lim_{x \to c^-} f(x) = L^- = L = L^+ = \lim_{x \to c^+} f(x)$.

2.4 ALGEBRA OF LIMITS

Suppose α is a constant. If $\lim_{x \to c} f(x) = L_1$ & $\lim_{x \to c} g(x) = L_2$, then

1. $\lim_{x \to c} [f(x) + g(x)] = L_1 + L_2$.
2. $\lim_{x \to c} [f(x) - g(x)] = L_1 - L_2$.
3. $\lim_{x \to c} [f(x) g(x)] = L_1 L_2$.
4. $\lim_{x \to c} [\alpha f(x)] = \alpha L_1$.
5. $\lim_{x \to c} \left[\frac{f(x)}{g(x)} \right] = \frac{L_1}{L_2}, \lim_{x \to c} g(x) = L_2 \neq 0$.
6. $\lim_{x \to c} [f(x)]^n = L_1^n$, where *n* is a positive integer. (Power Law)
7. $\lim_{x \to c} \sqrt[n]{f(x)} = \sqrt[n]{\lim_{x \to c} f(x)}$, where *n* is a positive integer. (Root Law)

Following theorems give important additional properties of limits.

Theorem 2.4.1: If $f(x) \le g(x)$ when x is near c (except possibly at c) and the limits of $f(x)$ and $g(x)$ both exist as x approaches c, then $\lim_{x \to c} f(x) \le \lim_{x \to c} g(x)$.

Theorem 2.4.2: (The Squeeze Theorem) If $f(x) \le g(x) \le h(x)$ when x is near c (except possibly at c) and $\lim_{x \to c} f(x) = \lim_{x \to c} h(x) = L$, then $\lim_{x \to c} g(x) = L$.

2.5 INFINITE LIMITS AND LIMIT OF INFINITY

In the previous section, we have discussed the limit at finite point and value of limit is a finite value. We now need to explore the possibilities of extending the definition by using ∞ for both the values of c and L.

Let $f(x)$ be a function defined in open interval containing c. Then f is said to tend to $+\infty$ as x tends to c, if for each real number $M > 0$, there exists a $\delta > 0$ such that

$$f(x) > M \text{ whenever } 0 < |x - c| < \delta.$$

Symbolically, it is written as

$$\lim_{x \to c} f(x) = +\infty.$$

Function f is said to tend to $-\infty$ as x tends to c, if for each real number $M < 0$, there exists $\delta > 0$ such that

$$f(x) < M \text{ whenever } 0 < |x - c| < \delta.$$

Symbolically, it is denoted as

$$\lim_{x \to c} f(x) = -\infty.$$

If $f(x)$ is a function defined in open interval $[c, \infty)$, then f is said to tend to a limit L as x tends to ∞, if for each real number $\in > 0$, there exists a $k \in [c, \infty)$ such that

$$|f(x) - L| < \in \text{ whenever } x < k.$$

Symbolically, we can denote as

$$\lim_{x \to \infty} f(x) = L.$$

If $f(x)$ is a function defined in open interval $(-\infty, c]$. Then f is said to tend to a limit L as x tends to $-\infty$, if for each real number $\in > 0$, there exists a $k \in (-\infty, c]$ such that

$$|f(x) - L| < \in \text{ whenever } x < k.$$

Symbolically, we can denote as

$$\lim_{x \to -\infty} f(x) = L.$$

2.6 SOME IMPORTANT LIMITS

Here, we give some important limits without proof.

1. $\lim\limits_{x\to 0}\dfrac{\sin(x)}{x}=1.$

2. $\lim\limits_{x\to 0}\dfrac{\tan(x)}{x}=1.$

3. $\lim\limits_{x\to 0}(1+x)^{1/x}=e.$

4. $\lim\limits_{x\to 0}\dfrac{log(1+x)}{x}=1.$

5. $\lim\limits_{x\to 0}\dfrac{a^x-1}{x}=\log a\ ,\ \forall a>0.$

6. $\lim\limits_{h\to 0}\left(1+\dfrac{x}{h}\right)^h=e^x.$

7. $\lim\limits_{x\to 0}\dfrac{x^n-a^n}{x-a}=na^{n-1},\ \forall\, n\neq 0\,\text{and}\,a\neq 0.$

Some Solved Examples Based on Limit

EXAMPLE 4 Evaluate the limit of the following functions:

(a) $\lim\limits_{x\to 0}2^{-1/x^2}.$

(b) $\lim\limits_{x\to 2}\dfrac{3x^2+4x}{2x+2}.$

(c) $\lim\limits_{x\to 0}\dfrac{\sin 3x}{\sin 4x}.$

Solution:

(a) LHS: $\lim\limits_{x\to 0^-}2^{-1/x^2}=\lim\limits_{x\to 0}2^{-1/x^2}=\dfrac{1}{2^\infty}=\dfrac{1}{\infty}=0.$

RHS: $\lim\limits_{x\to 0^+}2^{-1/x^2}=\lim\limits_{x\to 0}2^{-1/x^2}=\dfrac{1}{2^\infty}=\dfrac{1}{\infty}=0.$

$LHS=RHS\Rightarrow$ limit exists and value is 0.

(b) First, $\lim\limits_{x\to 2}3x^2+4x=3.2^2+4.2=12+8=20.$

Next, $\lim\limits_{x\to 2}2x+2=6\neq 0.$

Then by the quotient rule of limits, we obtain

$$\lim_{x\to 2}\frac{3x^2+4x}{2x+2}=\frac{\lim_{x\to 2}3x^2+4x}{\lim_{x\to 2}2x+2}=\frac{20}{6}=\frac{10}{3}.$$

(c) $$\lim_{x\to 0}\frac{\sin 3x}{\sin 4x}=\lim_{x\to 0}\frac{3}{4}\frac{\frac{\sin 3x}{3x}}{\frac{\sin 4x}{4x}}=\frac{3}{4}\frac{\lim_{x\to 0}\frac{\sin 3x}{3x}}{\lim_{x\to 0}\frac{\sin 4x}{4x}}=\frac{3}{4}\frac{1}{1}=\frac{3}{4}.$$

EXAMPLE 5 Using $\in-\delta$ definition, show that $\lim_{x\to 0}x\sin\frac{1}{x}=0$.

Solution: Let $\in>0$ be given. To show that the $\lim_{x\to 0}x\sin\frac{1}{x}=0$ using $\in-\delta$ definition. We have to show that for a given $\in>0$, there exists a positive $\delta>0$ such that

$$\left|x\sin\frac{1}{x}-0\right|<\in \text{ whenever } 0<|x-0|<\delta.$$

Here, $\left|x\sin\frac{1}{x}-0\right|=|x|\,|\sin\frac{1}{x}|\le|x|$, because $|\sin\frac{1}{x}|\le 1$.

$$\therefore \quad \left|x\sin\frac{1}{x}-0\right|<\in \text{ whenever} |x|<\in.$$

If we choose a number δ such that $0<\delta\le\in$, we have

$$\left|x\sin\frac{1}{x}-0\right|<\in \text{ whenever } 0<|x|<\delta.$$

Hence, $\lim_{x\to 0}x\sin\frac{1}{x}=0$.

EXAMPLE 6 Given that $|f(x)-2|\le 3(x+1)^2\ \forall x\in R$, evaluate $\lim_{x\to -1}f(x)$.

Solution: We know that from absolute function,

$$-3(x+1)^2\le f(x)-2\le 3(x+1)^2.$$

$$\Rightarrow -3(x+1)^2+2\le f(x)\le 3(x+1)^2+2.$$

Now, $\lim_{x\to -1}-3(x+1)^2+2=2$ and $\lim_{x\to -1}3(x+1)^2+2=2$.

By the sandwich theorem,

$$\lim_{x\to -1}-3(x+1)^2+2=2=\lim_{x\to -1}3(x+1)^2+2.$$

Thus, $$\lim_{x\to -1}f(x)=2.$$

2.7 CONTINUITY OF FUNCTIONS

Generally, if the graph of the function $y = f(x)$ is a continuous curve (without break), then it is called a continuous function. But sometimes it is difficult to draw the graph of the function. Here, we discuss the analytical definition of continuity.

Cauchy's definition of continuity: Let $f(x)$ be a real valued function defined on an open interval I and a point $c \in I$. The function $f(x)$ is said to be continuous at point c, if for a given positive number $\in > 0$ (however small) there exists a positive number $\delta > 0$ such that

$$|f(x) - f(c)| < \in \text{ whenever } |x - c| < \delta.$$

Geometrically, as illustrated in Figure 2.2, if the function is continuous then the points $(x, f(x))$ on the graph of f approach the point $(c, f(c))$ on the graph. So, there is no gap in the curve.

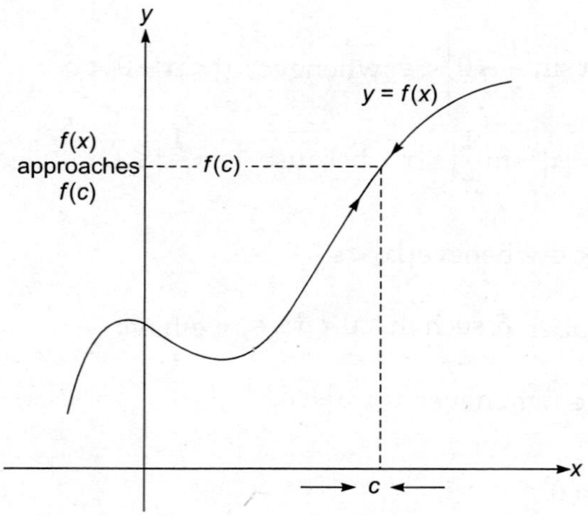

Figure 2.3

A function $f(x)$ is said to continuous at point c, if

(a) $f(c)$ is defined (that is, c is in the domain of f),

(b) $\lim_{x \to c} f(x)$ exists.

(c) $\lim_{x \to c} f(x) = f(c)$.

If a function f is continuous at each point in an open interval (a, b), then, we can say that f is continuous on (a, b). Similarly, we can also define continuity of a function on some closed interval as:

"A function f is said to be continuous on a closed interval $[a,b]$, if f is continuous on open interval (a,b) and $\lim\limits_{x \to b^-} f(x) = f(b)$ *(f is continuous from left at b) and* $\lim\limits_{x \to a^+} f(x) = f(a)$ *(f is continuous from right at a)".*

Remark: In general, to show that a function is continuous everywhere. We shall show that it is continuous at an *arbitrary* real number.

2.8 DISCONTINUITY

Definition: A function f is said to be discontinuous at c if it is not continuous at that point. This point is called the point of discontinuity.

2.9 TYPE OF DISCONTINUITY

(a) Removable discontinuity: A function f is said to have a removable discontinuity at a point c. If $\lim\limits_{x \to c} f(x)$ exists and is equal to function value at c,

i.e.,
$$\lim_{x \to c^-} f(x) = \lim_{x \to c^+} f(x) \neq f(c).$$

This type of discontinuity can easily be removed by extension of definition.

For example, the function $f(x) = \dfrac{\sin(x)}{x}$ is not defined at $x = 0$. However, the limit of this function at $x = 0$ is equal to 1. Hence, this function is not continuous at $x = 0$. This type of discontinuity can be removed by defining a new function,

$$g(x) = \begin{cases} \dfrac{\sin(x)}{x}, & x \neq 0 \\ 1, & x = 0 \end{cases}.$$

Now, the function $g(x)$ is continuous everywhere and is the continuous extension of $f(x)$.

(b) Discontinuity of the first kind: A function $f(x)$ is said to have a discontinuity of first kind at the point c. If

$$\lim_{x \to c^-} f(x) \neq f(c) = \lim_{x \to c^+} f(x),$$

$$\lim_{x \to c^-} f(x) = f(c) \neq \lim_{x \to c^+} f(x),$$

i.e., both the limits exist but are not equal.

For illustration, consider a function $f(x) = \begin{cases} 2, & x \geq 1 \\ 3, & x < 1 \end{cases}$. Here, both the limits $\lim\limits_{x \to 1^-} f(x) = 3$ and $\lim\limits_{x \to 1^+} f(x) = 2$ exist but are not equal to function value $f(1) = 2$. Therefore, the above function is discontinuous at $x=1$ and has discontinuity of the first kind (jump discontinuity) at $x=1$.

(c) Discontinuity of the Second Kind: A function $f(x)$ is said to have a discontinuity of the second kind at the point c, if none of the limits $\lim_{x\to c^-} f(x)$ and $\lim_{x\to c^+} f(x)$ exist at c.

For example, consider the function $f(x)=\sqrt{(1-x)}+\sqrt{(x-1)}$. In this case both the limits do not exist, therefore the function has discontinuity of the second kind at $x=1$.

(d) Mixed Discontinuity: A function $f(x)$ is said to have a mixed discontinuity of the first kind on one side ($\lim_{x\to c^-} f(x)$) and a discontinuity of the second kind on the other side ($\lim_{x\to c^+} f(x)$) and vice versa.

(e) Infinite Discontinuity: A function $f(x)$ is said to have an infinite discontinuity at a point $x=c$. If one or more of the functional limits are $+\infty$ or $-\infty$.

2.10 SOME IMPORTANT PROPERTIES OF CONTINUOUS FUNCTIONS

Let f and g be two functions defined on an interval I. If f and g are continuous at a point $c \in I$, then

(a) $f \pm g$ is also continuous at c.

(b) fg is continuous at c.

(c) f/g is continuous at c if $g(c) \neq 0$ and has a discontinuity at c if $g(c)=0$.

The following theorems give additional properties of continuous functions.

Theorem 2.10.1: Polynomial function and rational function (denominator of rational function is non-zero) are continuous everywhere.

Theorem 2.10.2: If the function f is continuous at $\in I$, then $|f|$ is also continuous at c.

Theorem 2.10.3: If the function g is continuous at $c \in I$ and the function f is continuous at c, then fog is also continuous at c.

Theorem 2.10.4: (Intermediate Value Theorem) Let f be defined on closed interval $[a,b]$. If f is continuous on interval $[a,b]$ and k is any real number between $f(a)$ and $f(b)$, interval [a, b] by (a, b), then there exists at least one real number $x \in [a,b]$ such that $f(x)=k$.

Solved Example Based on Continuity

EXAMPLE 7 Examine the continuity at point $x=2$ for the function $f(x)$ defined as

$$f(x)=\begin{cases} \dfrac{1}{x-2}\operatorname{cosec}\left(\dfrac{1}{x-2}\right), & x \neq 2 \\ 0, & x=2 \end{cases}.$$

Solution: RHL:

$$\lim_{x\to 2^+}\frac{1}{x-2}\operatorname{cosec}\left(\frac{1}{x-2}\right)=\lim_{h\to 0}\frac{1}{(h+2)-2}\operatorname{cosec}\left(\frac{1}{(h+2)-2}\right)$$

$$=\lim_{h\to 0}\frac{1}{h\sin\left(\frac{1}{h}\right)}=\frac{1}{\lim\limits_{h\to 0}h\sin\left(\frac{1}{h}\right)}=\frac{1}{0}=+\infty.$$

LHL:

$$\lim_{x\to 2^-}\frac{1}{x-2}\operatorname{cosec}\left(\frac{1}{x-2}\right)=\lim_{h\to 0}\frac{1}{(2-h)-2}\operatorname{cosec}\left(\frac{1}{(2-h)-2}\right)$$

$$=\lim_{h\to 0}\frac{1}{-h\sin\left(-\frac{1}{h}\right)}=\frac{1}{\lim\limits_{h\to 0}h\sin\left(\frac{1}{h}\right)}=\frac{1}{0}=+\infty.$$

$$\lim_{x\to 2}f(x)=\infty\neq f(2).$$

So, the function is not continuous at the point $x=2$ and has an infinite discontinuity of the second kind.

EXAMPLE 8 Show that $f:\mathbb{R}\to\mathbb{R}$ s.t. $f(x)=\left(x^2+1\right)^2$ is continuous on $\mathbb{R}$, using composition of two continuous functions is also continuous.

Solution: If we assume that $g(x)=x^2$ and $h(x)=x^2+1$. Because g and h are polynomial functions and polynomial functions are continuous everywhere. Therefore, g and h are continuous on $\mathbb{R}$.

By the composition of two functions,

$$f(x)=(goh)(x)=\left(x^2+1\right)^2.$$

By Theorem 2.10.3, $f(x)$ is also continuous on $\mathbb{R}$.

EXAMPLE 9 Examine the type of discontinuity at the point $x=1$ for the following function

$$f(x)=\begin{cases}\dfrac{x^2-4x+3}{x^2-1}, & x\neq 1\\ 2, & x=1.\end{cases}$$

Solution:

$$\lim_{x\to 1}\frac{x^2-4x+3}{x^2-1}=\lim_{x\to 1}\frac{(x-1)(x-3)}{(x+1)(x-1)}=\lim_{x\to 1}\frac{x-3}{x+1}=-1$$

$$\lim_{x\to 1}f(x)=-1\neq 2=f(1).$$

So, the function is not continuous at point $x=1$. The discontinuity is of the first kind and can be removed by defining function as $f(x)=-1$.

EXAMPLE 10 Show that the function f defined as

$$f(x)=\begin{cases} 1, \text{ if } x \text{ is rational} \\ -1, \text{ if } x \text{ is irrational} \end{cases}$$

is discontinuous everywhere.

Solution: To show that the function $f(x)$ is discontinuous at every point of $\mathbb{R}$, we assume that $x=c$ is an arbitrary point of R. If we take any positive number $\delta>0$, then $c-\delta$ and $c+\delta$ are two distinct real numbers. Since there are infinite many real numbers lies in between two distinct real numbers, therefore, there exists infinite many rational and irrational numbers in the open interval $(c-\delta, c+\delta)$. For every $\delta>0$, there exists a point $x\in(c-\delta, c+\delta)$ such that if we take x as a rational number and c as an irrational number, then

$$|f(x)-f(c)|=|1-(-1)|=2$$

and if we take x as an irrational number and c as a rational number, then

$$|f(x)-f(c)|=|-1-1|=2.$$

Thus, whatever the point c in R, for every $\delta>0$ there exists a point x in $(c-\delta, c+\delta)$ such that

$$|f(x)-f(c)|=2.$$

If we choose $\in=2>0$, there exists no $\delta>0$ such that

$$|f(x)-f(c)|<\in, \text{ whenever } |x-c|<\delta.$$

By the $\in-\delta$ definition, $f(x)$ is not continuous at $x=c$. But $x=c$ is an arbitrary point of R. Thus, the function f is discontinuous at every point of R.

EXAMPLE 11 Show that if f is continuous at $x=c$, then $|f|$ is also continuous at $x=c$.

Solution: If f is continuous at point $x=c$, then for a given $\in>0$, there exists a $\delta>0$ such that

$$|f(x)-f(c)|<\in, \text{ whenever } |x-c|<\delta.$$

As we know that for any two real numbers x_1 and x_2

$$\big||x_1|-|x_2|\big|\le|x_1-x_2|$$

Now,

$$\big||f(x)|-|f(c)|\big|\le|f(x)-f(c)|<\in, \text{ whenever } |x-c|<\delta.$$

By the $\in-\delta$ definition, $|f|$ is also continuous at $x=c$.

Exercise

1. Show that $\lim_{x\to -1}\sqrt{x+1}$ does not exist and $\lim_{x\to -1}\sqrt{x^2+1}$ exists.
2. Find the value of δ such that $\left|\sqrt{x}-2\right|<0.05$ if $0<|x-4|<\delta$.

   In problems 3-10, a positive number $\in$ and limit L of a function $f(x)$ at $x=c$ are given. Find the number $\delta>0$ for which

$$|f(x)-L|<\in \text{ if } |x-c|<\delta.$$

3. $\lim_{x\to 4} 2x=8;\ \in=0.1.$
4. $\lim_{x\to -2}\frac{1}{2}x=-1;\ \in=0.01.$
5. $\lim_{x\to 2} 2x-1=3;\ \in=0.1.$
6. $\lim_{x\to 3} 5x-2=13;\ \in=0.1.$
7. $\lim_{x\to 2}\frac{x^2-1}{x-2}=4;\ \in=0.5.$
8. $\lim_{x\to -1}\frac{x^2-1}{x+1}=-2;\ \in=0.05.$
9. $\lim_{x\to 2} 3x+1=7;\ \in=0.01.$
10. $\lim_{x\to 2} x^2=4;\ \in=0.01.$
11. Evaluate $\lim_{x\to 1} 5x^2+3x^2-6x^2-3x+5$.
12. Evaluate $\lim_{x\to 2}\left[x^3+\frac{x^2-4}{x-2}\right]$.
13. Evaluate $\lim_{x\to 0}\left[\frac{x\left(x^2+1\right)}{2}+x^5+5\right]$.
14. Show that $\lim_{x\to 4}\sqrt{x}\left(1+\sqrt{x}\right)=6$.
15. Show that $\lim_{x\to -1}\frac{x^3+5x^2+3x}{2x^5+3x^4-2x^2-1}=-\frac{1}{2}$.

16. Show that $\lim_{x\to0}\frac{\sin 3x}{x}=3$.

17. Prove that $\lim_{x\to0}\frac{\sin 3x}{\sin 2x}=\frac{3}{2}$.

18. Evaluate $\lim_{x\to0}\frac{1-\cos 2x}{x^2}$.

19. Show that $\lim_{x\to0}\frac{x-|x|}{x}$ does not exist.

20. Let $f(x)$ be defined as

$$f(x)=\begin{cases}2-x, \text{ if } 1\le x\le 2\\ x^2-2x, \text{ if } x>2\end{cases}.$$

Find the $\lim_{x\to2}f(x)$.

21. Prove that $\lim_{x\to0}\frac{3x+2|x|}{5x-3|x|}$ does not exist.

22. Prove that the $\lim_{x\to2}\frac{|x-2|}{x-2}$ does not exist.

23. A function $f(x)$ is defined as $f(x)=\begin{cases}1, & x\neq0\\ 2, & x\neq0\end{cases}$. Does the $\lim_{x\to0}f(x)$ exist?

In problems 24-28, a positive number $\in$ and limit L of a function $f(x)$ at $x=-\infty$ are given. Find a negative number N for which $|f(x)-L|<\in$ if $x<N$.

24. $\lim_{x\to-\infty}\frac{1}{x+2}=0; \in=0.05.$

25. $\lim_{x\to-\infty}\frac{1}{x^2}=0; \in=0.01.$

26. $\lim_{x\to-\infty}\frac{4x-1}{2x+3}=2; \in=0.01.$

27. $\lim_{x\to-\infty}\frac{1}{x}=0; \in=0.001.$

28. $\lim_{x\to-\infty}\frac{3x+1}{x+2}=3; \in=0.01.$

In problems 29-31, a positive number $\in$ and limit L of a function $f(x)$ at $x=\infty$ are given. Find the positive number N for which $|f(x)-L|<\in$ if $x>N$,

29. $\lim\limits_{x\to\infty}\dfrac{3x+1}{x+2}=3; \in=0.01.$

30. $\lim\limits_{x\to\infty}\dfrac{1}{x+2}=0; \in=0.05.$

31. $\lim\limits_{x\to\infty}\dfrac{1}{x^2}=0; \in=0.01.$

32. $\lim\limits_{x\to\infty}\dfrac{4x-1}{2x+3}=2; \in=0.01.$

33. $\lim\limits_{x\to\infty}\dfrac{1}{x}=0; \in=0.001.$

34. Evaluate $\lim\limits_{x\to\infty}\dfrac{5x^2+1}{10x^3+8000x}$.

35. Find the value of the constants a,b and c for which the function

$$f(x)=\begin{cases}\dfrac{\sin(a+1)x+\sin x}{x} & \text{when } x<0\\ c, & \text{when } x=0\\ \dfrac{(x+bx)^{\frac{1}{2}}-x^{\frac{1}{2}}}{bx^{\frac{3}{2}}}, & \text{when } x>0\end{cases}$$

is continuous at $x=0$.

36. Discuss the continuity of the following functions $f(x)$ and find the point x (if any) at which f is not continuous

(a) $f(x)=x^3-2x+3$

(b) $f(x)=(x-4)^{16}$

(c) $f(x)=\dfrac{x}{x^2+1}$

(d) $f(x)=\dfrac{x-4}{x^2-16}$

(e) $f(x)=\dfrac{x}{|x|-3}$

(f) $f(x)=\frac{|x|}{x}$

37. Discuss the continuity of the function $f(x)=\begin{cases}\frac{\sin x}{x}, & x\neq 0\\ 1, & x=0\end{cases}$ at $x=0$ and determine whether such value is a removable discontinuity.

38. Discuss the continuity of the function $f(x)=\sqrt{c-x}+\sqrt{x-c}$ at point $x=c$.

39. Show that $f(x)=\begin{cases}\frac{|x-1|}{x-1}, & x\neq 1\\ 1, & x=1\end{cases}$ is discontinuous at $x=1$.

40. Discuss the continuity of the function $f(x)=\begin{cases}\frac{xe^{\frac{1}{x}}}{1+e^{\frac{1}{x}}}, & x\neq 0\\ 0, & x=0\end{cases}$ at $x=0$.

41. Show that the function $f(x)=\begin{cases}\frac{e^{\frac{1}{x}}-1}{e^{\frac{1}{x}}+1}, & x\neq 0\\ 0, & x=0\end{cases}$ is discontinuous $x=0$.

42. Examine the continuity of $f(x)=|x|+|x-1|$ at the point $x=0$ and $x=1$.

43. Find the points of discontinuity (if any) of the function

$$f(x)=\begin{cases}2, & x=0\\ x+1, & 0<x<1\\ 1, & x=1\\ x-1, & 1<x<2\\ 0, & x=2\end{cases}$$

Also, discuss the kind of each discontinuity.

44. Find the points of discontinuity (if any) of the function

$$f(x)=\begin{cases}1, & x\leq 0\\ 3-x, & 0<x\leq 1\\ \frac{4}{1+x}, & x>1\end{cases}$$

Also, discuss the kind of each discontinuity.

45. Show that the function $f:\mathbb{R}\to\mathbb{R}$ defined as

$$f(x)=\begin{cases}\dfrac{1}{2}, \text{ if } x \text{ is rational}\\ \dfrac{1}{3}, \text{ if } x \text{ is irrational}\end{cases}$$

is discontinuous everywhere.

46. Show that the function $f:\mathbb{R}\to\mathbb{R}$ defined as

$$f(x)=\begin{cases}x, \text{ if } x \text{ is rational}\\ -x, \text{ if } x \text{ is irrational}\end{cases}$$

is discontinuous everywhere except at $x=0$.

47. Show that the function $f:\mathbb{R}\to\mathbb{R}$ defined as

$$f(x)=\begin{cases}x^2\sin\dfrac{1}{x}, \text{ if } x\neq 0\\ 0, \text{ if } x=0\end{cases}$$

is continuous at $x=0$.

48. Examine the continuity of the function $f:\mathbb{R}\to\mathbb{R}$ defined as

$$f(x)=\begin{cases}\dfrac{|x|}{x}, \text{ if } x\neq 0\\ 0, \text{ if } x=0\end{cases}$$

3 CHAPTER

Derivatives and Their Applications

The study of derivatives of functions is the most important topic of differential calculus because quantities changes in the real world, i.e., number of bacteria in culture, inflation of currency, and speed of rocket. In this chapter, we will discuss the concept of derivative which is used to study the velocity of a body, to the slope of tangent line and any instantaneous rate of change.

3.1 DERIVATIVE

Let $f(x)$ be a real valued function defined on an interval $I=[a,b]\subseteq R$. The function is said to be differentiable at a point $c\ (a<c<b)$ if

$$\lim_{x\to c}\frac{f(x)-f(c)}{x-c}$$

exists finitely. The value of limit is called the derivative or differential coefficient of function f at c and is denoted by $f'(c)$. That is,

$$f'(c)=\lim_{x\to c}\frac{f(x)-f(c)}{x-c}.$$

3.1.1 Left Hand and Right Hand Derivatives

The left hand derivative of f at $x=c$ is defined as

$$Lf'(c)=\lim_{x\to c^-}\frac{f(x)-f(c)}{x-c}$$

provided the limit exists.

Similarly, the right hand derivative of f at $x=c$ is defined as

$$Rf'(c)=\lim_{x\to c^+}\frac{f(x)-f(c)}{x-c}$$

provided the limit exists.

Result: If a function f is differentiable at a point c, then $Lf'(c)=Rf'(c)$.

EXAMPLE 1 Discuss the differentiability of function $f(x)=|x-1|+|x-2|$ at point $x=1$ and $x=2$.

Solution: We can rewrite the given function as

$$f(x)=\begin{cases} 3-2x, & \text{when } x<1 \\ 1, & \text{when } 1\le x<2. \\ 2x-3, & \text{when } x\ge 2 \end{cases}$$

For $x=1$, the right hand and left hand derivatives are

$$Rf'(1)=\lim_{x\to 1^+}\frac{f(x)-f(1)}{x-1}=\lim_{x\to 1}\frac{1-1}{x-1}=0.$$

$$Lf'(1)=\lim_{x\to 1^-}\frac{f(x)-f(1)}{x-1}=\lim_{x\to 1}\frac{-2x+2}{x-1}=-2.$$

Since, $Lf'(1)\neq Rf'(1)$, so the function $f(x)$ is not differentiable at $x=1$.

Next, for $x=2$, the right hand and left hand derivatives are

$$Rf'(2)=\lim_{x\to 2^+}\frac{f(x)-f(2)}{x-2}=\lim_{x\to 2}\frac{2x-3-1}{x-2}=\lim_{x\to 2}\frac{2(x-2)}{x-2}=2.$$

$$Lf'(2)=\lim_{x\to 2^-}\frac{f(x)-f(2)}{x-2}=\lim_{x\to 1}\frac{1-1}{x-2}=0.$$

Since, $Lf'(2)\neq Rf'(2)$, so the function $f(x)$ is also not differentiable at $x=2$.

EXAMPLE 2 Show that the function $f(x)=\begin{cases} x^2\sin\left(\frac{1}{x}\right), & \text{if } x\neq 0 \\ 0, & \text{if } x=0 \end{cases}$ is differentiable at point $x=0$.

Solution: The right hand and left hand derivatives are

$$Rf'(0)=\lim_{x\to 0^+}\frac{f(x)-f(0)}{x-0}=\lim_{x\to 0^+}\frac{f(x)}{x}=\lim_{x\to 0^+}\frac{x^2\sin\left(\frac{1}{x}\right)}{x}=\lim_{x\to 0}x\sin\left(\frac{1}{x}\right)=0.$$

$$Lf'(0)=\lim_{x\to 0^-}\frac{f(x)-f(0)}{x-0}=\lim_{x\to 0^-}x\sin\left(\frac{1}{x}\right)=0$$

Since, $Lf'(0)=Rf'(0)$, so the function $f(x)$ is differentiable at $x=0$.

EXAMPLE 3 Show that the following function

$$f(x)=\begin{cases} \dfrac{x}{1+e^{\frac{1}{x}}}, & \text{if } x\neq 0 \\ 0, & \text{if } x=0 \end{cases}$$

is not differentiable at point $x=0$.

Solution: The right hand and left hand derivatives are

$$Rf'(0)=\lim_{x\to 0^+}\frac{f(x)-f(0)}{x-0}=\lim_{x\to 0^+}\frac{f(x)}{x}=\lim_{x\to 0^+}\frac{1}{1+e^{\frac{1}{x}}}=\lim_{x\to 0^+}\frac{e^{-\frac{1}{x}}}{e^{-\frac{1}{x}}+1}$$

$$=\frac{0}{0+1}=0.$$

$$Lf'(0)=\lim_{x\to 0^-}\frac{f(x)-f(0)}{x-0}=\lim_{x\to 0^-}\frac{1}{1+e^{\frac{1}{x}}}=\frac{1}{1+0}=1.$$

Since, $Lf'(0)\neq Rf'(0)$, so the function $f(x)$ is not differentiable at $x=0$, i.e., $f'(0)\neq 0$.

3.1.2 Differentiability in an Interval

If a function $f(x)$ is defined on interval $[a,b]$ and differentiable at each point of an interval $[a,b]$ except the end points, then the function is differentiable on open interval (a,b).

If the function is derivable at each point of the interval $[a,b]$, then the function is derivable on closed interval $[a,b]$.

3.2 GEOMETRICAL INTERPRETATION OF DERIVATIVE

If $P[x,f(x)]$ and $Q[x+\Delta x,f(x+\Delta x)]$ are the two neighbouring points on the curve $y=f(x)$. Let the chord PQ, joining points P and Q, meet the x-axis at the point R and tangent at P meet the x-axis at S. Let $\angle QRU=\alpha$ and $\angle PSU=\psi$. Now, draw the perpendicular PT and QU on x-axis and perpendicular PV on QU. Then,

$$PV=TU=OU-OT$$
$$=a+h-a=h$$

and

$$QV=QU-VU=QU-PT$$
$$=f(a+h)-f(a)$$

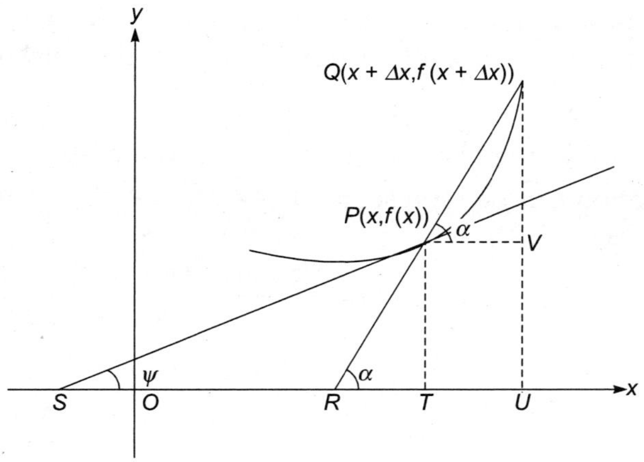

Therefore,

$$\tan\alpha = \frac{QV}{PV} = \frac{f(a+h)-f(a)}{h} \tag{3.1}$$

As $h \to 0$, the point Q approaches to the point P moving along the curve, then chord PQ approaches the tangent line PS and the angle α tends to angle ψ. If we take limit $h \to 0$ the equation (3.1) gives

$$\tan(\psi) = \lim_{h\to 0}\frac{f(a+h)-f(a)}{h} = f'(x)$$

Now, we see that $f'(x)$ is the slope of the tangent line to the curve $y = f(x)$ at the point P makes with x-axis.

3.3 VELOCITY PROBLEM

Suppose an object moves along a straight line according to an equation of motion $s = f(t)$, where s is the displacement of the object from the origin at time t and f defines the motion, called the position function of the object. Then, the average velocity over the time interval $[t, t+\Delta t]$ is

$$\text{Average velocity} = \frac{\text{Displacement}}{\text{Time}} = \frac{f(t+\Delta t)-f(t)}{\Delta t}.$$

If we want to compute the average velocities over short time intervals, i.e., $\Delta t \to 0$, we define the instantaneous velocity $v(t)$ at time t as limit of these average velocities.

$$v(t) = \lim_{\Delta t\to 0}\frac{f(t+\Delta t)-f(t)}{\Delta t}.$$

EXAMPLE 4 Suppose that a ball is dropped from 420 m height above the ground. Find (i) the velocity of the ball after 3 seconds, and (ii) the velocity of the ball after it hits the ground?

Solution: We know that the distance (in metres) covered after t seconds is $4.9t^2$. So, the equation of motion can be represented by $= f(t) = 4.9t^2$. Then the velocity of the ball after t seconds is

$$v(t) = \lim_{\Delta t \to 0} \frac{f(t+\Delta t) - f(t)}{\Delta t}$$

$$= \lim_{\Delta t \to 0} \frac{4.9(t+\Delta t)^2 - 4.9t^2}{\Delta t}$$

$$= \lim_{\Delta t \to 0} \frac{4.9\left((\Delta t)^2 + 2t\Delta t\right)}{\Delta t}$$

$$= \lim_{\Delta t \to 0} 4.9(\Delta t + 2t) = 9.8t.$$

(i) The velocity after 3 seconds is $v(3) = (9.8)(3) = 29.4$.

(ii) Suppose that ball hits the ground at time t_1, then

$$s(t_1) = 4.9t_1^2 = \text{height above the ground} = 420.$$

Now, we have

$$t_1 = \sqrt{\frac{420}{4.9}} \approx 9.3 \text{seconds}.$$

Then, the velocity of ball when it hits the ground is

$$v(t_1) = 9.8t = 9.8 * 9.3 = 91.14.$$

3.4 SOME IMPORTANT THEOREMS AND RESULTS ON DERIVATIVES

3.4.1 Theorem

(Necessary condition for the existence of derivative) Continuity is a necessary condition for the existence of derivatives but not sufficient condition.

$$\textit{Differentiability} \Rightarrow \textit{continuity}\ \textbf{but}\ \textit{Continuity} \nRightarrow \textit{Differentiability}.$$

3.4.2 Theorem

Let function f and g be defined on an interval $[a,b]$ and are differentiable at a point $c \in [a,b]$. Then, $f \pm g, fg$ and $\frac{f}{g}$ are also differentiable at c and:

(a) $(f \pm g)'(c) = f'(c) \pm g'(c)$,

(b) $(f.g)'(c) = f'(c)g(c) + f(c)g'(c)$,

(c) $\left(\frac{f}{g}\right)'(c) = \frac{g(c)f'(c) - f(c)g'(c)}{[g(c)]^2}$, *if* $g'(c) \neq 0$.

3.4.3 Theorem

(The chain rule) Let f and g be the two differentiable functions and $F = fog$ is the composite function defined by $(x) = f(g(x))$, then F is a differentiable function and

$$F'(x) = f'(g(x)).g'(x).$$

In Leibnitz notation, if $y = f(u)$ is a differentiable function of u and $u = g(x)$ is a differentiable function of x, then y is a differentiable function and

$$\frac{dy}{dx} = \frac{dy}{du}\frac{du}{dx}.$$

3.4.4 Theorem

If $x = f(t)$ and $y = g(t)$ are differentiable functions of some parameters $t, t \in [a,b]$, i.e.,

$$\frac{dx}{dt} = f'(t) \text{ and } \frac{dy}{dt} = g'(t),$$

then

$$\frac{dy}{dx} = \frac{\frac{dy}{dt}}{\frac{dx}{dt}} = \frac{g'(t)}{f'(t)}, \text{ provided } g'(t) \neq 0.$$

3.4.5 Some Standard Results on Derivatives

There are some standard derivatives of their functions shown in Table 3.1.

Table 3.1

Functions	Derivatives	Functions	Derivatives	Functions	Derivatives
c = *constant*	0	$\text{cosec}^{-1}x$	$-\frac{1}{x\sqrt{x^2-1}}$	$\log_e x$	$\frac{1}{x}$
x^n	nx^{n-1}	$\sec^{-1}x$	$\frac{1}{x\sqrt{x^2-1}}$	$\log_a x$	$\frac{1}{x}\log a$

(Contd.)

$\sin x$	$\cos x$	$\cot^{-1} x$	$-\dfrac{1}{1+x^2}$	$\sin^{-1} x$	$\dfrac{1}{\sqrt{1-x^2}}$
$\cos x$	$-\sin x$	e^x	e^x	$\cos^{-1} x$	$-\dfrac{1}{\sqrt{1-x^2}}$
$\tan x$	$\sec^2 x$	c^x	$c^x \log c$	$\tan^{-1} x$	$\dfrac{1}{1+x^2}$
$\operatorname{cosec} x$	$-\operatorname{cosec} x \cot x$	$\sinh x$	$\cosh x$	$\operatorname{cosech} x$	$-\operatorname{cosec} x \coth x$
$\sec x$	$\sec x \tan x$	$\cosh x$	$\sinh x$	$\operatorname{sech} x$	$\operatorname{sech} x \tanh x$
$\cot x$	$-\operatorname{cosec}^2 x$	$\tanh x$	$\operatorname{sech}^2 x$	$\coth x$	$-\operatorname{cosech}^2 x$

Solved examples based on section 3.4

EXAMPLE 5 Find the derivatives of the following functions:

(i) $x + \cos(x)$ (ii) $\dfrac{\sin(x)}{x}, x \neq 0$ (iii) $x^3 e^x$.

Solution:

(i) $[x+\cos(x)]' = [x]' + [\cos(x)]' = 1 - \sin(x)$.

(ii) $[\sin(x)/x]' = \dfrac{[\sin x]'x - [x]'\sin x)}{x^2} = \dfrac{\cos x\, x - \sin x}{x^2}, x \neq 0$.

(iii) $[x^3 e^x]' = [x^3]' e^x + x^3 [e^x]' = 3x^2 e^x + x^3 e^x = (x^3 + 3x^2) e^x$.

EXAMPLE 6 Find the derivatives of the following functions:

(i) $[\sin(x)]^{x^2}$.

(ii) $x^{x^{x^{x^{\cdot^{\cdot^{\cdot}}}}}}$.

Solution:

(i) Let $y = [\sin(x)]^{x^2}$.

Taking logarithm of both the sides, we have

$$\log y = x^2 \log[\sin x].$$

Now, differentiating both the sides with respect to x, we have

$$\frac{d}{dx}(\log y) = \frac{d}{dx}(x^2 \log[\sin x])$$

$$\Rightarrow \quad \frac{1}{y}\frac{dy}{dx} = \frac{d}{dx}\left(x^2\right)\log[\sin x] + x^2 \frac{d}{dx}[\log[\sin x]]$$

$$= 2x\log[\sin x] + x^2 \frac{\cos x}{\sin x}$$

$$= 2x\log[\sin x] + x^2 \cot x.$$

Hence,

$$\frac{dy}{dx} = y[2x\log\sin x + x^2 \cot x]$$

$$= [\sin x]^{x^2}[2x\log\sin x + x^2 \cot x].$$

(ii) Let $y = x^{x^{x^{x^{\cdot^{\cdot^{\cdots\infty}}}}}}$.

We can rewrite the function as

$$y = x^y.$$

Taking logarithm of both the sides, we have

$$\log y = y \log x.$$

Now, differentiating both the sides with respect to x. Then, we have

$$\frac{d}{dx}(\log y) = \frac{d}{dx}(y \log x)$$

$$\Rightarrow \quad \frac{1}{y}\frac{dy}{dx} = \frac{dy}{dx}\log x + y\frac{1}{x}$$

$$\Rightarrow \quad \left\{\frac{1}{y} - \log x\right\}\frac{dy}{dx} = \frac{y}{x}$$

$$\Rightarrow \quad \frac{dy}{dx} = \frac{y^2}{x(1 - y\log x)}, \text{ where} = x^{x^{x^{x^{\cdot^{\cdot^{\cdots\infty}}}}}}.$$

EXAMPLE 7 Find the derivatives of $f(x) = \sqrt{mx^2 + 2nx + l}$, where m, n and l are constants.

Solution: Let $= f(u) = \sqrt{u}$, where, $u = mx^2 + 2nx + l$. By chain rule, we have

$$\frac{dy}{dx} = \frac{dy}{du}\frac{du}{dx},$$

where, $\dfrac{dy}{du}=\dfrac{1}{2\sqrt{u}}$ and $\dfrac{du}{dx}=2(mx+n)$.

Hence,

$$\frac{dy}{dx}=\frac{1}{2\sqrt{u}}2(mx+n)=\frac{mx+n}{\sqrt{mx^2+2nx+l}}.$$

EXAMPLE 8 Find the derivative $\dfrac{dy}{dx}$ for the following parametric functions:

(i) $x(t)=a(t+\sin t),\ y(t)=t^2$.

(ii) $x(t)=r(\cos t+t\sin t), y(t)=r(\sin t-t\cos t)$.

(iii) $x(t)=e^t(\cos t-\sin t), y(t)=e^t(\cos t+\sin t)$.

***Solution*:**

(i) Since $x(t)$ and $y(t)$ both are differentiable of t. Therefore,

$$\frac{dx}{dt}=a(1+\cos t) \text{ and } \frac{dy}{dt}=2t$$

Thus,

$$\frac{dy}{dx}=\frac{\frac{dy}{dt}}{\frac{dx}{dt}}=\frac{2t}{a(1+\cos t)}.$$

(ii) Here, $x(t)=r(\cos t+t\sin t)$, $y(t)=r(\sin t-t\cos t)$.

This gives,

$$\frac{dx}{dt}=r(-\sin t+t\cos t+\sin t)=rt\cos t,$$

$$\frac{dy}{dt}=r(\cos t-\cos t+t\sin t)=rt\sin t.$$

$$\therefore \qquad \frac{dy}{dx}=\frac{dy}{dt}\Big/\frac{dx}{dt}=\frac{rt\sin t}{rt\cos t}=\tan t.$$

(iii) Here, $x(t)=e^t(\cos t-\sin t), y(t)=e^t(\cos t+\sin t)$.

This gives,

$$\frac{dx}{dt}=e^t(\cos t-\sin t)+e^t(-\sin t-\cos t)=-2e^t\sin t,$$

$$\frac{dy}{dt}=e^t(-\sin t+\cos t)+e^t(\cos t+\sin t)=2e^t\cos t.$$

Therefore,

$$\frac{dy}{dx} = \frac{\frac{dy}{dt}}{\frac{dx}{dt}} = \frac{2e^t \cos t}{-2e^t \sin t} = -\cot t.$$

3.5 APPLICATIONS OF DERIVATIVES

One of the most important applications of derivatives is the optimization problem. In the optimization problem, the maximum or minimum value of the function depends on real world problem. In this section, we locate maximum and minimum values. First we define the meaning of maximum and minimum.

3.5.1 Absolute Maximum and Minimum

Definition: Let f be a function defined on an interval I and $c \in I$. Then, function f has:

(i) an absolute maximum at c if $f(c) \geq f(x)$ for all x in I

(ii) an absolute minimum at c if $f(c) \leq f(x)$ for all x in I

and values $f(c)$ are called maximum and minimum values respectively. The maximum and minimum values of f are called the extreme values.

3.5.2 Local Maximum and Minimum (Relative Maximum and Minimum)

Definition: Let f be a function defined on an interval I and $c \in I$. Then function f has:

(i) an local (relative) maximum at c if $f(c) \geq f(x)$ for all x in near c.

(ii) an local (relative) minimum at c if $f(c) \geq f(x)$ for all x in near c.

Figure 3.1 shows the graph of function f with absolute maximum and minimum and local maximum and minimum.

We can make the following observations from Figure 3.1.

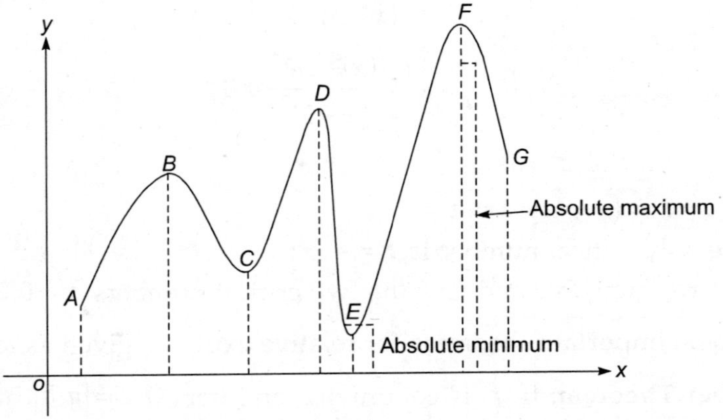

Figure 3.1

1. Local maximum values of the function $f(x)$ are at the points B and D.
2. Local minimum values are at the points A, C and G.
3. Absolute maximum value of function at the point F.
4. Absolute minimum value of function at the point E.

3.5.3 Critical Number and Critical points

Definition: A number c in domain of a function $f(x)$ is said to be a **critical number** if $f'(c)=0$ or $f'(c)$ does not exist. The point $(c, f(c))$ on the graph is called a **critical point**.

EXAMPLE 9 Find the critical numbers for the following functions:

(i) $f(x)=4x^2-4x+2$.

(ii) $f(x)=\dfrac{e^x}{x-3}, x\neq 3$.

(iii) $f(x)=2\sqrt{x}(6-x), x\geq 0$.

Solution:

(i) $f'(x)=8x-4$ is defined for all values of x. After solving the equation $8x-4=0$, we have $x=\frac{1}{2}$. Therefore, $x=\frac{1}{2}$ is a critical number for the given function.

(ii) Here,

$$f'(x)=\frac{(x-3)e^x-e^x}{(x-3)^2}.$$

The derivative is well defined because $x=3$ does not belong to the domain of function. Then, the critical number is found by solving the following equation:

$$\frac{(x-3)e^x-e^x}{(x-3)^2}=0$$

$$\Rightarrow \quad \frac{(x-4)e^x}{(x-3)^2}=0$$

$$\Rightarrow \quad (x-4)e^x=0.$$

But $e^x>0\neq 0 \Rightarrow (x-4)=0 \Rightarrow x=4$.

Thus, the only critical number is $x=4$.

(iii) Similarly, we can solve and find the two critical numbers $x=0,2$.

There are some important theorems for relative extrema, given as follows:

Critical Number Theorem: If f is continuous on interval $I=[a,b]$ and has a relative extremum at point $c\in[a,b]$, then c must be a critical number of f.

The Extreme Value theorem: Let f be defined on closed interval $I=[a,b]$. If f is continuous on I, then f attains an absolute maximum value $f(c)$ and absolute minimum value $f(d)$ at some numbers c and d in $[a,b]$.

Monotone Function Theorem: Let f be a differentiable function on the open interval (a,b).

If $f'(x)>0$ on (a,b), then f is strictly an increasing function on (a,b).

If $f'(x)<0$ on (a,b), then f is strictly a decreasing function on (a,b).

Fermat's Theorem: If f has a relative maximum or minimum at c and if f is differentiable at c, then $f'(c)=0$.

3.5.4 First Derivative Test

From the critical number theorem, we know that every relative extremum is a critical point. However, every critical point of continuous function is not necessarily a relative extremum. The few observations that are summarized in a procedure are called the *first-derivative test for relative extrema*.

First-Derivative Test for Relative Extrema

Step 1. Determine the critical numbers of a continuous function f, i.e., find all the c numbers such that either $f'(c)=0$ or $f(c)$ does not exist.

Step 2. Classify each critical point $(c,f(c))$ as follows:

(a) The point $(c,f(c))$ has a **relative maximum** if $f'(x)>0$ (graph rising) for all x in an open interval (a,c) to the left of c and $f'(x)<0$ (graph falling) for all x in an open interval (c,b) to the right of c (Figure 3.2).

(b) The point $(c,f(c))$ has a **relative minimum** if $f'(x)<0$ (graph falling) for all x in an open interval (a,c) to the left of c and $f'(x)>0$ (graph rising) for all x in an open interval (c,b) to the right of c (Figure 3.3).

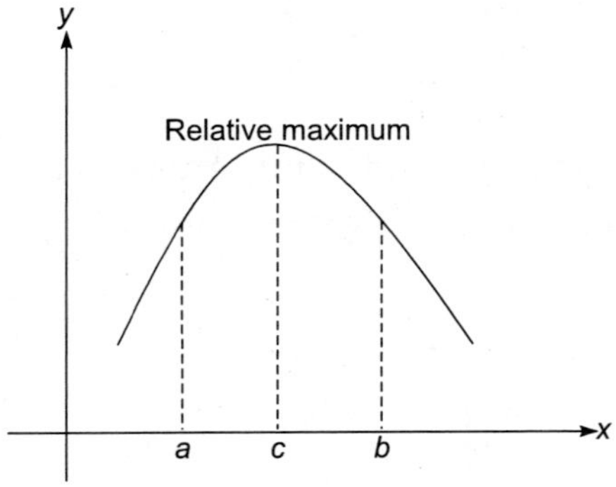

Figure 3.2

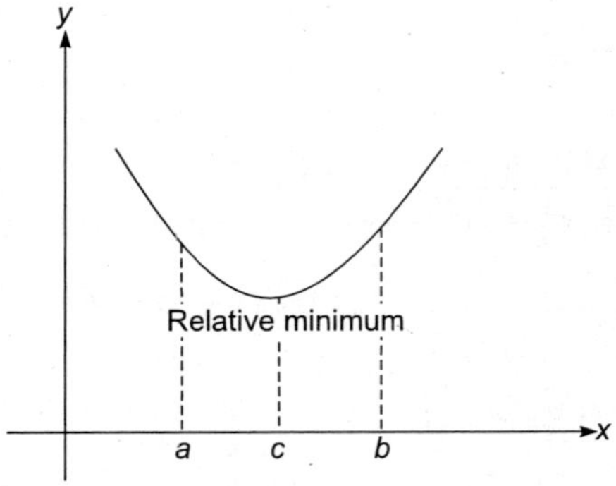

Figure 3.3

(c) The point $(c, f(c))$ is not an **extremum** if $f'(x)$ has the same sign for all x in an open interval (a, c) and (c, b) on each side of c.

EXAMPLE 10 Use the first derivative test to find the relative extrema for the function $f(x) = x^3 - 3x^2 - 9x + 1$.

Solution: Here, $f'(x) = 3x^2 - 6x - 9.$

The critical number is found by solving the following equation

$$3x^2 - 6x - 9 = 3(x^2 - 2x - 3) = 0$$

$$\Rightarrow \quad x^2 - 2x - 3 = 0$$

$$\Rightarrow \quad (x + 1)(x - 3) = 0 \Rightarrow x = -1 \text{ and } x = 3$$

There are two critical numbers $x = -1$ and $x = 3$. We see that $f'(x) > 0$, when $x < -1$ and $x > 3$ and $f'(x) < 0$ when $-1 < x < 3$. Then the first derivative test tells us that there is a relative maximum at $x = -1$ and a relative minimum at $x = 3$.

3.5.5 Second Derivative Test

Let f be a function such that $f'(c) = 0$ and the second derivative exists on an open interval containing c.

If $f''(c) > 0$, then there exists a **relative minimum** at $x = c$.

If $f''(c) < 0$, then there exists a **relative maximum** at $x = c$.

If $f''(c) = 0$, then the second–derivative test **fails** (either a maximum or a minimum, or neither may occur).

The above theorem can be generalized as

Let $f'(c) = f''(c) = \ldots\ldots\ldots\ldots = f^{n-1}(c) = 0$ and $f^n(c) \neq 0$,

1. **If *n* is odd,** then the function has neither maximum nor minimum at $x = c$.
2. **If *n* is even**:
   (i) If $f^n(c) < 0$, then function is relative maximum at $x = c$.
   (ii) If $f^n(c) > 0$, then function is relative minimum at $x = c$.

EXAMPLE 11 Find the relative maximum and minimum values of $f(x) = x^3 - 6x^2 + 9x - 4$ using both the first and second derivative tests.

Solution: Here, $f(x) = x^3 - 6x^2 + 9x - 4$

$\Rightarrow f'(x) = 3x^2 - 12x + 9$

$\Rightarrow f''(x) = 6x - 12.$

First derivative test: To find the critical numbers, we set

$$f'(x) = 3x^2 - 12x + 9 = 3(x-1)(x-3) = 0.$$

From the above equation, we obtain two critical points $x=1$ and $x=3$.

We see that $f'(x)>0$, when $x<1$ and $x>3$ and $f'(x)<0$, when $1<x<3$. Then, the first derivative test tells us that there is a relative maximum at $x=1$ and a relative minimum at $x=3$.

Second Derivative Test:

$$f''(x) = 6(x-2).$$

At $x=1$, $f''(1)=6(1-2)=-6<0$, so the test tells us that there is a relative maximum at $x=1$.

At $x=3$, $f''(3)=6(3-2)=6>0$, so the test tells us that there is a relative minimum at $x=3$.

3.5.6 Concavity and Inflection Points

In Figure 3.4, we see that both the graphs are increasing functions on interval (a,b) but their behaviour is different. They look different because they bend in different directions. The bending of the graphs is measured by its concavity.

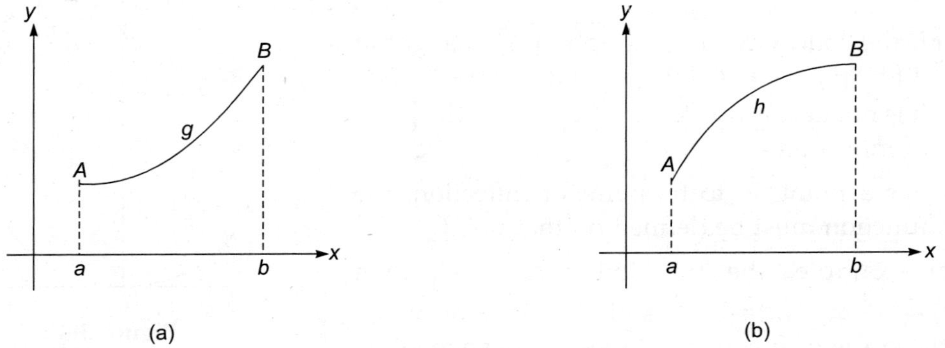

Figure 3.4

Definition: A graph of a function f is said to be **concave upward** on interval I if the graph lies above all its tangents on I. If the graph of f lies below all of its tangents on I, then it is called **concave downward** on I.

Definition: A point $P(c, f(c))$ on a curve $y = f(x)$ is said to be an *inflection point,* if f is continuous there and the curve changes from *concave upward* to *concave downward* while passing through the point P or vice versa.

Figure 3.5 shows the graph of function that is concave upward on the intervals $(b,c),(d,e)$ and (e,f) and concave downward on the interval $(a,b),(c,d)$ and (f,g). And points B,C,D and F are the points of inflection.

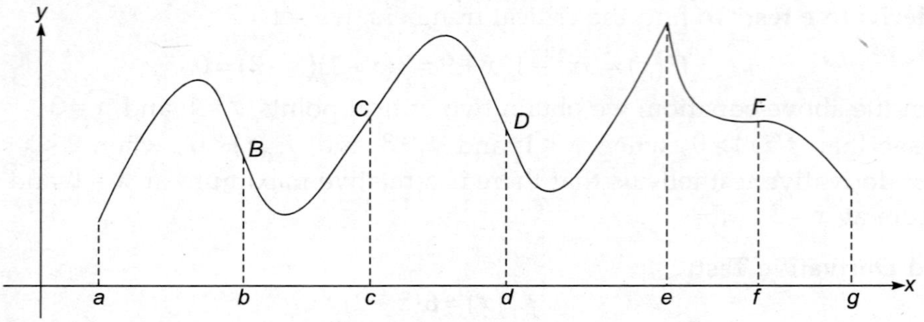

Figure 3.5

From Figure 3.4, we observe that graph of function will be concave up on interval (a,b), where the slope of the tangent to the curve increases and concave down where the slope decreases. This yields the following concavity test:

Concavity Test: Let f be defined on interval I.

(a) If $f''(x)>0$ for all x in I, then the graph of f is concave upward on I.

(b) If $f''(x)<0$ for all x in I, then the graph of f is concave downward on I.

Remark:

1. If the concavity of the graph of f changes at $P(c,f(c))$ then $f''(c)=0$ but if $f''(c)=0$, then it is not necessary, that the point P is the point of inflection.

2. For a point P to be point of inflection, the function must be defined on that point.

For example, the function $=f(x)=x^4$, then $f'(x)=4x^3$ & $f''(x)=12x^3$ and $f''(0)=0$ at $x=0$. But the curve of function f does not change while passing through the origin (Figure 3.6).

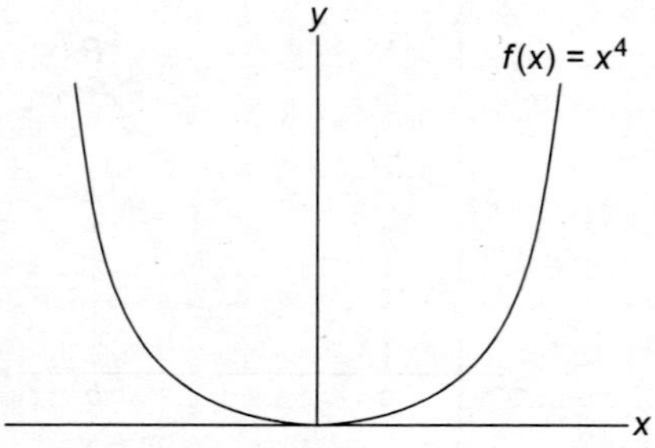

Figure 3.6

EXAMPLE 12 Find where the graph of the function $f(x)=x^4-6x^3+12x^2+12x-60$ is concave up and where concave down.

Solution: Here, we find that

$$f(x)=x^4-6x^3+12x^2+12x-60$$

$$\Rightarrow f'(x)=4x^3-18x^2+24x+12$$

$$\Rightarrow f''(x)=12x^2-36x+24=12\left(x^2+3x+2\right)=12(x-1)(x-2).$$

Therefore, $f''(x) > 0$, when $-\infty < x < 1$, so the graph of the function is concave up.

$f''(x) < 0$, when $1 < x < 2$, so the graph of function is concave down.

$f''(x) > 0$, when $2 < x < \infty$, so the graph of function is concave up.

EXAMPLE 13 Discuss concavity, point of inflection and relative maximum and minimum for the function $f(x) = x^4 - 4x^3$. Also, sketch the graph of the function.

Solution: **For critical points and Inflection point**:

To find the critical points and point of inflection, we calculate $f'(x) = 0$ & $f''(x) = 0$,

$$f'(x) = 4x^3 - 12x^2 = 4x^2(x-3) = 0 \Rightarrow x = 0 \text{ and } x = 3.$$

There are two critical points at $x = 0$ & $x = 3$. To find the point of inflection, we look at the second order derivative critical numbers,

$$f''(x) = 12x^2 - 24x = 12x(x-2) = 0$$

Since $f''(x) = 0$, when x = 0 or 2, we divide the real line into intervals with these numbers as end points and have the following information as shown in following table.

Interval	$f''(x) = 12x(x-2)$	**Concavity**
$(-\infty, 0)$	+	upward
$(0, 2)$	–	downward
$(2, \infty)$	+	upward

From the above table, point $(0,0)$ and $(2,-16)$ are inflection points, since the graph of function changes from concave upward to concave downward and concave downward to concave upward respectively.

For Maximum and Minimum:

Consider, $f''(x) = 12x(x-2)$. At $x = 3$,

$$f''(3) = 12 * 3(3-2) = 36 > 0,$$

so the second derivative test tells us that there is a relative minimum at $x = 3$.

At $x = 0$, $f''(0) = 0$, so the second order derivative gives no information about the critical point. But since $f'(x) < 0$ for $x < 0$ and also for $0 < x < 3$, the first order derivative test tells us that there are no relative maximum and minimum at point $x = 0$.

Using the concavity, point of inflection and relative maximum and minimum, we sketch the graph of the function shown in Figure 3.7.

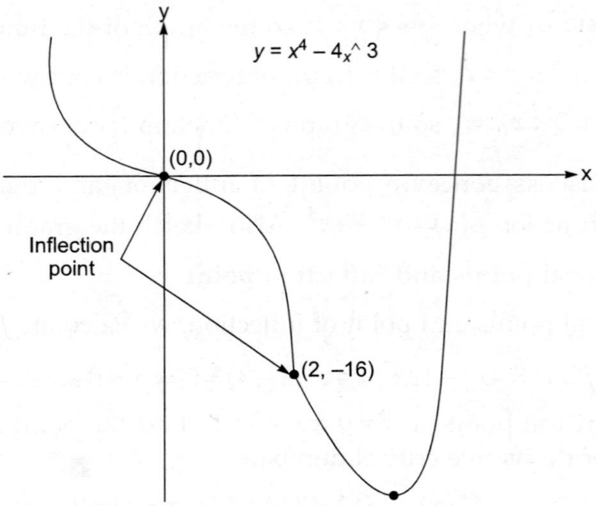

Figure 3.7

3.5.7 Rolle's Theorem and Mean Value Theorem

In this section, we discuss the Rolle's theorem and mean value theorem:

Rolle's Theorem: Let f be defined on closed interval $[a,b]$. If f is

(i) continuous on $[a,b]$,

(ii) differentiable on open interval (a,b), and

(iii) $f(a)=f(b)$,

then there exists at least one value c between a and b such that $f'(c)=0$.

Proof. There are three cases:

Case 1. When $f(x)=k$, where k is a constant.

Then, $f'(x)=0$, so the number c can be taken to be any number in the open interval (a,b).

Case 2. When $f(x)>f(a)$ for some x in (a,b).

If f is defined and continuous on closed interval, by the extreme value theorem, f has a maximum value somewhere in $[a,b]$. Since $f(a)=f(b)$, it must attain this maximum value at a number c in open interval (a,b). Then, f has a relative maximum at c and f is differentiable at c. Therefore, $f'(c)=0$ by Fermat's theorem.

Case 3. When $f(x)<f(a)$ for some x in (a,b).

Again, if f is defined and continuous on closed interval, by the extreme value theorem, we can say that f has a minimum value somewhere in $[a,b]$. Since $f(a)=f(b)$, it must attain this minimum value at a number c in the open interval (a,b). Again, $f'(c)=0$ by Fermat's theorem.

EXAMPLE 14 Verify Rolle' theorem for the function

$$f(x)=x(x+3)e^{-x/2} \text{ on } [-3,0].$$

Solution: Here, the function $f(x)=x(x+3)e^{-x/2}$ is continuous on closed $[-3,0]$ and differentiable on interval $(-3,0)$.

$$\therefore \quad f(x)=\left(x^2+3x\right)e^{-\frac{x}{2}}.$$

$$\Rightarrow f'(x)=(2x+3)e^{-\frac{x}{2}}+\left(x^2+3x\right)e^{-\frac{x}{2}}\left(-\frac{1}{2}\right)=-\frac{1}{2}e^{-\frac{x}{2}}\left(x^2-x-6\right)$$

which exists for every value of x in interval $[-3,0]$. Thus, $f(x)$ is differentiable in the interval $[-3,0]$ and therefore it is also continuous in the same interval.

Also,

$$f(-3)=0=f(0).$$

Thus, all the three conditions of Rolle's theorem are satisfied. So, there exists at least one value of c in between -3 and 0.

Now,

$$f'(c)=0$$

$$\Rightarrow -\frac{1}{2}e^{-\frac{c}{2}}\left(c^2-c-6\right)=0$$

$$\Rightarrow \quad (c^2-c-6)=0$$

$$\Rightarrow \quad (c-3)(c+2)=0$$

$$\Rightarrow \quad c=3,-2.$$

Since the value $x=-2$ lies between -3 and 0, the Rolle's theorem is verified.

Mean Value Theorem: Let f be defined on closed interval $[a,b]$. If f is

(i) continuous on $[a,b]$, and

(ii) differentiable on open interval (a,b) and then there exists at least one value c between a and b such that

$$f'(c)=\frac{f(b)-f(a)}{b-a}.$$

Proof. We consider a function $\psi(x)=f(x)+Ax$, where A is a constant such that $\psi(a)=\psi(b)$, i.e.,

$$\psi(a)=\psi(b)$$

$$\Rightarrow \quad f(a)+Aa=f(b)+Ab$$

$$\Rightarrow \qquad -A = \frac{f(b) - f(a)}{b - a}.$$

Since the function $\psi(x)$ is the sum of two continuous and differential functions, it is also

(i) continuous on closed interval $[a, b]$,

(ii) differentiable on open interval (a, b), and

(iii) $\psi(a) = \psi(b)$

Thus, the function satisfying all the conditions of Rolle's theorem, there exists a point c in between a and b such that $\psi'(c) = 0$. But

$$\psi'(x) = f'(x) + A \Rightarrow \psi'(c) = 0 = f'(c) + A$$

$$f'(c) = -A = \frac{f(b) - f(a)}{b - a}$$

This means that

$$f'(c) = \frac{f(b) - f(a)}{b - a}.$$

EXAMPLE 15 Verify the mean value theorem for the function $f(x) = x^3 + x^2$ on $[1, 2]$.

Solution: Because $f(x) = x^3 + x^2$ is a polynomial function, it is differentiable and also continuous on the closed interval $[1, 2]$. Thus, both the hypotheses of the mean value theorem are satisfied.

By differentiating f, we have

$$f'(x) = 3x^2 + 2x,$$

for all x. Therefore, by the mean value theorem, there exists at least one value c between 1 and 2 such that

$$f'(c) = \frac{f(2) - f(1)}{2 - 1}$$

is satisfied when

$$3c^2 + 2c = \frac{12 - 2}{2 - 1} = 10.$$

Solving the equation $3c^2 + 2c - 10 = 0$. We have

$$c = -\frac{1 \pm \sqrt{31}}{3}.$$

The negative value is not in the open interval $(1, 2)$. Since the positive value $= -\frac{1 + \sqrt{31}}{3} \approx 1.52$ lies between 1 and 2, the mean value theorem is verified.

3.5.8 Curve Sketching with Asymptotes

In this section, we discuss and sketch the graph of the function with asymptotes. We define asymptotes with the help of limits to infinity. In chapter 2, we have already discussed the infinite limit.

Vertical and Horizontal Asymptotes: A line $x = c$ is said to be a *vertical asymptote* of the graph of f if either of the one-sided limits

$$\lim_{x\to c^-} f(x) \quad \text{or} \lim_{x\to c^+} f(x)$$

is infinite.

A line $y = L$ is said to be *horizontal asymptote* of the graph of f if

$$\lim_{x\to+\infty} f(x) = L \quad \text{or} \lim_{x\to-\infty} f(x) = L.$$

EXAMPLE 16 Sketch the graph of $f(x) = \dfrac{3x-4}{x-2}$ with asymptotes.

Solution:

Vertical Asymptotes: Because vertical asymptotes for the function $f(x) = \dfrac{3x-4}{x-2}$ occur at values of c for which the limits $\lim\limits_{x\to c^-} f(x)$or $\lim\limits_{x\to c^+} f(x)$ is infinite. We find the value for which the denominator of rational function $f(x)$ is equal to zero. In this case $x = 2$ is the value that causes division by zero, so we find

$$\lim_{x\to 2^-} \frac{3x-4}{x-2} = -\infty \quad \text{and} \lim_{x\to 2^+} \frac{3x-4}{x-2} = \infty.$$

This means that $x = 2$ is a vertical asymptote and that the graph is moving downward as $x \to 2$ from the left and upward as $x \to 2$ from the right.

Horizontal Asymptotes: For the horizontal asymptotes, we compute

$$\lim_{x\to+\infty} \frac{3x-4}{x-2} = \lim_{x\to+\infty} \frac{3-\frac{4}{x}}{1-\frac{2}{x}} = 3,$$

$$\lim_{x\to-\infty} \frac{3x-4}{x-2} = \lim_{x\to-\infty} \frac{3-\frac{4}{x}}{1-\frac{2}{x}} = 3.$$

This means that $y = 3$ is a horizontal asymptote.

Using the concavity, point of inflection and relative maximum and minimum, we sketch the graph of the function shown in Figure 3.8.

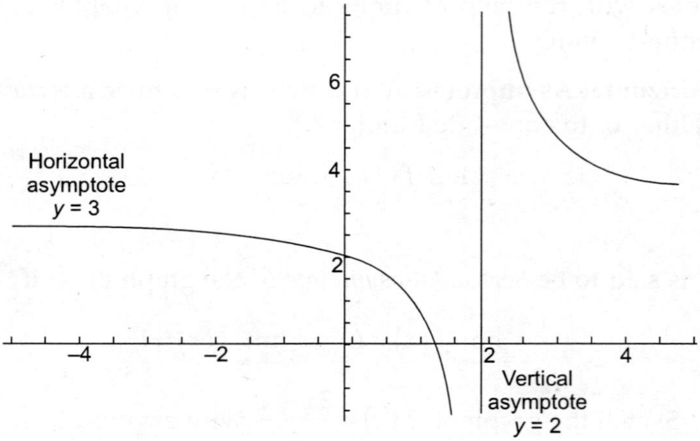

Figure 3.8

Vertical Tangents and Cusps

If the function f is continuous at the point $P(c, f(c))$, then the graph of f has a vertical tangent at P, if

$$\lim_{x\to c^-} f'(x) \text{ and } \lim_{x\to c^+} f'(x)$$

are either both $+\infty$ or $-\infty$;

And the graph of f has a cusp at P, if

$$\lim_{x\to c^-} f'(x) \text{ and } \lim_{x\to c^+} f'(x)$$

are infinite with opposite signs (one $+\infty$ and $-\infty$).

Some possibilities of vertical tangents and cusps are shown in Figure 3.9.

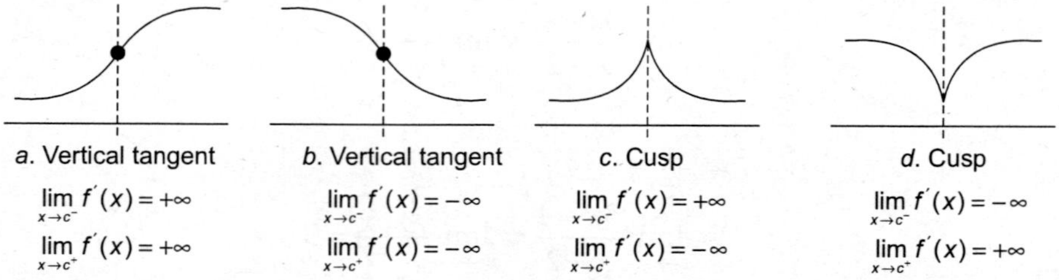

Figure 3.9 Vertical tangents and cusps

EXAMPLE 17 Determine whether the graph of function $f(x) = x^{\frac{2}{3}}(2x+5)$ has a vertical tangent or a cusp.

Solution: Here, $f(x) = x^{\frac{2}{3}}(2x+5) = 2x^{\frac{5}{3}} + 5x^{\frac{2}{3}}$

$$f'(x) = \frac{10}{3}x^{\frac{2}{3}} + \frac{10}{3}x^{-\frac{1}{3}} = \frac{10}{3}x^{-\frac{1}{3}}(x+1).$$

From the above expression, we see that $f'(x)$ becomes infinity at point $x=0$. Then, we have

$$\lim_{x\to 0^-}\frac{10}{3}x^{-\frac{1}{3}}(x+1) = -\infty \text{ and } \lim_{x\to 0^+}\frac{10}{3}x^{-\frac{1}{3}}(x+1) = +\infty.$$

So, there is a cusp at point $(0,0)$ and the graph of function shown in Figure 3.10.

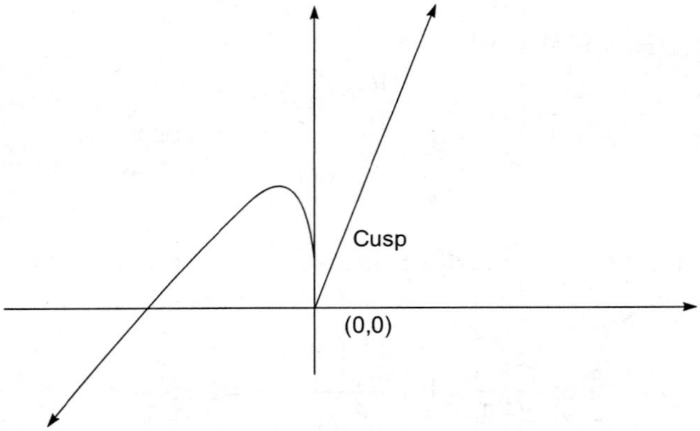

Figure 3.10

3.5.9 L' Hospital's Rule

To evaluate the indeterminate forms $\frac{0}{0}, \frac{\infty}{\infty}$ and other indeterminate forms, a rule was developed to deal with such forms of limit expression by a French mathematician L'Hospital and this rule is known as L'Hospital's rule.

L'Hospital's Rule: Let f and g be differentiable function and $g'(c) = 0$ on an open interval containing c (except possibly at c). If

$$\lim_{x\to c} f(x) = 0 \text{ and } \lim_{x\to c} g(x) = 0,$$

or that

$$\lim_{x\to c} f(x) = \pm\infty \text{ and } \lim_{x\to c} g(x) = \pm\infty.$$

(in other words, if the $\lim_{x\to c}\frac{f(x)}{g(x)}$ has indeterminate form of $\frac{0}{0},\frac{\infty}{\infty}$).

Then,

$$\lim_{x\to c}\frac{f(x)}{g(x)}=\lim_{x\to c}\frac{f'(x)}{g'(x)}$$

if the limit on right side exists (or is ∞ or $-\infty$).

EXAMPLE 18 Evaluate the following:

(i) $\lim_{x\to 0}\frac{\sin x}{x}$. (ii) $\lim_{x\to 1}\frac{x^2-1}{x-1}$.

Solution:

(i) Here, $\lim_{x\to 0}\sin x=0$ and $\lim_{x\to 0}x=0$, so that $\lim_{x\to 0}\frac{\sin x}{x}$ is of the indeterminate form $\frac{0}{0}$.

Now, apply L'Hospital's rule

$$\lim_{x\to 0}\frac{\sin x}{x}=\lim_{x\to 0}\frac{\frac{d}{dx}(\sin x)}{\frac{d}{dx}(x)}=\lim_{x\to 0}\cos x=1.$$

(ii) Again, we have $\lim_{x\to 0}\frac{x^2-1}{x-1}$ is of the indeterminate form $\frac{0}{0}$. Now apply L'Hospital's rule

$$\lim_{x\to 1}\frac{x^2-1}{x-1}=\lim_{x\to 1}\frac{\frac{d}{dx}\left(x^2-1\right)}{\frac{d}{dx}(x-1)}=\lim_{x\to 1}\frac{2x}{1}=2.$$

EXAMPLE 19 Evaluate the $\lim_{x\to 0}\frac{x\cos x-\log(1+x)}{x^2}$.

Solution: In this case also it is observed that $\lim_{x\to 0}\frac{x\cos x-\log(1+x)}{x^2}$ is of the indeterminate form $\frac{0}{0}$.

Now, apply L'Hospital's rule,

$$\lim_{x\to 0}\frac{x\cos x-\log(1+x)}{x^2}=\lim_{x\to 1}\frac{\frac{d}{dx}(x\cos x-\log(1+x))}{\frac{d}{dx}\left(x^2\right)}$$

$$=\lim_{x\to 1}\frac{\cos x-x\sin x-\frac{1}{1+x}}{2x}$$

Still $\frac{0}{0}$ form exists, apply the rule once again. This gives,

$$\lim_{x\to 1}\frac{\cos x - x\sin x - \frac{1}{1+x}}{2x} = \lim_{x\to 1}\frac{2\sin x - x\cos x + 1/(1+x)^2}{2} = \frac{1}{2}.$$

EXAMPLE 20 Evaluate $\lim_{x\to\infty}\frac{2x^2-3x+1}{3x^2+5x-2}$.

Solution: In this case also it is observed that $\lim_{x\to\infty}\frac{2x^2-3x+1}{3x^2+5x-2}$ is of the indeterminate form $\frac{\infty}{\infty}$.

Now, apply L'Hospital's rule. This gives,

$$\lim_{x\to\infty}\frac{2x^2-3x+1}{3x^2+5x-2} = \lim_{x\to\infty}\frac{\frac{d}{dx}(2x^2-3x+1)}{\frac{d}{dx}(3x^2+5x-2)} = \lim_{x\to 1}\frac{4x-3}{6x+5} = \frac{2}{3}.$$

Other Indeterminate Forms

In other than standard indeterminate form $\frac{0}{0}$ and $\frac{\infty}{\infty}$ such as $1^\infty, 0^0, \infty^0, \infty-\infty$ and $0.\infty$ can often be manipulated algebraically into standard forms $\frac{0}{0}$ and $\frac{\infty}{\infty}$ and then evaluated using L'Hospital's rule.

EXAMPLE 21 Evaluate the following limits:

(i) $\lim_{x\to 0} x\log x$

(ii) $\lim_{x\to\infty}\left[1+\frac{1}{x}\right]^x$

(iii) $\lim_{x\to 1}\left[1-x^2\right]^{1/\log(1-x)}$

(iv) $\lim_{x\to\infty}[x+e^x]^{\frac{2}{x}}$

(v) $\lim_{x\to 0^+}\left(\frac{1}{x}-\frac{1}{\sin x}\right)$

Solution:

(i) Here, $\lim_{x\to 0} x\log x$ is of $0.\infty$ type, therefore

$$\lim_{x\to 0} x\log x = \lim_{x\to 0}\left(\frac{\log x}{\frac{1}{x}}\right) \quad \left[\frac{\infty}{\infty}\right]$$

$$= \lim_{x\to 0}\left(\frac{\frac{1}{x}}{\frac{1}{x^2}}\right) \quad \text{(Apply L – Rule)}$$

$$= \lim_{x\to 0}(-x) = 0.$$

(ii) Here, $\lim_{x\to\infty}\left(1+\frac{1}{x}\right)^x$ is of 1^∞ type.

Suppose, $L = \lim_{x\to\infty}\left(1+\frac{1}{x}\right)^x$

Taking the logarithm on both the sides, we get

$$\log L = \log\left[\lim_{x\to+\infty}\left(1+\frac{1}{x}\right)^x\right]$$

$$= \lim_{x\to+\infty} x\log\left(1+\frac{1}{x}\right)$$

$$= \lim_{x\to+\infty}\frac{\log\left(1+\frac{1}{x}\right)}{1/x} \quad \left(\text{From}\left(\frac{0}{0}\right)\right)$$

$$= \lim_{x\to+\infty}\frac{\frac{1}{\left(1+\frac{1}{x}\right)}\left(-\frac{1}{x^2}\right)}{-\frac{1}{x^2}} \quad \text{(L'Hospital's rule)}$$

$$= \lim_{x\to+\infty}\frac{1}{1+\frac{1}{x}} = \frac{1}{1+0} = 1.$$

Therefore, $\ln L = 1$, i.e., $L = e$.

(iii) $\lim_{x\to 1}\left[1-x^2\right]^{1/\log(1-x)}$ is of 0^0 type.

Let $L = \lim_{x\to 1}\left(1-x^2\right)^{\frac{1}{\log(1-x)}}$.

Taking the logarithm on both the sides, we get

$$\log L = \lim_{x\to 1}\left(\frac{1}{\log(1-x)}\right)\log\left(1-x^2\right) \qquad (0.\infty \text{ form})$$

$$= \lim_{x\to 1}\left(\frac{\log(1-x^2)}{\log(1-x)}\right) \qquad \left(\frac{\infty}{\infty}\text{ form}\right)$$

$$= \lim_{x\to 1}\frac{{}^{-2x}\!/\!_{(1-x^2)}}{{}^{-1}\!/\!_{(1-x)}} \qquad \text{[applying L'Hopsital rule]}$$

$$= \lim_{x\to 1}\left(\frac{2x}{1+x}\right) = 1.$$

Therefore, $L = e^1 = 1$.

(iv) $\lim\limits_{x\to\infty}[x+e^x]^{\frac{2}{x}}$ is of the type ∞^0.

Let $L = \lim\limits_{x\to\infty}\left(x+e^x\right)^{\frac{2}{x}}$.

Taking logarithm on both sides, we get

$$\log \text{L} = \lim_{x\to\infty}\left(\frac{2}{x}\right)\log\left(x+e^x\right) \qquad [0.\infty \text{ form}]$$

$$=2\lim_{x\to\infty}\frac{\log\left(x+e^x\right)}{x} \qquad [\infty/\infty \text{ form}]$$

Applying L'Hospital's rule, we get

$$= 2\lim_{x\to\infty}\left\{\frac{{}^{1+e^x}\!/\!_{x+e^x}}{1}\right\} = 2\lim_{x\to\infty}\left\{{}^{1+e^x}\!/\!_{x+e^x}\right\} \quad \left(\text{still } \frac{\infty}{\infty}\text{ form}\right)$$

Again, applying L'Hospital's rule, we get

$$= 2\lim_{x\to\infty}\left\{{}^{e^x}\!/\!_{1+e^x}\right\} \qquad \left(\text{still } \frac{\infty}{\infty}\text{ form}\right)$$

Again, applying L'Hospital's rule, we get

$$= 2\lim_{x\to\infty}\left\{{}^{e^x}\!/\!_{e^x}\right\} = 2.1 = 1.$$

(v) $\lim\limits_{x\to 0^+}\left(\frac{1}{x}-\frac{1}{\sin x}\right)$ is of the type $\infty-\infty$.

Rewriting the above expression, we have

$$\lim_{x\to 0}\left(\frac{1}{x}-\frac{1}{\sin x}\right)=\lim_{x\to 0}\left(\frac{\sin x - x}{x\sin x}\right) \qquad \left(\frac{0}{0}\text{form}\right)$$

Applying L'Hospital's rule, we get

$$\lim_{x\to 0}\left(\frac{\sin x - x}{x\sin x}\right)=\lim_{x\to 0}\left(\frac{\cos x - 1}{x\cos x+\sin x}\right) \left(\text{still}\frac{0}{0}\text{form}\right)$$

$$=\lim_{x\to 0}\left(\frac{\sin x}{x\sin x + 2\cos x}\right)=\frac{0}{2}=0.$$

3.5.10 Optimization in Business, Economics and Life Sciences

Mathematical model of system is a symbolic representation involving mathematical concepts and symbols. Thus, there is a translation of the given real world problem to a mathematical formulation. Once the given problem is transformed to mathematical symbols, it gets divorced from the real world and is treated in mathematical terms only. Mathematical modeling is the process by which a problem, as it appears in the real world, is interpreted and represented in mathematical terms.

In this section, we will use mathematical modelling and calculus to solve optimization problems.

3.5.10.1 Economics

Maximizing Profit

If $C(x)$ (*cost function*) is the cost of producing x units of a certain product, then the **marginal cost** is the rate of change of C with respect to x.

$$\text{Marginal cost} = \frac{d}{dx}C(x).$$

Let $p(x)$ be the price per unit that the manufacturing firm can charge if it sells x units. Then, p is called the **demand function** (or **price function**) and we would expect it to be a decreasing function of x.

If x units of the product are sold and the price per unit is $p(x)$, then the total revenue is

$$R(x)=x\,p(x)$$

and R is called the **revenue function**. The derivative of the revenue function R with respect to x is called the marginal revenue function. In other words, marginal revenue function is the rate of change of R with respect to number of units sold.

$$\text{Marginal revenue} = \frac{d}{dx}R(x).$$

If we assume that x units are sold, then the total profit is

$$P(x) = R(x) - C(x)$$

and P is called the profit function. The marginal profit function is the derivative of profit function.

$$\text{Marginal profit function} = \frac{d}{dx}P(x).$$

If $P'(x) = 0$, when $R'(x) = C'(x)$, and by using economics arguments we can show that the maximum occurs at corresponding critical point.

Marginal profit = Marginal Revenue–Marginal cost,

i.e., $$P'(x) = R'(x) - C'(x).$$

"Profit is maximized when marginal revenue equals marginal cost."

EXAMPLE 22 Suppose the total cost (in dollars) of manufacturing x units of a certain commodity are

$$C(x) = \frac{1}{8}x^2 + 5x + 98$$

and all units can be sold at a price of $p(x) = \frac{1}{2}(75 - x)$ dollar per unit. Determine the price that corresponds to the maximum profit.

Solution: Here, the marginal cost is

$$C'(x) = \frac{1}{4}x + 5.$$

The revenue is

$$R(x) = x\,p(x) = \frac{1}{2}\left(75x - x^2\right).$$

The marginal revenue is

$$R'(x) = \frac{75}{2} - x.$$

The marginal profit is

$$P'(x) = R'(x) - C'(x) = \left(\frac{75}{2} - x\right) - \left(\frac{1}{4}x + 5\right) = \frac{65}{2} - \frac{5}{4}x.$$

Profit is maximum, when marginal profit is zero, i.e., $P'(x) = 0$

$$\Rightarrow \qquad \frac{65}{2} - \frac{5}{4}x = 0 \Rightarrow x = 26.$$

Thus, the price that corresponds to the maximum profit is

$$p(26)=\frac{1}{2}(75-26)=24.5.$$

3.5.10.2 Business Management

EXAMPLE 23 Suppose you own a rare book whose value in t years from now is modeled as $300e^{\sqrt{3t}}$. If the prevailing rate of interest remains constant 8% compounded, when will be the most advantageous time to sell it?

Solution: If P is the sum of money that will be available in future t years from now and the interest is compounded continuously at the rate $100r\%$, then its present value will be

$$P=Ve^{-rt},$$

where $V=$ the value after t years.

But $V=300e^{\sqrt{3t}}$ and $r=0.08$, then

$$P(t)=300e^{\sqrt{3t}}e^{-0.08t}.$$

For the maximum value, $P'(t)=0$, we have

$$300e^{\sqrt{3t}-0.08t}\left(-0.08+\frac{\sqrt{3}}{2\sqrt{t}}\right)=0$$

$$\Rightarrow \qquad \left(-0.08+\frac{\sqrt{3}}{2\sqrt{t}}\right)=0\Rightarrow t=\frac{3}{4}\left(\frac{1}{0.08}\right)^2=117.1875.$$

3.5.10.3 Physiology

EXAMPLE 24 If $c(t)$ is the concentration in blood at time t of drug injected in blood stream intramuscularly and concentration of drug is modeled by the following equation:

$$c(t)=\frac{k}{b-a}\left(e^{-at}-e^{-bt}\right),\quad t\geq 0 \quad (\text{from paper written by E. Heinz,1949})$$

where a,b $(b>a)$ and k are positive constants that depend on drugs. Find out when the largest concentration occurs in blood stream? Also, determine the concentration of drug as $t\to\infty$.

Solution: The concentration of the drug is given by the following equation:

$$c(t)=\frac{k}{b-a}\left(e^{-at}-e^{-bt}\right),\quad t\geq 0.$$

Then,

$$c'(t) = \frac{k}{b-a}\left(-ae^{-at} + be^{-bt}\right) = \frac{k}{b-a}\left(be^{-bt} - ae^{-at}\right).$$

Let t^* be the time when maximum concentration occurs. For extremum $c'\left(t^*\right) = 0$

$$\frac{k}{b-a}\left(be^{-bt^*} - ae^{-at^*}\right) = 0$$

$$\Rightarrow \qquad be^{-bt^*} = ae^{-at^*}$$

$$\Rightarrow \qquad \frac{b}{a} = e^{(b-a)t^*}$$

$$\Rightarrow \qquad t^* = \frac{1}{b-a}\log\left(\frac{b}{a}\right).$$

Now, we need to show that $c''(t) < 0$ for maximum value at t^*.

$$\Rightarrow \qquad c''(t) = \frac{k}{b-a}\left(-b^2e^{-bt} + a^2e^{-at}\right) = \frac{k}{b-a}\left(a^2e^{-at} - b^2e^{-bt}\right).$$

For $t = t^* = \frac{1}{b-a}\log\left(\frac{b}{a}\right)$, we have

$$c''\left(\frac{1}{b-a}\log\left(\frac{b}{a}\right)\right) = \frac{k}{b-a}\left(a^2e^{-a\left(\frac{1}{b-a}\log\left(\frac{b}{a}\right)\right)} - b^2e^{-b\left(\frac{1}{b-a}\log\left(\frac{b}{a}\right)\right)}\right)$$

$$= \frac{k}{b-a}\left(a^2\left(\frac{b}{a}\right)^{-\left(\frac{a}{b-a}\right)} - b^2\left(\frac{b}{a}\right)^{-\left(\frac{b}{b-a}\right)}\right)$$

$$= \frac{ka^2}{b-a}\left(\left(\frac{b}{a}\right)^{-\left(\frac{a}{b-a}\right)} - \left(\frac{b}{a}\right)^2\left(\frac{b}{a}\right)^{-\left(\frac{b}{b-a}\right)}\right)$$

$$= \frac{ka^2}{b-a}\left(\left(\frac{b}{a}\right)^{-\left(\frac{a}{b-a}\right)} - \left(\frac{b}{a}\right)\left(\frac{b}{a}\right)^{-\left(\frac{a}{b-a}\right)}\right)$$

$$= -ka\left(\frac{b}{a}\right)^{-\left(\frac{a}{b-a}\right)}.$$

Since $b > a$, $\frac{b}{a} > 1$, then $c''\left(t^*\right) < 0$. This means at $t^* = \frac{1}{b-a}\log\left(\frac{b}{a}\right)$, largest concentration occurs. To know the behaviour of concentration as $t \to \infty$, we compute the limit as

$$\lim_{t\to\infty} c(t) = \lim_{t\to\infty} \frac{k}{b-a}\left(e^{-at} - e^{-bt}\right) = 0$$

This tell us that concentration of drugs will decrease to zero as time tends to infinity (in long run). The graph of the concentration of drug shown in Figure 3.11.

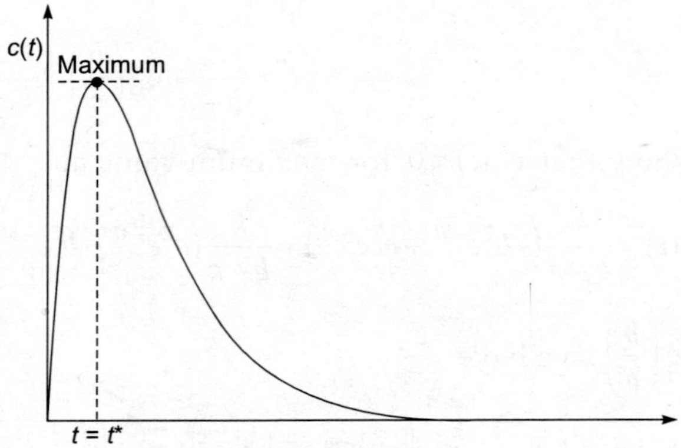

Figure 3.11

Exercise

1. Show that the function $f : R \to R$ defined as

$$f(x) = \begin{cases} x^2 \sin\frac{1}{x}, & \text{if } x \neq 0 \\ 0, & \text{if } x = 0 \end{cases}$$

is differentiable at $x = 0$.

2. Examine the differentiability of the following function at the point indicated in front of each function:

(a) $f(x) = \begin{cases} 2 + x, \text{if } x \leq 2 \\ 6 - x, \text{if } x > 2 \end{cases}$ at $x = 2$

(b) $f(x) = (x-3)^{\frac{1}{3}}$ at $x = 3$

$$f(x)=\begin{cases}\dfrac{x-1}{2x^2-7x+5}, & \text{if } x\neq 1\\ -\dfrac{1}{3}, & \text{if } x=1\end{cases}$$

3. Find the slope of the graph of the function $f(x)=x^{\frac{1}{3}}$ at point $(1,1)$.
4. If the distance travelled by a particle in time t is given by $s(t)=64t-8t^2$, then find the velocity and acceleration at any time t.
5. An object moving on the x-axis has position $x(t)=t^3-9t^2+24t+20$ after t seconds. What is the total distance traveled by the object during the first 8 seconds?
6. Find the derivative of the following functions from the definition.
   (a) x^n
   (b) $\cos x$
   (c) $x^2 \cos x$
   (d) $\sin\left(x^2+1\right)$
7. Differentiate the following functions with respect to x.
   (a) $\cos^4 x \cos x^4$
   (b) $e^{x+x\log x}$
   (c) $\sin\sin x$
   (d) $\left(x^2-x+1\right)^2$
   (e) $\cos\left(a^3+x^3\right)$
   (f) $\tan\tan\tan x$
   (g) $\exp\exp\exp x$
   (h) $\log\log x$
   (i) $\dfrac{1}{\left(x^4+1\right)^3}$
8. Give geometrical meaning and physical significance of the derivative.
9. Find the derivative of the function $f(x)=\left(x^2+1\right)e^x \tan x$ with respect to x.
10. Find the derivative of the following function with respect to x:
   (a) $(\cos x)^{\sin x}$
   (b) $(x^2)^{(x^2)^{\left(x^2\right)^{\left(x^2\right)^{\cdots\cdots\infty}}}}$

(c) $(\sin x)^{(\sin x)^{(\sin x)^{(\sin x)^{\cdots\cdots\infty}}}}$

(d) $x^{\sin^{-1} x}$

(e) $e^{x \sin x^2} + (\cosh x)^{\cosh x}$

11. Find the derivatives of the following parametric curves:
    (a) $x = a\cos t, y = a\sin t$
    (b) $x = at, y = a^3 t^3$
    (c) $x = t^2 + t + 1, y = \sin\frac{\pi}{2}t + \cos\frac{\pi}{2}t$

12. If $\sqrt{x} + \sqrt{y} = 4$, find $\frac{dy}{dx}$.

13. If $(a^x)^{(a^x)^{(a^x)^{(a^x)^{\cdots\cdots\infty}}}}$, then show that

$$\frac{dy}{dx} = \frac{y^2 \log_e x}{x(1 - y\log_e x \log_e y)}.$$

14. Show that the function $f(x) = x^3 - 27x$ satisfies the Rolle's theorem on interval $\left[-3\sqrt{x}, 3\sqrt{3}\right]$.

15. Show that the function $f(x) = x^2 - 3x + 6$ satisfies the Rolle's theorem on interval $[1,2]$. Find the point at which the derivative is zero.

16. Show that the following function satisfies all the requirements of the Rolle's theorem on the interval indicated in front of each function. Also, find the points at which the derivatives are zero.
    (a) $f(x) = x(x-1), \quad [0,1]$.
    (b) $f(x) = x^2 - 5x + 11, \ [1,4]$.
    (c) $f(x) = (x-1)^2 (x-2)^2, [1,2]$.

17. Show that the following functions *f* satisfy all the requirements of the mean value theorem on the interval indicated in front of each function. Also, find the points c in between interval $[a,b]$ such that $f'(c) = \frac{f(b) - f(a)}{b - a}$.

    (a) $f(x) = 2x^2 + 1, \quad [0,2]$.

    (b) $f(x) = x^5 + 3, \ [2,4]$.

(c) $f(x)=\frac{1}{x+1}, [0,2]$.

(d) $f(x)=e^{x}, [0,1]$.

(e) $f(x)=\log x, \left[\frac{1}{2},2\right]$.

18. Show that the function $f(x)=x^3-3x^2+2x$ satisfies the mean value theorem in $\left[0,\frac{1}{2}\right]$.

19. Using the mean value theorem, prove the following inequalities:

(a) $x-1>\log x>\frac{x-1}{x}$ for $x>1$.

(b) $x>\tan^{-1} x>\frac{x}{1+x^2}$ for $x>0$.

(c) $\frac{\tan x}{x}>\frac{x}{\sin x}$ for $0<x<\frac{\pi}{2}$.

20. Find the critical numbers for the following functions:

(a) $f(x)=(x-1)^2(x+2)$.

(b) $f(x)=x^3-3x^2$.

(c) $f(x)=x^5-x^4$.

(d) $f(x)=|x-3|$.

21. Whether the following functions f defined on the interval indicated in front of each function yield a minimum, maximum or neither.

(a) $f(x)=(x-1)^2(x+2),[-1,1]$.

(b) $f(x)=x^3-3x^2,[-1,3]$.

(c) $f(x)=x^5-x^4,[-1,1]$.

(d) $f(x)=|x-3|,[-4,4]$.

(e) $f(x)=x^4-8x^3+18x^2+60x-8,[1,5]$.

22. Examine the following functions for concavity (where the graph is concave up or concave down) and find the inflexion points:

(a) $f(x)=4+3x-x^3$

(b) $f(x)=\left(x^3+x\right)^2$

(c) $f(x)=x^3+3x^2$

(d) $f(x)=x+\frac{4}{x}$

(e) $f(x)=\frac{x}{x^2+1}$

23. Use the first-derivative test to classify each of the critical numbers of the following functions as a relative minimum, a relative maximum or neither

(a) $f(x)=x^3-3x^2-9x+1$

(b) $f(x)=x-2\sin x$, for $0\le x\le 2\pi$

(c) $f(x)=\left(x^3-3x+1\right)$

(d) $f(x)=\frac{e^{-x^2}}{3-2x}$

24. Find the critical points of the following functions and use the second derivative test to classify each of the critical numbers as a relative minimum, a relative maximum or neither.

(a) $f(x)=2x^3+3x^2-12x+11$

(b) $f(x)=(x-1)(x-2)(x-3)$

(c) $f(x)=4\cos x+3\sin x$

(d) $f(x)=\left(x^2-3x+1\right)e^{-x}$

(e) $f(x)=x^5-5x^4+5x^3-10$

(f) $f(x)=(x+1)^4(x-2)^5$

(g) $f(x)=(x-a)^4(x-b)^3$

25. Show that $f(x)=\frac{x}{1+x\tan x}$ is maximum when $x=\cos x$.

26. Use calculus to prove that for constants a, b and c, the vertex (relative extremum) of the quadratic function $y=ax^2+bx+c\,(a\ne 0)$ occurs at $x=-\frac{b}{2a}$.

27. Show that if $f(x)=a\log x+bx^2+x$ has its relative extremum values at $x=-1$ and $x=2$, then $a=2$ and $b=-\frac{1}{2}$.

28. Evaluate $\lim_{x \to -\infty} \frac{95x^2 + 57x + 30}{x^5 - 1000}$

29. Sketch the graph of the following functions using vertical and horizontal asymptotes:

(a) $f(x) = \frac{3x-5}{x-2}$

(b) $f(x) = \frac{x^2 - x - 2}{x-3}$

(c) $f(x) = \frac{3x+5}{7-x}$

(d) $f(x) = \frac{x^2 - 4x - 5}{2x^2 + x - 10}$

30. Evaluate the following limits:

(a) $\lim_{x \to 0} \frac{1 - \cos x^2}{x^2 \sin x^2}$

(b) $\lim_{x \to 2} \frac{x^7 - 128}{x^3 - 8}$

(c) $\lim_{x \to 0} \frac{1 - \cos x}{\sec x}$

(d) $\lim_{x \to 0} \frac{1 - \cos x^2}{\sin^3 x}$

(e) $\lim_{x \to 0} \frac{\sin ax}{\cos bx}, ab \neq 0$

(f) $\lim_{x \to \frac{\pi}{2}} \frac{\sin 2x \cos x}{x \sin 4x}$

(g) $\lim_{x \to 0} x^2 \sin \frac{1}{x}$

(h) $\lim_{x \to 0} \frac{e^x \sin x - x - x^2}{x^2 + x \log(1-x)}$

(i) $\lim_{x \to 0} \left(\frac{1}{e^x - 1} - \frac{1}{x} \right)$

(j) $\lim_{x \to 0} \left(\frac{\tan x}{x} \right)^{\frac{1}{x^2}}$

(k) $\lim_{x \to 0} (1+x)^{\frac{1}{x}}$

(l) $\lim_{x \to 1-} \left(\frac{\log(1-x)}{\cot \pi x} \right)$

(m) $\lim_{x \to \frac{\pi}{2}+} \left(\frac{\log\left(x - \frac{\pi}{2}\right)}{\tan x} \right)$

31. A manufacturer estimates that when x units of a particular commodity are produced each month, the total cost will be $c(x) = \frac{1}{8}x^2 + 4x + 200$ (in dollars) and all units can be sold at a price of $p(x) = 49 - x$ dollars per unit. Determine the price that corresponds to the maximum profit.
32. A manufacturer estimates that when x units of a particular commodity are produced each month, the total cost will be $c(x) = \frac{1}{8}x^2 + 4x + 200$ (in dollars) and all units can be sold at a price of $p(x) = 49 - x$ dollar per unit. Determine the level of production at which the average cost is minimized.
33. A manufacturer estimates that when x units of a particular commodity are produced each month, the total cost will be $c(x) = \frac{1}{8}x^2 + 5x + 98$ (in dollars) and all units can be sold at a price of $p(x) = \frac{1}{2}(75 - x)$ dollar per unit. Determine the price that corresponds to the maximum profit and find the level of production at which the average cost is minimized.
34. A manufacturer estimates that when x units of a particular commodity are produced each month, the total cost will be $c(x) = \frac{1}{5}(x + 30)$ (in dollars) and all units can be sold at a price of $p(x) = \frac{70 - x}{x + 30}$ dollar per unit. Determine the price that corresponds to the maximum profit and find the level of production at which the average cost is minimized.

35. It is estimated that t years from now the population of a certain country will be

$$p(t) = \frac{160}{1 + 8e^{-0.01t}}$$

million. When will the population be growing most rapidly?

36. In example, we showed that the Heinz concentration function

$$c(t) = \frac{k}{b-a}\left(e^{-at} - e^{-bt}\right)$$

for $b > a$ has exactly one critical number, namely, $t_c = \frac{1}{b-a}\log\left(\frac{b}{a}\right)$. Find the $c''(t)$ and show that a relative maximum for concentration $c(t)$ occurs at time $t = t_c$ by using second-derivative test.

37. Find the area of the largest rectangle that can be inscribed in a semicircle of radius r.

38. A store has been selling 200 unit of a product a week at $\$350$ each. A market survey shows that for each $\$10$ rebate offered to buyers, the numbers of units will increase by 20 a week. Find the demand function and revenue function. How large a rebate should offer to maximize its revenue?

4 CHAPTER

Successive Differentiation

Derivatives of the second order and generally, of higher orders play an important role in mathematics and its applications in different areas of mechanics, physics and engineering. This shows the existence and importance of higher order derivatives than the first order derivatives. Suppose that a function $y = f(x)$ defined on an interval has the derivatives $f'(x)$ with respect to x. If the function $f'(x)$ is differentiable, then its derivative is called the second order derivative. Similarly, third and higher order derivatives can be determined. In this chapter, we discuss the higher order derivatives of several types of functions.

4.1 DEFINITION AND EXAMPLES

The derivative of the nth order $f^n(x)$ is the derivative of the derivative of the $(n-1)^{th}$ order:

$$f^n(x) = \left[f^{n-1}(x)\right]' = \lim_{\Delta x \to 0} \frac{f^{n-1}(x+\Delta x) - f^{n-1}(x)}{\Delta x}$$

provided the limit exists. The higher order derivatives of $y = f(x)$ denoeted by $f'(x), f''(x), f'''(x), \ldots, f^n(x), ..$ are called its successive derivatives with respect to x. We can use various notations to denote the successive derivatives as

$$y^{(n)}, f^n(x), D^n(y), \frac{d^n y}{d^n x} \text{ or } y_n(x).$$

EXAMPLE 1 Find the fifth order derivatives of $y = f(x) = x^4 + 3x + 5$ at $x = 1$.

Solution: We are given that,

$$y = x^4 + 3x + 5.$$

$$y' = 4x^3 + 3 \Rightarrow y'(1) = 41^3 + 3 = 7$$

$$y'' = 12x^2 \Rightarrow y''(1) = 12$$

$$y''' = 24x \Rightarrow y'''(1) = 24$$

$$y^{iv} = 24 \Rightarrow y^{\{iv\}}(1) = 24$$
$$y^{v} = 0 \Rightarrow y^{\{v\}}(1) = 0$$

EXAMPLE 2 Find the first four derivatives of $y = f(x) = e^{4x} + \cos 3x + x^4$.

Solution: We are given that,

$$y = e^{4x} + \cos 3x + x^3.$$
$$y' = 4e^{4x} - 3\sin 3x + 3x^2$$
$$y'' = 16e^{4x} - 9\cos 3x + 6x$$
$$y''' = 64e^{4x} + 27\sin 3x + 6$$
$$y^{iv} = 256e^{4x} + 81\cos x.$$

EXAMPLE 3 Find the first and third derivatives of $y = f(x) = x^2 \sin x$.

Solution: We have $y = x^2 \sin x$

$$y' = 2x\sin x + x^2 \cos x$$
$$y'' = 2\sin x + 2x\cos x + 2x\cos x - x^2 \sin x = 2\sin x + 4x\cos x - x^2 \sin x$$
$$y''' = 2\cos x - 4x\sin x + 4\cos x - 2x\sin x - x^2 \cos x = 6\cos x - 6x\sin x - x^2 \cos x$$

EXAMPLE 4 If $y = A\sin ax + B\cos ax$, then prove that $y'' + a^2 y = 0$.

Solution: If $y = A\sin ax + B\cos ax$, then we have

$$y' = Aa\cos ax - Ba\sin ax.$$
$$\Rightarrow y'' = -Aa^2 \sin ax - Ba^2 \cos ax = -a^2(A\sin ax + B\cos ax) = -a^2 y$$
$$\Rightarrow y'' + a^2 y = 0.$$

EXAMPLE 5 If $y = \cos\{m\sin^{-1} x\}$, then prove that

$$\left(1 - x^2\right)y'' - xy' + m^2 y = 0.$$

Solution: Differentiating y with respect to x, we have

$$y' = -\sin\{m\sin^{-1} x\}m\frac{1}{\sqrt{1-x^2}}.$$
$$\Rightarrow y'\sqrt{1-x^2} = -m\sin\{m\sin^{-1} x\}.$$

Again differentiate w.r.t. x, we have

$$y''\sqrt{1-x^2} - \frac{xy'}{\sqrt{1-x^2}} = -m^2 \cos\{m\sin^{-1} x\}\frac{1}{\sqrt{1-x^2}}$$

$$\Rightarrow y''\left(1-x^2\right)-xy'=-m^2\cos\{m\sin^{-1}x\}$$

$$\Rightarrow y''\left(1-x^2\right)-xy'+m^2y=0.$$

EXAMPLE 6 If $y=\left(\tan^{-1}x\right)^2$, then prove that

$$\left(1+x^2\right)^2y''+2x\left(1+x^2\right)y'-2=0.$$

Solution: Here, $y=\left(\tan^{-1}x\right)^2$, then

$$y'=2\tan^{-1}x\frac{1}{1+x^2}\Rightarrow\left(1+x^2\right)y'=2\tan^{-1}x.$$

Differentiating again w.r.t. x, we have

$$\left(1+x^2\right)y''+2xy'=2\frac{1}{1+x^2}.$$

$$\Rightarrow\left(1+x^2\right)^2y''+2x\left(1+x^2\right)y'=2$$

$$\Rightarrow\left(1+x^2\right)^2y''+2x\left(1+x^2\right)y'-2=0.$$

4.2 HYPERBOLIC FUNCTION

Function e^x can be expressed as

$$e^x=\underbrace{\frac{e^x+e^{-x}}{2}}+\underbrace{\frac{e^x-e^{-x}}{2}}.$$

The first part of the expression is an even function and the second part of the expression is an odd function. These even and odd functions are called hyperbolic functions. The even function is called the *hyperbolic cosine* of x and the odd function is called the *hyperbolic sine* of x. They are denoted by

$$\sinh x=\frac{e^x-e^{-x}}{2}\quad\text{and}\quad\cosh x=\frac{e^x+e^{-x}}{2}.$$

From above two *hyperbolic* functions, we can define four more functions as

Hyperbolic tangent $\quad\tanh x=\dfrac{\sinh x}{\cosh x}=\dfrac{e^x-e^{-x}}{e^x+e^{-x}}.$

Hyperbolic cotangent $\quad\coth x=\dfrac{\cosh x}{\sinh x}=\dfrac{e^x+e^{-x}}{e^x-e^{-x}}.$

Hyperbolic secant $\operatorname{sech}x = \dfrac{1}{\cosh x} = \dfrac{2}{e^x + e^{-x}}.$

Hyperbolic cosecant $\operatorname{csch}x = \dfrac{1}{\sinh x} = \dfrac{2}{e^x - e^{-x}}.$

4.2.1 Hyperbolic Identities

The hyperbolic functions satisfy various identities that are similar to identities of trigonometric functions. There are some useful hyperbolic identities as follows:

1. $\cosh^2 x - \sinh^2 x = 1.$

Proof $\cosh^2 x - \sinh^2 x = \left(\dfrac{e^x + e^{-x}}{2}\right)^2 - \left(\dfrac{e^x - e^{-x}}{2}\right)^2$

$$= \frac{1}{4}\left(e^{2x} + e^{-2x} + 2\right) - \frac{1}{4}\left(e^{2x} + e^{-2x} - 2\right)$$

$$= 1.$$

2. $1 - \tanh^2 x = \operatorname{sech}^2 x.$

Proof $1 - \tanh^2 x = 1 - \left(\dfrac{e^x - e^{-x}}{e^x + e^{-x}}\right)^2$

$$= \left(1 + \frac{e^x - e^{-x}}{e^x + e^{-x}}\right)\left(1 - \frac{e^x - e^{-x}}{e^x + e^{-x}}\right)$$

$$= \left(\frac{e^x + e^{-x} + e^x - e^{-x}}{e^x + e^{-x}}\right)\left(\frac{e^x + e^{-x} - e^x + e^{-x}}{e^x + e^{-x}}\right)$$

$$= \left(\frac{2e^x}{e^x + e^{-x}}\right)\left(\frac{2e^{-x}}{e^x + e^{-x}}\right) = \left(\frac{2}{e^x + e^{-x}}\right)^2$$

$$= (\operatorname{sech}x)^2$$

$$= \operatorname{sech}^2 x.$$

3. $\coth^2 x - 1 = \operatorname{csch}^2 x.$

Proof If we divide $\cosh^2 x - \sinh^2 x = 1$ by $\sinh^2 x$, we get

$$\coth^2 x - 1 = \operatorname{csch}^2 x.$$

Similarly, we can also prove the following identities that are shown in Table 4.1.

Table 4.1 Some hyperbolic identities

$\cosh x + \sinh x = e^x$
$\cosh x - \sinh x = e^{-x}$
$\cosh(-x) = \cosh x$
$\sinh(-x) = -\sinh x$
$\sinh 2x = 2\sinh x \cosh x$
$\cosh 2x = \cosh^2 x + \sinh^2 x$
$\cosh 2x = 2\cosh^2 x - 1$
$\cosh 2x = 2\sinh^2 x + 1$
$\sinh(x+y) = \sinh x \cosh y + \cosh x \sinh y$
$\cosh(x+y) = \cosh x \cosh y + \sinh x \sinh y$
$\sinh(x-y) = \sinh x \cosh y - \cosh x \sinh y$
$\cosh(x-y) = \cosh x \cosh y - \sinh x \sinh y$

4.2.2 Graphs of Hyperbolic Functions

The graphs of the hyperbolic functions can be generated by combining the graphs of the functions $y = \frac{e^x}{2}$ and $y = \frac{e^{-x}}{2}$. Figure 4.1 shows the graphs of hyperbolic functions.

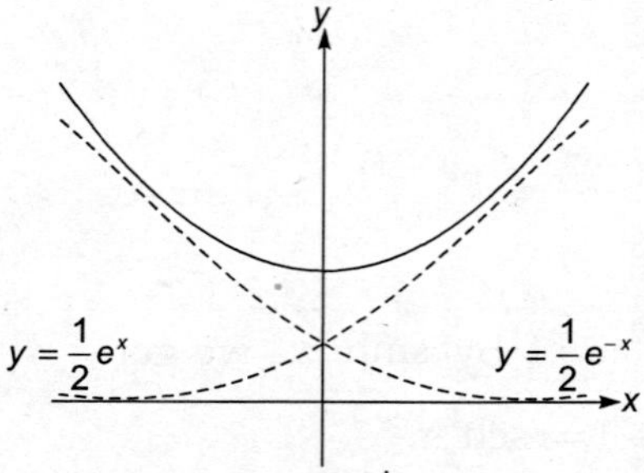

$y = \cosh x$

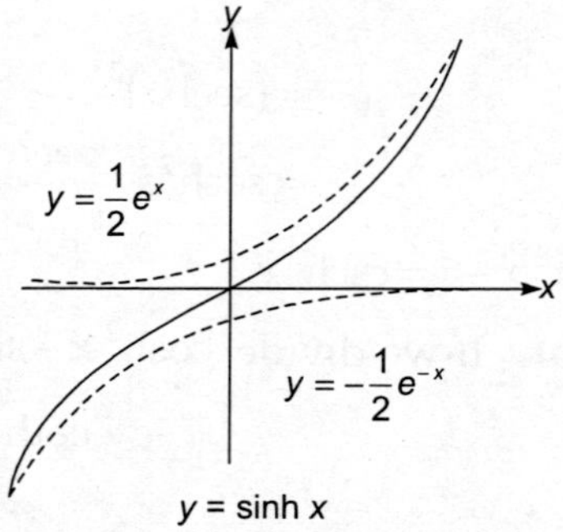

$y = \sinh x$

(Contd.)

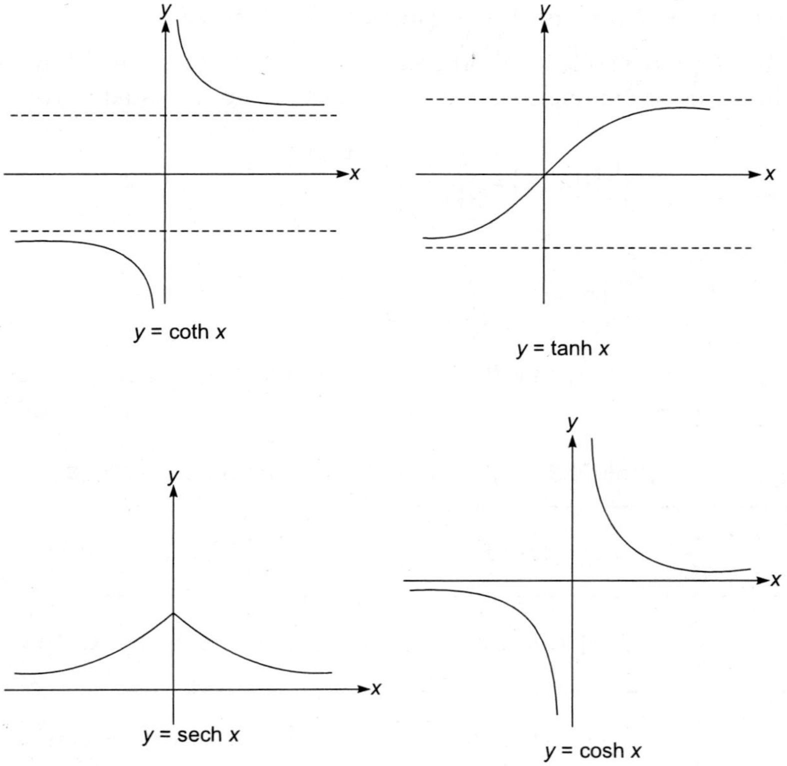

Figure 4.1 Graphs of hyperbolic functions

The parametric equations

$$x = \cosh t, \qquad y = \sinh t \qquad t \in (-\infty, \infty)$$

represent a portion of the curve $x^2 - y^2 = 1$, as may be satisfied by

$$x^2 - y^2 = \cosh^2 t - \sinh^2 t = 1 \text{ and } x = \cosh t > 0 .$$

Then the curve $x^2 - y^2 = 1, x > 0$ is the right half of a larger curve called unit hyperbola (Figure 4.2). ***"This is the reason why the functions are called hyperbolic functions"***.

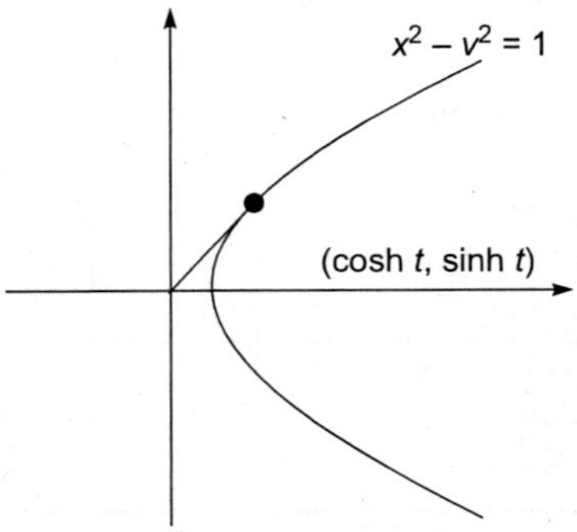

Figure 4.2

4.2.3 Derivatives Formula for Hyperbolic Functions

The derivative of hyperbolic sine and cosine function can be obtained by expressing these functions in terms of exponential function and then apply usual derivative formulas:

$$\frac{d}{dx}[\sinh x]=\frac{d}{dx}\left(\frac{e^x-e^{-x}}{2}\right)=\frac{e^x+e^{-x}}{2}=\cosh x.$$

$$\frac{d}{dx}[\cosh x]=\frac{d}{dx}\left(\frac{e^x+e^{-x}}{2}\right)=\frac{e^x-e^{-x}}{2}=\sinh x.$$

Similarly, we can obtain the derivatives of remaining hyperbolic functions that are shown in Table 4.2.

Table 4.2 Derivatives of the hyperbolic functions

$\frac{d}{dx}[\sinh x]=\cosh x$	$\frac{d}{dx}[\operatorname{sech} x]=-\operatorname{sech} x\tanh x$
$\frac{d}{dx}[\cosh x]=\sinh x$	$\frac{d}{dx}[\operatorname{csch} x]=-\operatorname{csch} x\coth x$
$\frac{d}{dx}[\tanh x]=\operatorname{sech}^2 x$	$\frac{d}{dx}[\coth x]=-\operatorname{csch}^2 x$

If $\sinh y = x$, then y is called the *inverse hyperbolic* sine function of x and is denoted as

$y=\sinh^{-1} x$.

The other hyperbolic functions, namely $\cosh^{-1} x$, $\tanh^{-1} x$, $\coth^{-1} x$, $\operatorname{sech}^{-1} x$ and $\operatorname{csch}^{-1} x$ are defined in similar manner. Because the hyperbolic functions are expressible in terms of exponential functions, then the inverse hyperbolic functions are expressible in terms of natural logarithms. Table 4.3 shows that the following relationships hold for x in the domain of inverse hyperbolic functions and derivatives of inverse hyperbolic functions:

Table 4.3 Inverse hyperbolic functions and their derivatives

$\sinh^{-1} x=\log\left(x+\sqrt{x^2+1}\right)$	$\frac{d}{dx}[\sinh^{-1} x]=\frac{1}{\sqrt{1+x^2}}$
$\tanh^{-1} x=\frac{1}{2}\log\left(\frac{1+x}{1-x}\right)$	$\frac{d}{dx}[\cosh^{-1} x]=\frac{1}{\sqrt{x^2-1}}, x>1$

(Contd.)

$\operatorname{sech}^{-1}x=\log\left(\frac{1+\sqrt{1-x^2}}{x}\right)$	$\frac{d}{dx}[\tanh^{-1}x]=\frac{1}{1-x^2},\lvert x\rvert<1$
$\cosh^{-1}x=\log\left(x+\sqrt{x^2-1}\right)$	$\frac{d}{dx}[\coth^{-1}x]=\frac{1}{1-x^2},\lvert x\rvert>1$
$\coth^{-1}x=\frac{1}{2}\log\left(\frac{x+1}{x-1}\right)$	$\frac{d}{dx}[\operatorname{sech}^{-1}x]=\frac{1}{x\sqrt{1-x^2}},0<x<1$
$\operatorname{csch}^{-1}x=\log\left(\frac{1}{x}+\frac{\sqrt{1-x^2}}{\lvert x\rvert}\right)$	$\frac{d}{dx}[\operatorname{csch}^{-1}x]=\frac{1}{\lvert x\rvert\sqrt{1+x^2}},x\neq 0$

Solved Examples Based on Hyperbolic Functions

EXAMPLE 7 Find the derivative of the function $y=\cosh 2x$.

Solution: We are given that, $y=\cosh 2x$.

Differentiating with respect to x, we obtain

$$\frac{dy}{dx}=\sinh 2x\frac{d}{dx}(2x)=\sinh 2x\ 2=2\sinh 2x.$$

EXAMPLE 8 Find the derivative of the function $y=\sinh^{-1}(ax+b)$.

Solution: We are given that, $y=\sinh^{-1}(ax+b)$.

Differentiating with respect to x, we obtain

$$\frac{dy}{dx}=\frac{1}{\sqrt{1+(ax+b)^2}}\frac{d}{dx}(ax+b)$$

$$=\frac{1}{\sqrt{1+(ax+b)^2}}(a)$$

$$=\frac{a}{\sqrt{1+(ax+b)^2}}.$$

EXAMPLE 9 Find the derivative of the function $y=\cosh^{-1}(\cosh x)$

Solution: We are given that, $y=\cosh^{-1}(\cosh x)$.

Differentiating with respect to x, we obtain

$$\frac{dy}{dx}=\frac{1}{\sqrt{\cosh^2 x-1}}\frac{d}{dx}(\cosh x)$$

$$=\frac{1}{\sqrt{\cosh^2 x-1}}\sinh x$$

$$=\frac{1}{\sqrt{\sinh^2 x}}\sinh x, |\sinh x|>1$$

$$=\frac{\sinh x}{|\sinh x|}, |\sinh x|>1.$$

4.3 n^{th} DERIVATIVE OF FUNCTIONS

Let us now calculate the nth order derivative of some elementary functions.

4.3.1 $y = (ax + b)^m$, $a \neq 0$ [Power Function]

If $y=(ax+b)^m$ where m is any real number. Using differentiation, we get

$$y'=am(ax+b)^{m-1}=m(ax+b)^{m-1}.a.$$

$$\Rightarrow y''=am(m-1)(ax+b)^{m-2}a=m(m-1)(ax+b)^{m-2}.a^2$$

$$\Rightarrow y'''=m(m-1)(m-2)(ax+b)^{m-3}.a^3.$$

After n^{th} differentiations, we have

$$y^{(n)}=m(m-1)(m-2)\ldots\ldots.(m-(n-1))(ax+b)^{m-n}.a^n.$$

Case 1: When m is a positive integer, then n^{th} order derivative is given by

$$y^{(n)}=\frac{m(m-1)(m-2)\ldots\ldots.(m-(n-1))(m-n)\ldots\ldots3.2.1}{(m-n)\ldots.3.2.1}a^n(ax+b)^{m-n}.$$

$$\Rightarrow y^{(n)}=\frac{m!}{(m-n)!}a^n(ax+b)^{m-n}.$$

Thus,

$$\boxed{D^n\left[(ax+b)^m\right]=\frac{m!}{(m-n)!}a^n(ax+b)^{m-n}, \ m \text{ is a positive integer.}}$$

Case 2: When $m=n=$ a positive integer, then

$$y^{(n)}=\frac{n!}{(0)!}a^n=n!a^n=\text{constant.}$$

Thus,

$$\boxed{\boldsymbol{D^n\left[(ax+b)^m\right] = n!a^n = \text{constant}, m = n \text{ and is a positive integer.}}}$$

Case 3: When $n > m$, then the all derivatives of order $m + 1$, $m + 2$,...etc. are zero, i.e.,

$$y^{(n)} = 0.$$

Thus,

$$\boxed{\boldsymbol{D^n\left[(ax+b)^m\right] = 0, n > m \text{ and is a positive integer.}}}$$

Case 4: When $m = -r$ and $r > 0$, then

$$y^{(n)} = -r(-r-1)(-r-2)........(-r-(-r-1))(ax+b)^{-r-n}.a^n$$

$$\Rightarrow \qquad y^{(n)} = \frac{a^n(-1)^n r(r+1)(r+2)........(r+n-1)}{(ax+b)^{r+n}}$$

$$\Rightarrow \qquad y^{(n)} = \frac{a^n(-1)^n(r+n-1)!}{(r-1)!(ax+b)^{n+r}}.$$

For $r = 1$,

$$y^{(n)} = \frac{a^n(-1)^n n!}{(ax+b)^{n+1}}.$$

Thus,

$$\boxed{\boldsymbol{D^n[(ax+b)^m] = \begin{cases} \dfrac{a^n(-1)^n(r+n-1)!}{(r-1)!(ax+b)^{n+r}}, & \text{when } m = -r, \quad r > 0 \\[2ex] \dfrac{a^n(-1)^n n!}{(ax+b)^{n+1}}, & \text{when } m = -r, \quad r = 1 \end{cases}}}$$

4.3.2 $y = e^{ax}$, a is a Real Constant and $y = a^x$, $a > 0$ [Exponential Function]

Differentiating the given function $y = e^{ax}$ w.r.t. x successively, we have

$$y' = ae^{ax}$$

$$y'' = a^2e^{ax}$$

$$.........$$

$$\boldsymbol{y^{(n)} = a^n e^{ax}}.$$

Again differentiating $y = a^x, a > 0$ w.r.t. x successively, we get

$$y' = a^x \log_e a$$

$$y'' = a^x (\log_e a)^2$$

$$.........$$

$$y^{(n)} = a^x (\log_e a)^n.$$

Thus,

$$\boldsymbol{D^n\left[e^{ax}\right] = a^n e^{ax}, a \text{ is a real constant},}$$

$$\boldsymbol{D^n\left[a^x, a > 0\right] = a^x (\log_e a)^n.}$$

4.3.3 y = log (ax + b), (ax + b) > 0 [Logarithm Function]

The first order derivative of the given function is

$$y' = \frac{a}{(ax+b)}.$$

Using section 4.3.1 case (4) when $m = -r$ and $r > 0$ for $y' = \dfrac{a}{(ax+b)}$. Here, $r = 1$ and differentiating $(n-1)$ times, we get

$$y^{(n)} = \frac{a^n (-1)^{n-1} (n-1)!}{(ax+b)^n}.$$

Thus,

$$\boldsymbol{D^n\left[\log(ax+b), (ax+b) > 0\right] = \frac{a^n (-1)^{n-1} (n-1)!}{(ax+b)^n}.}$$

4.3.4 y = sin(ax + b) or y = cos(ax + b) a ≠ 0 [Trigonometric Function]

Differentiating the function $y = \sin(ax+b), a \neq 0$ w.r.t. x successively, we have

$$y' = a\cos(ax+b) = a\sin\left(\frac{\pi}{2} + ax + b\right)$$

$$y'' = -a^2 \sin(ax+b) = a^2 \sin\left(2\frac{\pi}{2} + ax + b\right)$$

$$y''' = -a^3 \cos(ax+b) = a^3 \sin\left(3\frac{\pi}{2} + ax + b\right)$$

$$\vdots$$

$$\boldsymbol{y^{(n)} = a^n \sin\left(\frac{n\pi}{2} + ax + b\right).}$$

Similarly, we get the nth derivative of the function $y = \cos(ax+b), a > 0$ as

$$\boldsymbol{y^{(n)} = a^n \cos\left(\frac{n\pi}{2} + ax + b\right).}$$

Thus,

$$D^n\left[\sin(ax+b), a \neq 0\right] = a^n \sin\left(\frac{n\pi}{2} + ax + b\right),$$

$$D^n\left[\cos(ax+b), a \neq 0\right] = a^n \cos\left(\frac{n\pi}{2} + ax + b\right).$$

4.3.5 $y = e^{ax}\sin(bx + c)$ or $y = e^{ax}\cos(bx + c)$ [Product Functions]

Here, differentiating the given function $y = e^{ax}\sin(bx+c)$ w.r.t. x successively, we get

$$\begin{aligned} y' &= ae^{ax}\sin(bx+c) + be^{ax}\cos(bx+c) \\ &= e^{ax}\left[a\sin(bx+c) + b\cos(bx+c)\right]. \end{aligned}$$

To rewrite in the form of *cosine* function, we put $a = r\cos\theta$ and $b = r\sin\theta$ so that

$$r = \sqrt{a^2+b^2} \text{ and } \theta = \tan^{-1}\left(\frac{b}{a}\right).$$

Then,

$$y' = e^{ax}[r\cos\theta\sin(bx+c) + r\sin\theta\cos(bx+c)] = re^{ax}\sin(bx+c+\theta)$$

$$\begin{aligned} \Rightarrow \quad y'' &= r\left[e^{ax}\sin(bx+c+\theta)\right]' \\ &= re^{ax}[a\sin(bx+c+\theta) + b\cos(bx+c+\theta)] \\ &= re^{ax}[r\cos\theta\sin(bx+c+\theta) + r\sin\theta\cos(bx+c+\theta)] \\ &= r^2e^{ax}\sin(bx+c+2\theta) \\ &\vdots \\ y^{(n)} &= r^ne^{ax}\sin(bx+c+n\theta) \end{aligned}$$

$$\Rightarrow \quad y^{(n)} = \left(a^2+b^2\right)^{\frac{n}{2}} e^{ax}\sin\left(bx+c+n\tan^{-1}\left(\frac{b}{a}\right)\right).$$

Similarly, n^{th} order derivative of the function $= e^{ax}\cos(bx+c)$ as

$$y^{(n)} = \left(a^2+b^2\right)^{\frac{n}{2}} e^{ax}\cos\left(bx+c+n\tan^{-1}\left(\frac{b}{a}\right)\right).$$

Thus,

$$D^n[e^{ax}\sin(bx+c)] = \left(a^2+b^2\right)^{\frac{n}{2}} e^{ax}\sin\left(bx+c+n\tan^{-1}\left(\frac{b}{a}\right)\right),$$

$$D^n[e^{ax}\cos(bx+c)] = \left(a^2+b^2\right)^{\frac{n}{2}} e^{ax}\cos\left(bx+c+n\tan^{-1}\left(\frac{b}{a}\right)\right)$$

Solved Examples for n^{th} Derivative of functions

EXAMPLE 10 Find the nth order derivative of function $y=(3x+1)^3$.

Solution: The successive derivatives of the given function are:

$$y' = 3(3x+1)^2.3 = 6(3x+1)^2$$

$$y'' = 6.2.(3x+1).3 = 36(3x+1)$$

$$y''' = 36.3 = 108$$

and $$y^{(n)} = 0\text{, for } n = 4 > m = 3.$$

EXAMPLE 11 Find the nth order derivative of function $y=\sin^3 x$.

Solution: We know that, $\sin 3x = 3\sin x - 4\sin^3 x$, rewrite the above formula as

$$\sin^3 x = \frac{3}{4}\sin x - \frac{1}{4}\sin 3x.$$

Now, $$y = \frac{3}{4}\sin x - \frac{1}{4}\sin 3x.$$

The nth derivative of y is $y^{(n)} = \frac{d^n}{dx^n}\left(\frac{3}{4}\sin x\right) - \frac{d^n}{dx^n}\left(\frac{1}{4}\sin 3x\right)$.

Using formula for n^{th} derivatives for trigonometric function, we have

$$y^{(n)} = \frac{3}{4}\sin\left(n\frac{\pi}{2}+x\right) - \frac{1}{4}\sin\left(\frac{n\pi}{2}+3x\right).$$

EXAMPLE 12 Find the nth order derivative of function $y = e^{2x}\sin\left(2x-\frac{\pi}{4}\right)$.

Solution: We are given that, $y = e^{2x}\sin\left(2x-\frac{\pi}{4}\right)$.

Using product of function rule, we have

$$D^n[e^{ax}\sin(bx+c)] = \left(a^2+b^2\right)^{\frac{n}{2}} e^{ax}\sin\left(bx+c+n\tan^{-1}\left(\frac{b}{a}\right)\right).$$

Here, $r = \sqrt{2^2+2^2} = \sqrt{8}$ and $\theta = \tan^{-1}(1) = \frac{\pi}{4}$.

Now, the nth order derivative of given function is

$$y^{(n)} = (8)^{\frac{n}{2}} e^{2x}\sin\left(2x-\frac{\pi}{4}+n\frac{\pi}{4}\right).$$

EXAMPLE 13 Find the nth order derivative of function $y = \log(2x+3)$.

Solution: We are given that, $y = \log(2x+3)$.

By using, the formula $D^n\left[\log(ax+b),(ax+b)>0\right] = \dfrac{a^n(-1)^{n-1}(n-1)!}{(ax+b)^n}$, we get

$$y^{(n)} = \frac{2^n(-1)^{n-1}(n-1)!}{(2x+3)^n}.$$

EXAMPLE 14 If $y = \dfrac{1}{a^2-x^2}$, where a is a real constant, find $y^{(n)}$.

Solution: We are given that, $y = \dfrac{1}{a^2-x^2}$ is not in any of the above standard form.

Therefore, we rewrite the given function in the following form

$$y = \frac{1}{2a}\left[\frac{1}{x+a} - \frac{1}{x-a}\right]$$

Now, the function is in standard form and by using $D^n\left[\dfrac{1}{(ax+b)}\right] = \dfrac{a^n(-1)^n n!}{(ax+b)^{n+1}}$, we get

$$y^{(n)} = \frac{1}{2a}D^n\left[\frac{1}{x+a}\right] - \frac{1}{2a}D^n\left[\frac{1}{x-a}\right]$$

$$= \frac{1}{2a}\left(\frac{(-1)^n n!}{(x+a)^{n+1}}\right) - \frac{1}{2a}\left(\frac{(-1)^n n!}{(x-a)^{n+1}}\right)$$

$$= \frac{(-1)^n n!}{2a}\left(\frac{1}{(x+a)^{n+1}} - \frac{1}{(x-a)^{n+1}}\right).$$

EXAMPLE 15 If $y = \dfrac{1}{6x^2-5x+1}$, where a is a real constant, find $y^{(n)}$.

Solution: Rewrite the given function in the following form

$$y = \left[\frac{2}{2x-1} - \frac{3}{3x-1}\right].$$

Now, we use $D^n\left[\dfrac{1}{(ax+b)}\right] = \dfrac{a^n(-1)^n n!}{(ax+b)^{n+1}}$, we get

$$y^{(n)} = D^n\left[\frac{2}{2x-1}\right] - D^n\left[\frac{3}{3x-1}\right]$$

$$= \left(\frac{2^{n+1}(-1)^n\, n!}{(2x-1)^{n+1}}\right) - \left(\frac{3^{n+1}(-1)^n\, n!}{(3x-1)^{n+1}}\right)$$

$$= (-1)^n\, n!\left(\frac{2^{n+1}}{(2x-1)^{n+1}} - \frac{3^{n+1}}{(3x-1)^{n+1}}\right)$$

$$= n!\left[\left(\frac{-3}{(3x-1)}\right)^{n+1} - \left(\frac{-2}{(2x-1)}\right)^{n+1}\right]$$

$$= n!\left[\left(\frac{3}{(1-3x)}\right)^{n+1} - \left(\frac{2}{(1-2x)}\right)^{n+1}\right].$$

EXAMPLE 16 If $y = \dfrac{1}{x^2 + a^2}$, where a is a real constant, find $y^{(n)}$.

Solution: Rewrite the given function in following form

$$y = \frac{1}{2ai}\left[\frac{1}{x-ai} - \frac{1}{x+ai}\right].$$

Now, we use $D^n\left[\dfrac{1}{(ax+b)}\right] = \dfrac{a^n(-1)^n\, n!}{(ax+b)^{n+1}}$, we get

$$y^{(n)} = \frac{1}{2ai}\left(D^n\left[\frac{1}{x-ai}\right] - D^n\left[\frac{1}{x+ai}\right]\right)$$

$$= \frac{1}{2ai}\left(\left(\frac{(-1)^n\, n!}{(x-ai)^{n+1}}\right) - \left(\frac{(-1)^n\, n!}{(x+ai)^{n+1}}\right)\right).$$

$$y^{(n)} = \frac{(-1)^n\, n!}{2ai}\left(\frac{1}{(x-ai)^{n+1}} - \frac{1}{(x+ai)^{n+1}}\right) \ldots\ldots (*)$$

We need the result in closed form. So we put $x = r\cos\theta$ and $a = r\sin\theta$,

So that $r=\sqrt{x^2+a^2}$ and $\theta=\tan^{-1}\dfrac{a}{x}$.

Then (*) can be written as

$$y^{(n)}=\frac{(-1)^n n!}{2ir\sin\theta}\left(\frac{1}{(r\cos\theta-ir\sin\theta)^{n+1}}-\frac{1}{(r\cos\theta+ir\sin\theta)^{n+1}}\right)$$

$$=\frac{(-1)^n n!r^{-(n+2)}}{2i\sin\theta}\left(\frac{1}{(\cos\theta-i\sin\theta)^{n+1}}-\frac{1}{(\cos\theta+i\sin\theta)^{n+1}}\right)$$

$$=\frac{(-1)^n n!r^{-(n+2)}}{2i\sin\theta}\left((\cos\theta-i\sin\theta)^{-(n+1)}-(\cos\theta+i\sin\theta)^{-(n+1)}\right)$$

$$=\frac{(-1)^n n!r^{-(n+2)}}{2i\sin\theta}\left((\cos(n+1)\theta+i\sin(n+1)\theta)-(\cos(n+1)\theta-i\sin(n+1)\theta)\right)$$

$$=\frac{(-1)^n n!r^{-(n+2)}}{2i\sin\theta}2i\sin(n+1)\theta$$

$$=\frac{(-1)^n n!r^{-(n+2)}}{\sin\theta}\sin(n+1)\theta$$

$$=\frac{(-1)^n n!r^{-(n+2)}}{\frac{a}{r}}\sin(n+1)\theta$$

$$=\frac{(-1)^n n!}{ar^{n+1}}\sin(n+1)\theta$$

$$=\frac{(-1)^n n!}{a^{n+2}}\sin^{n+1}\theta\sin(n+1)\theta$$

$$=\frac{(-1)^n n!}{a\left(x^2+a^2\right)^{\frac{n+1}{2}}}\sin\left((n+1)\tan^{-1}\frac{a}{x}\right).$$

EXAMPLE 17 If $y=\tan^{-1}\dfrac{2x}{1-x^2}$, then find $y^{(n)}$.

Solution: We are given that, $y = \tan^{-1}\dfrac{2x}{1-x^2}$, differentiating w.r.t. x, we have

$$y' = \frac{1}{1+\left(\frac{2x}{1-x^2}\right)^2}\,\frac{(1-x^2)2-2x(-2x)}{(1-x^2)^2} = \frac{2-2x^2+4x^2}{1+x^4-2x^2+4x^2} = \frac{2(1+x^2)}{(1+x^2)^2}$$

$$= \frac{2}{1+x^2}.$$

Differentiating y', $n-1$ times, we get

$$y^{(n)} = D^{n-1}\left[\frac{2}{(1+x^2)}\right] \qquad \text{(from Example 13)}$$

$$\Rightarrow \qquad y^{(n)} = \frac{2(-1)^{n-1}(n-1)!}{1(1+x^2)^{\frac{n}{2}}}\sin\left(n\tan^{-1}\frac{1}{x}\right),$$

or $$y^{(n)} = 2(-1)^{n-1}(n-1)!\sin^{n}\theta\,\sin(n\theta), \qquad \text{where } \theta = \tan^{-1}\frac{1}{x}.$$

4.4 LEIBNITZ THEOREM

Statement: Let $u(x)$ and $v(x)$ be the two functions of x having n^{th} order derivative. Then, the nth order derivative of product of these two functions is given by

$$D^n\left[u(x)v(x)\right] = (y = uv)^{(n)} = u^{(n)}v + {}^nC_1u^{(n-1)}v^{(1)} + {}^nC_2\,u^{(n-2)}v^{(2)} + \ldots\ldots\ldots{}^nC_ru^{(n-r)}v^{(r)} + \ldots$$
$$+ {}^nC_{n-1}u^{(1)}v^{(n-1)} + {}^nC_nv^{(n)}u$$

Proof: The proof of this theorem is based on mathematical induction. By direct differentiation, we have

$$(uv)^{(1)} = u^{(1)}v + uv^{(1)}.$$

$$\Rightarrow \qquad (uv)^{(2)} = \left(u^{(1)}v + uv^{(1)}\right)' = u^{(2)}v + 2u^{(1)}v^{(1)} + uv^{(2)}.$$

$$\Rightarrow \qquad (uv)^{(3)} = (u^{(2)}v + 2u^{(1)}v^{(1)} + uv^{(2)})'$$
$$= (u^{(2)}v)' + 2\left(u^{(1)}v^{(1)}\right)' + \left(uv^{(2)}\right)'$$
$$= u^{(3)}v + u^{(2)}v^{(1)} + 2\left(u^{(2)}v^{(1)} + u^{(1)}v^{(2)}\right) + uv^{(3)} + u^{(1)}v^{(2)}$$
$$= u^{(3)}v + 3u^{(2)}v^{(1)} + 3u^{(1)}v^{(2)} + uv^{(3)}.$$

So, the Leibnitz rule's holds for n = 1, 2 and 3.

Assume that the Leibnitz rule's holds for $n=k$, then

$$(uv)^{(k)} = u^k v + {}^kC_1 u^{(k-1)}v^{(1)} + {}^kC_2\, u^{(k-2)}v^{(2)} + \ldots\ldots\ldots\ldots {}^kC_r u^{(k-r)}v^{(r)} + \ldots$$
$$+ {}^kC_{k-1}u^{(1)}v^{(k-1)} + {}^kC_k v^{(k)}u$$

Differentiating both the sides, we have

$$(uv)^{(k+1)} = u^{k+1}v + (u^{(k)}v^{(1)} + {}^kC_1 u^{(k)}v^{(1)}) + ({}^kC_1 u^{(k-1)}v^{(2)} + {}^kC_2\, u^{(k-1)}v^{(2)})$$
$$+ {}^kC_2 u^{(k-2)}v^{(3)} + \ldots + \ldots {}^kC_{r-1}u^{(k-r-2)}v^{(r-1)}$$
$$+ ({}^kC_{r-1}u^{(k-r+1)}v^{(r)} + {}^kC_r u^{(k-r+1)}v^{(r)} + \ldots$$
$$+ {}^kC_k u^{(1)}v^{(k)} + {}^kC_k v^{(k+1)}u.$$

We know that

$${}^kC_{r-1} + {}^kC_r = {}^{k+1}C_r;\ {}^kC_k = {}^{k+1}C_{k+1} = 1;\ 1 + {}^kC_1 = 1 + k = {}^{k+1}C_1.$$

Therefore,

$$(uv)^{(k+1)} = u^{(k+1)}v + {}^{k+1}C_1\, u^{(k)}v^{(1)} + {}^{k+1}C_2 u^{(k-1)}v^{(2)}$$
$$+ {}^{k+1}C_r u^{(k-r+1)}v^{(r)} + \ldots + {}^{k+1}C_{k+1}v^{(k+1)}u.$$

So, the theorem is also true for $k+1$. By the principle of mathematical induction, the result is true for all n.

Solved Examples for n^{th} Derivative Using Leibnitz's Rule

EXAMPLE 18 If $y = \cos\{m\sin^{-1}x\}$, then prove that $(1-x^2)y'' - xy' + m^2y = 0$. Deduce that

$$(1-x^2)y^{(n+2)} - x(2n+1)y^{(n+1)} - (n^2 - m^2)y^{(n)} = 0.$$

Solution: Differentiating y with respect to x, we have

$$y' = -\sin\{m\sin^{-1}x\}m\frac{1}{\sqrt{1-x^2}}$$

$$\Rightarrow \quad y'\sqrt{1-x^2} = -m\sin\{m\sin^{-1}x\}.$$

Again differentiate w.r.t. x,

$$y''\sqrt{1-x^2} - \frac{xy'}{\sqrt{1-x^2}} = -m^2\cos\{m\sin^{-1}x\}\frac{1}{\sqrt{1-x^2}}$$

$$\Rightarrow \quad y''(1-x^2) - xy' = -m^2\cos\{m\sin^{-1}x\}$$

$$\Rightarrow \quad y''(1-x^2) - xy' + m^2y = 0 \qquad \text{(a)}$$

Now, differentiate (a) n times by Leibnitz rule, we get

$$D^n\left[y''\left(1-x^2\right)\right]-D^n\left[xy'\right]+D^n[m^2y]=0. \quad \text{(b)}$$

Applying Leibnitz rule for each part, we get

$$D^n\left[y''\left(1-x^2\right)\right]=\left(1-x^2\right)y^{n+2}+{}^nC_1y^{n+1}(-2x)++{}^nC_2y^n(-2)$$

$$=\left(1-x^2\right)y^{n+2}-2nx\,y^{(n+1)}-n(n-1)y^{(n)}$$

$$\Rightarrow \quad D^n\left[xy'\right]=y^{(n+1)}x+{}^nC_1y^{(n)}=xy^{(n+1)}+ny^{(n)}$$

$$\Rightarrow \quad D^n[m^2y]=m^2y^{(n)}.$$

Using above values in (b), we have

$$\left(1-x^2\right)y^{n+2}-2nx\,y^{(n+1)}-n(n-1)y^{(n)}-xy^{(n+1)}-ny^{(n)}+m^2y^{(n)}=0$$

$$\Rightarrow \quad \left(1-x^2\right)y^{n+2}-(2n+1)x\,y^{(n+1)}-\left(n^2-m^2\right)y^{(n)}=0.$$

EXAMPLE 19 Find the nth order derivative of function $y=e^x\log x$ using Leibnitz rule.

Solution: Here, $y=e^x\log x$.

Applying Leibnitz rule both the sides, we get

$$y^{(n)}=D^n\left(e^x\log x\right).$$

$$\Rightarrow \quad y^{(n)}=\left(e^x\right)^{(n)}\log x+{}^nC_1\left(e^x\right)^{(n-1)}(\log x)^{(1)}+\ldots\ldots\ldots\ldots+{}^nC_n\left(e^x\right)(\log x)^{(n)}.$$

Since,

$$\left(e^x\right)^{(n)}=e^x.$$

$$\Rightarrow (\log x)^{(n)}=\frac{(-1)^{n-1}(n-1)!}{x^n}.$$

Then,

$$y^{(n)}=e^x[\log x+{}^nC_1\frac{1}{x}-{}^nC_2\left(\frac{1}{x^2}\right)+2!\,{}^nC_3\left(\frac{1}{x^3}\right)+\ldots\ldots\ldots\ldots+\frac{(-1)^{n-1}(n-1)!}{x^n}.$$

EXAMPLE 20 Find the value $y^{(n)}(0)$, where $y=e^{m\cos^{-1}x}$ using Leibnitz rule.

Solution: Here $y=e^{m\cos^{-1}x}$.

Differentiating w.r.t., we get

$$y'=e^{m\cos^{-1}x}\,m\left(-\frac{1}{\sqrt{1-x^2}}\right)=-\frac{m}{\sqrt{1-x^2}}y.$$

Squaring both the sides, we have

$$y'^2\left(1-x^2\right)=m^2y^2.$$

Again differentiating w.r.t. x, we get

$$\left(1-x^2\right)2y'y''-2xy'^2=2m^2yy'.$$

$$\Rightarrow \qquad \left(1-x^2\right)y''-xy'-m^2y=0.$$

Apply Leibnitz rule n times both the sides, we have

$$\left[\left(1-x^2\right)y^{(n+2)}+{}^nC_1(-2x)y^{(n+1)}+{}^nC_2(-2)y^{(n)}\right]-\left[xy^{(n+1)}+{}^nC_1y^{(n)}\right]-m^2y^{(n)}=0.$$

$$\Rightarrow \qquad \left(1-x^2\right)y^{(n+2)}-(2n+1)xy^{(n+1)}-\left(n^2-m^2\right)y^{(n)}=0.$$

Putting $x=0$, we get

$$y^{(n+2)}(0)=\left(n^2+m^2\right)y^{(n)}(0)$$

$$y(0)=e^{m\frac{\pi}{2}}$$

$$y'(0)=-me^{m\frac{\pi}{2}}$$

For $n=0$,
$$y^{(2)}(0)=\left(0+m^2\right)e^{m\frac{\pi}{2}}=m^2e^{m\frac{\pi}{2}}$$

For $n=1$,
$$y^{(3)}(0)=\left(1^2+m^2\right)\left(-me^{m\frac{\pi}{2}}\right)=-m\left(1^2+m^2\right)e^{m\frac{\pi}{2}}$$

For $n=2$,
$$y^{(4)}(0)=\left(2^2+m^2\right)\left(m^2e^{m\frac{\pi}{2}}\right)=m^2\left(2^2+m^2\right)e^{m\frac{\pi}{2}}.$$

Similarly, we can find the values of higher order derivatives.

Exercise

1. Find the fifth order derivative of the functions given below by using Leibnitz rule:
   (a) x^3e^{ax} (b) $x^3\sin x$
   (c) $x^3\log x$ (d) $e^x\log x$
   (e) $x^2(2x+3)^{10}$ (f) $e^{4x}(3x+5)^2$
   (g) $e^x\cos 2x$ (h) $\sin x\sin 2x$
2. Find the nth order derivative of the following functions by using Leibnitz rule:
   (a) $\dfrac{\log x}{x}$ (b) $x^{n-1}\log x$

(c) $e^x \log x$ (d) $x^4 \sin(2x+3)$

3. If $y = a\cos(\log x) + b\sin(\log x)$, then prove that

$$x^2 y'' + xy' + y = 0.$$

Hence, show that, $x^2 y^{(n+2)} + (2n+1)xy^{(n+1)} + (n^2+1)y^{(n)} = 0$.

4. If $y = \log\left[x + \sqrt{(a^2 + x^2)}\right]$, then show that

$$(a^2 + x^2)y'' + xy' = 0.$$

Hence, deduce that

$$(a^2 + x^2)y^{(n+2)} + x(2n+1)y^{(n+1)} + n^2 y^{(n)} = 0.$$

5. If $y = \tan^{-1} x$, then prove that

$$(1+x^2)y'' + 2xy' = 0.$$

Use Leibnitz rule to deduce that

$$(1+x^2)y^{(n+2)} + 2x(n+1)y^{(n+1)} + n(n+1)y^{(n)} = 0.$$

6. If $y = \sin^{-1} x$, then prove that

$$(1-x^2)y'' - xy' = 0.$$

Use Leibnitz rule to deduce that

$$(1-x^2)y^{(n+2)} - (2n+1)xy^{(n+1)} - n^2 y^{(n)} = 0.$$

7. If $= \cos^{-1} x$, then prove that

$$(1-x^2)y'' - xy' - \alpha^2 y = 0.$$

Use Leibnitz rule to deduce that

$$(1-x^2)y^{(n+2)} - (2n+1)xy^{(n+1)} - (n^2+\alpha^2)y^{(n)} = 0.$$

8. If $y = e^{\propto \cos^{-1} x}$, then prove that

$$(1-x^2)y'' - xy' - \alpha^2 y = 0$$

Use Leibnitz rule to deduce that

$$(1-x^2)y^{(n+2)} - (2n+1)xy^{(n+1)} - (n^2+\alpha^2)y^{(n)} = 0.$$

9. If $y = e^{\propto \tan^{-1} x}$, then show that

$$\left(1+x^2\right)y'' + (2x-a)y' = 0.$$

Hence, deduce that

$$\left(1+x^2\right)y^{(n+2)} + (2nx+2x-a)y^{(n+1)} + n(n+1)y^{(n)} = 0.$$

10. If $y = \sin\left(a\sin^{-1} x\right)$,then prove that

$$\left(1-x^2\right)y'' - xy' + a^2 y = 0.$$

Deduce that

$$\left(1-x^2\right)y^{(n+2)} - (2n+1)xy^{(n+1)} - \left(n^2-a^2\right)y^{(n)} = 0.$$

Hence, or otherwise expand $\sin m\theta$ in the powers of $\sin\theta$. Find $\left(y_n\right)_0$.

11. If $y = \left(x + \sqrt{\left(1+x^2\right)}\right)^m$, then prove that

$$\left(1+x^2\right)y'' + xy' - m^2 y = 0.$$

Hence, deduce that

$$\left(1+x^2\right)y^{(n+2)} + (2n+1)xy^{(n+1)} + \left(n^2-m^2\right)y = 0.$$

12. If $P_n(x) = D^n\left[\left(x^2-1\right)^n\right] = \frac{d^n}{dx^n}\left(x^2-1\right)^n$,

Then prove that

$$\frac{d}{dx}\left[\left(1-x^2\right)\frac{dP_n}{dx^n}\right] + n(n+1)P_n = 0.$$

13. If $y^{1/m} + y^{-1/m} = 2x$, then show that

$$\left(x^3-1\right)y^{(n+2)} + (2n+1)xy^{(n+1)} + \left(n^3-m^2\right)y^{(n)} = 0$$

14. If $y = e^{p\log\left(ax^2+b\right)}$, then prove that

$$\left(ax^2+b\right)y'' + 2ax(1-p)y' - 2apy = 0.$$

Deduce from here that

$$\left(ax^3+b\right)y^{(n+2)} + 2ax(n+1-p)y^{(n+1)} + +an(n-1-2p)y^{(n)} = 0.$$

15. If $y = x^2 e^x$, show that

$$y^{(n)} = \frac{1}{2}n(n-1)y'' - n(n-2)y' + \frac{1}{2}(n-1)(n-2)y.$$

16. If $\cos^{-1}\left(\frac{y}{b}\right) = \log\left(\frac{x}{m}\right)^m$, prove that

$$x^2 y^{(n+2)} + (2n+1)xy^{(n+1)} + \left(n^2 + m^2\right)y^{(n)} = 0.$$

17. If $x + y = 1$, prove that

$$\frac{d^n}{dx^n}\left(x^n y^n\right) = n!\left\{y^n - \left(n_{C1}\right)^2 y^{n-1}x + \left(n_{C2}\right)^2 y^{n-2}.x^2 - \ldots + (-1)^n x^n\right\}.$$

18. If $y = x^n \log x$, prove that $xy^{(n+1)} = n\,!$.

19. If $y = e^{ax}\sin bx$, show that

$$y^{(n+1)} = 2ay^{(n)} - \left(a^2 + b^2\right)y^{(n-1)}.$$

20. If $x = \sin\left\{\frac{1}{a}\log y\right\}$, show that

$$\left(1 - x^2\right)y'' - xy' - a^2 y = 0.$$

Hence, prove that

$$\left(1 - x^2\right)y^{(n+2)} - (2n+1)xy^{(n+1)} - \left(n^2 + a^2\right)y^{(n)} = 0.$$

21. If $= \cosh\left(\frac{1}{m}\log y\right)$, prove that

$$\left(x^2 - 1\right)y'' + xy' - m^2 y = 0.$$

and $\left(x^2 - 1\right)y^{(n+2)} + (2n+1)xy^{(n+1)} + \left(n^2 - m^2\right)y^{(n)} = 0.$

22. If $= \left(x^2 - 1\right)^m$, prove that $y^{(2m)} = 2m!$.

5 CHAPTER

Parametric Curves, Polar Curves and Conic Sections

The curves in plane have been usually described by giving y as a function of $x\left(y=f(x)\right)$ or x as a function of $y\left(x=g(y)\right)$ or by giving a relation between x and y that defines y implicitly as a function of $x\left(f(x,y)=0\right)$.

In this chapter, we will study two new ways of expressing curves in the plane. We will begin by describing curves in parametric equations and then on polar coordinate system. Finally, we will study the equations of conics with axes parallel and tilted to the coordinate axes.

5.1 PARAMETRIC EQUATIONS

Suppose that a particle moves along the curve C in xy-plane as shown in Figure 5.1.

It is not possible to describe C by the equation of the form $y=f(x)$. But, if x-coordinate and y-coordinate of the particle are functions of time t, then the curve C can be expressed as $x=f(t)$ and $y=g(t)$. In such a way of describing a curve C gives rise to the following definition.

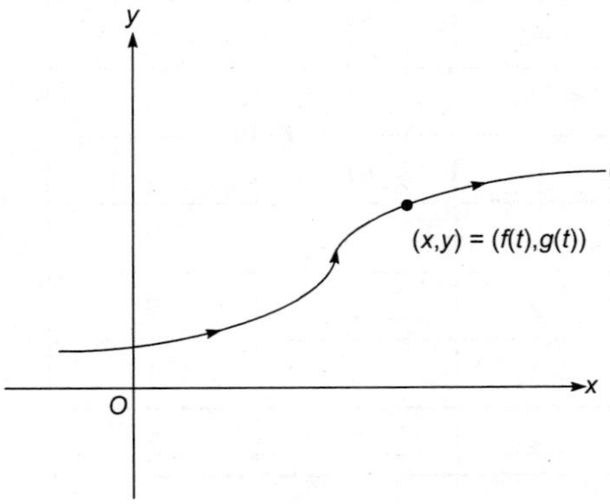

Figure 5.1

Definition: Suppose that a particle moves along the curve C in the xy-plane such that x and y are both given as functions of a third variable t by the equations

$$x=f(t) \text{ and } y=g(t) \tag{5.1}$$

The equations in (5.1) are known as *parametric equations* of motion for the particle and C is known as trajectory of the particle or graph of the equations. The variable t is known as the parameter for the equations.

Remarks:

1. For each value of t, a point (x,y) in xy-coordinate plane is determined.
2. As parameter t varies, the point $(x,y)=(f(t),g(t))$ varies and traces out a curve C.
3. The parameter t does not always represent time. But in many applications of parametric curves, t denotes time and so, $(x,y)=(f(t),g(t))$ is the position of a particle in xy-plane at time t and equations in (5.1) are known as parametric equations of motion of a particle with time t.
4. If there is no restriction on the parameter t are stated explicitly or implied by the equations, then it varies from $-\infty$ to ∞. Otherwise, if t is restricted to an interval $[a,b]$, then we will write $x=f(t)$, $y=g(t)(a\le t\le b)$.

EXAMPLE 1 Sketch the trajectory of the particle whose parametric equations of motion are

$$x=t^2-2t,\ y=t+1,\ -2\le t\le 4.$$

Solution: We sketch the trajectory by plotting the point (x,y) at each value of t in xy-plane and then connecting these points, we get a smooth curve C (as shown in Figure 5.2).

t	x	y
–2	8	–1
–1	3	0
0	0	1
1	–1	2
2	0	3
3	3	4
4	8	5

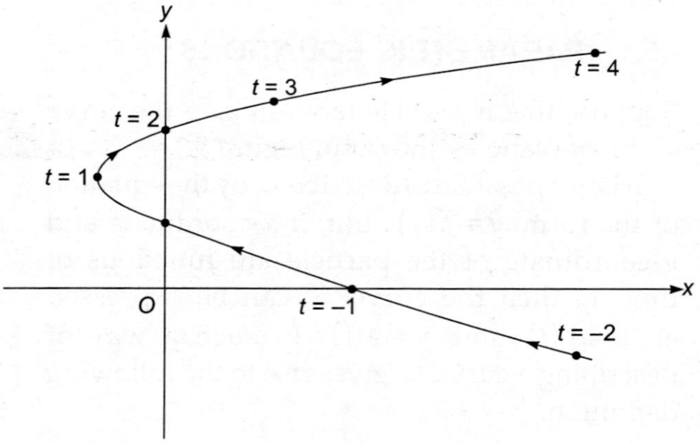

Figure 5.2

Notice that, a particle whose position is given by the parametric equations move along the curve C in the direction of the arrows as t increases.

EXAMPLE 2 Sketch the parametric curve of the parametric equations

$$x=t^2+t,\ y=2t-1,\ -1\le t\le 1.$$

Solution: We sketch the trajectory by plotting the point (x,y) at each value of t in xy-plane and then connecting these points, we get a smooth curve C (as shown in Figure 5.3).

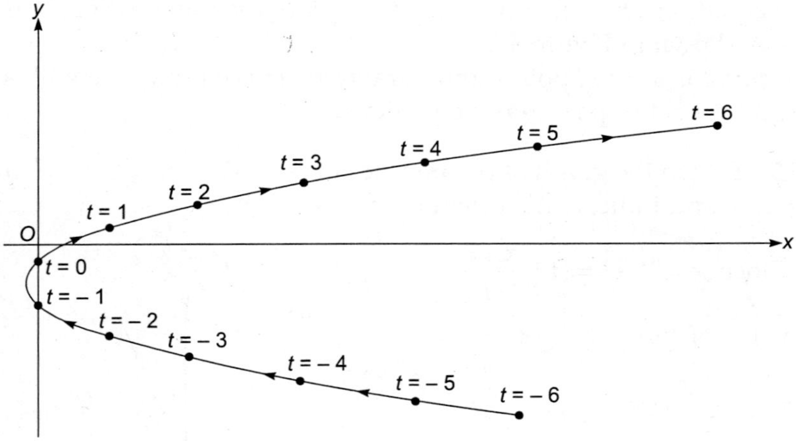

Figure 5.3

t	x	y
–6	30	–13
–5	20	–11
–4	12	–9
–3	6	–7
–2	2	–5
–1	0	–3
0	0	–1
1	2	1
2	6	3
3	12	5
4	20	7
5	30	9
6	42	11

Notice that a particle whose position is given by the parametric equations moves along the curve C in the direction of the arrows as t increases.

5.1.1 Orientation of Parametric Curves

Definition: The direction in which the parametric curve C is traced as the parameter increases is called the *direction* of increasing parameter or the *orientation* imposed on the curve by the equations. For example, the orientation of parametric curve is indicated by arrowheads as shown in Figure 5.2.

Thus, the curve is a set of points and parametric curve is a curve with an orientation imposed on it by a set of parametric equations.

EXAMPLE 3 Sketch the graph of parametric curve $x = 2t - 3$, $y = 6t - 7$ by eliminating the parameter t and indicate the orientation.

Solution: Since $x = 2t - 3 \Rightarrow t = \dfrac{x+3}{2}$

Put this value of t in y, we get

$$y = 6\left(\frac{x+3}{2}\right) - 7$$

$$\Rightarrow \qquad y = 3x + 2$$

It represents a straight line with slope of 3 and $y-$ intercept 2. Since as t increases, then x and y also increase. Thus, the orientation of the given parametric equations is indicated by the arrowheads as shown in Figure 5.4.

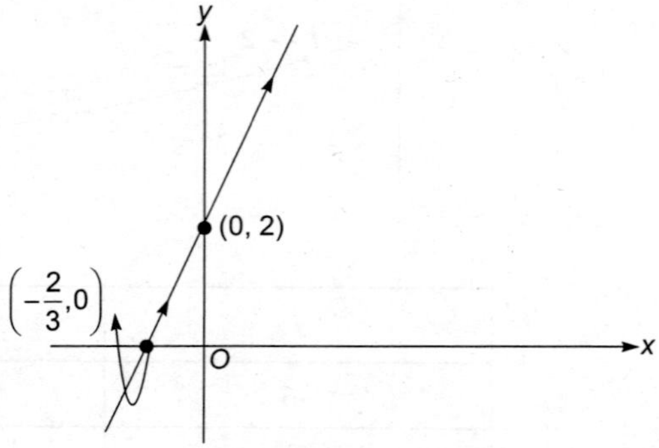

Figure 5.4

EXAMPLE 4 Find the graph of the curve described by the parametric equations $x = \cos t$, $y = \sin t$, $0 \le t \le 2\pi$ and also indicate the orientation on the graph.

Solution: By eliminating the parameter t, we get

$$x^2 + y^2 = \sin^2 t + \cos^2 t = 1.$$

Thus, the graph is contained in the unit circle $x^2 + y^2 = 1$ or the point (x, y) moves on the unit circle $x^2 + y^2 = 1$. The parameter t can be interpreted as the angle (in radians) swept out by the radial line from the origin to the point $(x, y) = (\cos t, \sin t)$ on the unit circle.

Since t increases from 0 to 2π, the point (x, y) traces the circle in counterclockwise direction starting at $(1,0)$ when $t = 0$ and then completing one full revolution when $t = 2\pi$. Thus, the orientation of the given parametric equations is indicated by arrowheads as shown in Figure 5.5.

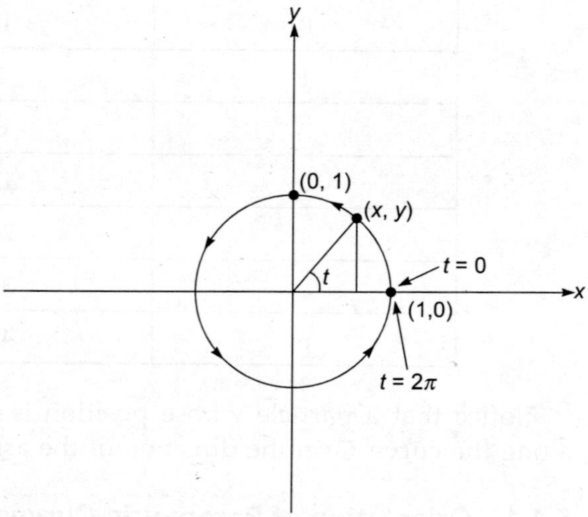

Figure 5.5

EXAMPLE 5 Find the graph of the parametric equations $x = \sin 2t$, $y = \cos 2t$, $0 \le t \le 2\pi$ and also indicate the orientation on the graph.

Solution: Again, by eliminating the parameter t, we have

$$x^2 + y^2 = 1.$$

Thus, the graph of the given parametric equations is same as in Example 4. But as t increases from 0 to 2π, the point $(x,y) = (\sin 2t, \cos 2t)$ moves twice around the unit circle $x^2 + y^2 = 1$ in clockwise direction starting from $(0,1)$.

Thus, the orientation of the given parametric equations is indicated by arrowheads as shown in Figure 5.6.

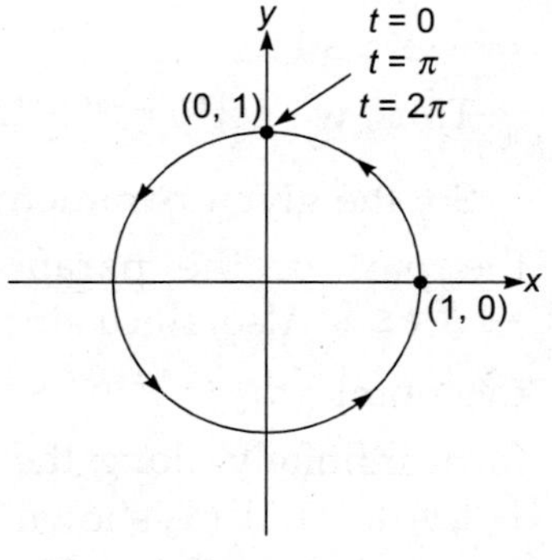

Figure 5.6

Remark: Examples 4 and 5 show different set of parametric equations which have the same curve.

EXAMPLE 6 Find the parametric equations of unit circle with centre (h,k) and radius R.

Solution: As in Example 4, the parametric equations of unit circle are

$$x_1 = \cos t,\ y_1 = \sin t,\ 0 \le t \le 2\pi$$

Multiply the expression for x_1 and y_1 by R, we get new parametric equations as

$$x = R\cos t,\ y = R\sin t,\ 0 \le t \le 2\pi \tag{5.2}$$

It is observed that the parametric equations in (5.2) represents a circle with centre $(0,0)$ and radius R, traces counterclockwise.

Now, shift the centre h units in the x-direction and k units in the y-direction, we get

$$x = h + R\cos t \text{ and } y = k + R\sin t,\ 0 \le t \le 2\pi. \tag{5.3}$$

Thus, the parametric equations in (5.3) represents the circle with centre (h,k) and radius R as shown in Figure 5.7.

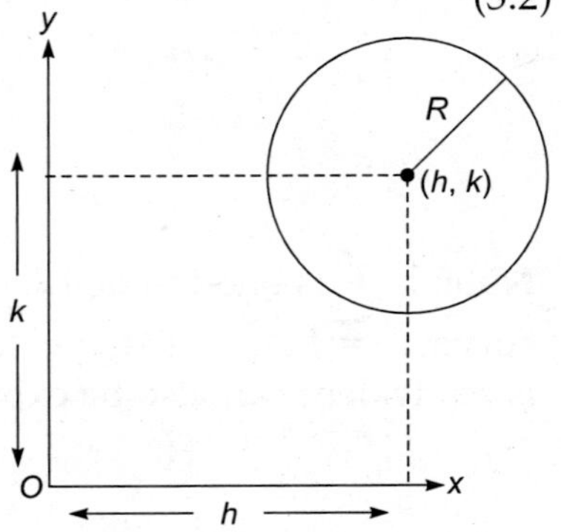

Figure 5.7

EXAMPLE 7 Find the graph of the parametric equations $x = \sin t$ and $y = \sin^2 t$ and also, indicate the orientation on the graph.

Solution: Here, $x = \sin t$ and $y = \sin^2 t$

$$\Rightarrow y = x^2.$$

Also, since $|\sin t| \le 1 \forall t$

$\Rightarrow -1 \le \sin t \le 1 \forall t$

$\Rightarrow -1 \le x \le 1$.

Thus, we get $y = x^2, -1 \le x \le 1$.

So, the given parametric equations represent the part of the parabola $y = x^2$ for which $-1 \le x \le 1$. Also, since $\sin t$ is a periodic function, the point $(x, y) = (\sin t, \sin^2 t)$ moves back and forth infinitely along the parabola $y = x^2$ from $(-1,1)$ to $(1,1)$ (as shown in Figure 5.8).

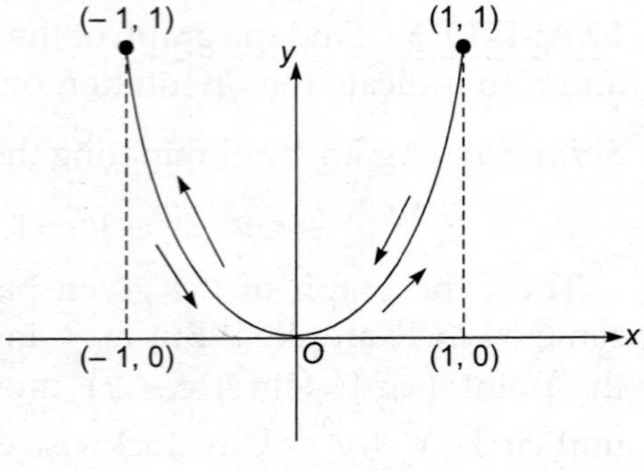

Figure 5.8

5.1.2 Expressing the Curves Parametrically or Parameterizing the Curves

An equation of the curves $y = f(x)$ can also be expressed as curves with parametric equations

$$x = t,\ y = f(t).$$

For example, the portion of the curve $y = \sin x$ over the interval $[-2\pi, 2\pi]$ can be expressed as parametric equations $x = t,\ y = \sin t,\ -2\pi \le t \le 2\pi$ (as shown in Figure 5.9).

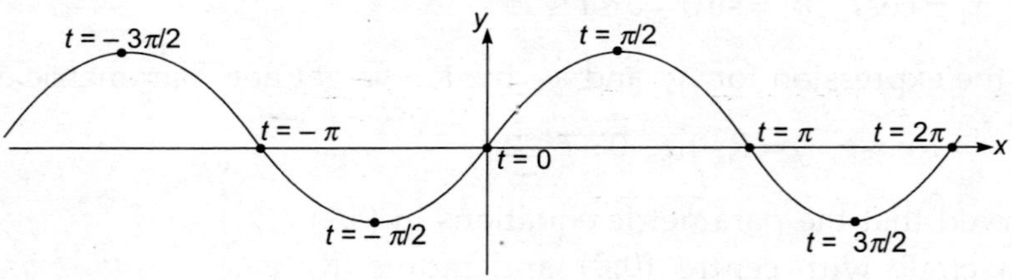

Figure 5.9

Note: If f is one-to-one function, then it has inverse function f^{-1}. The equation of the curves $y = f^{-1}(x)$ is equivalent to $x = f(y)$, i.e., the equation of the curves $y = f^{-1}(x)$ is equivalent can also be expressed as the curves with parametric equations

$$y = t,\ x = f(t).$$

EXAMPLE 8 Let the equation of the curve be $f(x) = x^5 + x + 1$. Find the parametric equations of f and f^{-1}.

Solution: $f(x) = x^5 + x + 1$. Let $x = t$ be the parameter, then $y = t^5 + t + 1$

So, the parametric equations of f are $x = t,\ y = t^5 + t + 1$.

Also, the parametric equations of f^{-1} are $x = t^5 + t + 1,\ y = t$.

EXAMPLE 9 Find the parametric equations of the curve $x = y^4 - 3y^2$.

Solution: Let $y = t$ be the parameter, then $x = t^4 - 3t^2$.

So, the parametric equations of the given curve are $y = t.\ x = t^4 = 3t^2$.

EXAMPLE 10 Find the parametric equations for the cycloid.

Solution:

Cycloid: The curve traced out by a point P on the circumference of a circle that rolls along a straight line is called a *cycloid* (Figure 5.10).

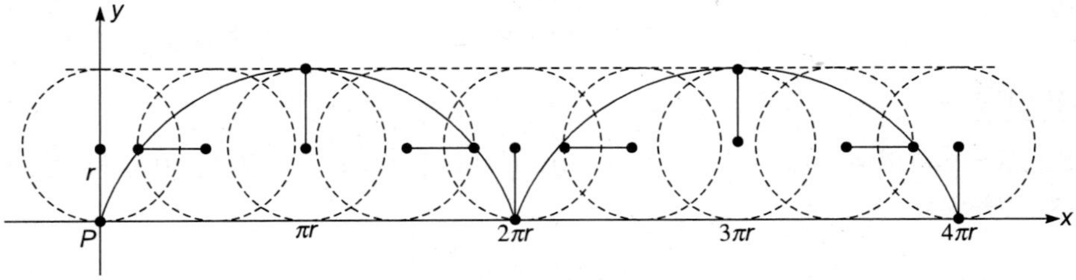

Figure 5.10

Let r be the radius of a circle that rolls along the x-axis and the position of a point P on the circle is at the origin and θ be the angle of rotation of the circle (i.e., when $\theta = 0$, then P is at the origin).

Let the circle be rotated through angle θ radians. Since the motion of P is a combination of the movement of circle's centre parallel to x-axis and the rotation of P about the centre as shown in Figure 5.11.

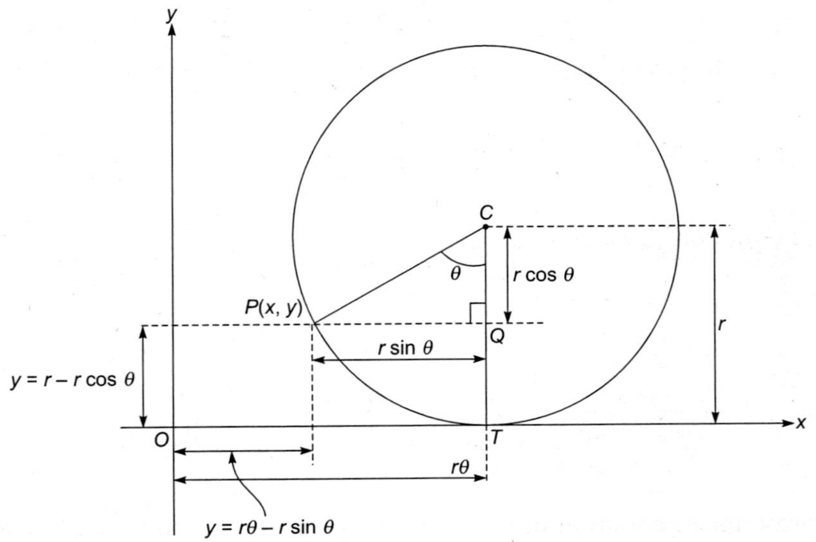

Figure 5.11

$\therefore |OT| = arcPT = r\theta$.

i.e., the circle moves a distance $r\theta$ from the origin along the x-axis.

Thus, the centre of the circle moves to a point $(r\theta, r)$.

Let (x, y) be the coordinates of P, then

$$x = |OT| - |PQ| = r\theta - r\sin\theta$$

and

$$y = |TC| - |QC| = r - r\cos\theta.$$

Hence, the parametric equations of the cycloid are:

$$x = r(\theta - \sin\theta) \text{ and } y = r(1 - \cos\theta)$$

EXAMPLE 11

(a) Show that the line segment joining the points (x_0, y_0) and (x_1, y_1) can be represented as

$$x = x_0 + (x_1 - x_0)\left(\frac{t - t_0}{t_1 - t_0}\right) \tag{5.4}$$

and
$$y = y_0 + (y_1 - y_0)\left(\frac{t - t_0}{t_1 - t_0}\right), \quad (t_0 \le t \le t_1) \tag{5.5}$$

(b) Which way is the line segment oriented?

(c) Find the parametric equations for the line segment traced from $(3, -1)$ to $(1, 4)$ varies from 1 to 2.

Solution:

(a) By eliminating the parameter, from equation (5.4), we get

$$\left(\frac{t - t_0}{t_1 - t_0}\right) = \frac{x - x_0}{x_1 - x_0}$$

Put in equation (5.5), then we get

$$y = y_0 + (y_1 - y_0)\left(\frac{x - x_0}{x_1 - x_0}\right)$$

$$\Rightarrow \quad y - y_0 = \left(\frac{y_1 - y_0}{x_1 - x_0}\right)(x - x_0) \tag{5.6}$$

It represents an equation of line segment joining the points (x_0, y_0) and (x_1, y_1).

Thus, equations (5.5) and (5.6) represent the parametric equations of the line segment joining the points (x_0, y_0) and (x_1, y_1).

(b) The line segment is oriented from (x_0, y_0) and (x_1, y_1) and it is indicated by arrowheads as shown in Figure 5.12.

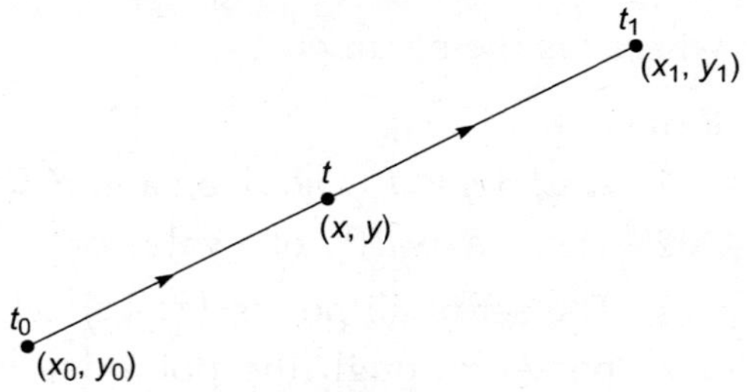

Figure 5.12

(c) From equations (5.4) and (5.5), we have

$$x = 3 + \frac{(1-3)(t-1)}{2-1} \Rightarrow x = 3 - 2(t-1) \Rightarrow x = 5 - 2t$$

and
$$y = (-1) + \frac{(t-1)}{2-1} \Rightarrow y = -1 + 5(t-1) \Rightarrow y = 5t - 2.$$

So, the parametric equations of line segment joining $(3,-1)$ to $(1,4)$ are

$$x = 5 - 2t \text{ and } y = 5t - 2,\ 1 \le t \le 2.$$

5.1.3 Parametric Form of a Line in $\mathbb{R}^3$

Since the direction of a line in the plane is measured by its slope. But, a line in the space, say in $\mathbb{R}^3$ can be measured when at least one point on a line and its direction are known.

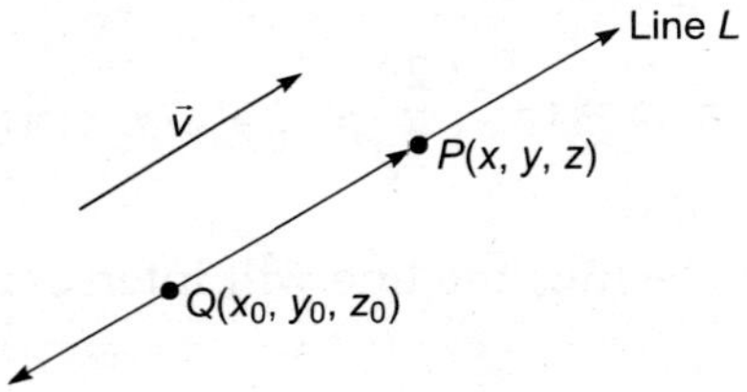

Figure 5.13

Let L be a line in $\mathbb{R}^3$ which is parallel to the vector $\vec{v} = v_1\hat{i} + v_2\hat{j} + v_3\hat{k}$ and also contains a point $Q(x_0, y_0, z_0)$ (as shown in Figure 5.13). So, line L has direction numbers v_1, v_2, v_3 and these direction numbers are denoted by $[v_1, v_2, v_3]$. The vector $\vec{v}$ is known as the direction vector of line L. Let $P(x, y, z)$ be any point on the line L such that $\overrightarrow{QP}$ is parallel to $\vec{v}$. Then, the vector equation is given by

$$\overrightarrow{QP} = t\vec{v}, \text{ for some } t.$$

$$\Rightarrow (x - x_0)\hat{i} + (y - y_0)\hat{j} + (z - z_0)\hat{k} = t\left(v_1\hat{i} + v_2\hat{j} + v_3\hat{k}\right)$$

$$\Rightarrow x - x_0 = tv_1, \quad y - y_0 = tv_2, \quad z - z_0 = tv_3.$$

Therefore, the parametric form of a line in $\mathbb{R}^3$ is defined as below.

Definition: A line L in $\mathbb{R}^3$ which is parallel to the vector $\vec{v} = v_1\hat{i} + v_2\hat{j} + v_3\hat{k}$ and containing a point (x_0, y_0, z_0), then the parametric equations of a line L in $\mathbb{R}^3$ are given by

$$x = x_0 + tv_1, \; y = y_0 + tv_2, \; z = z_0 + tv_3, \tag{5.7}$$

where t is the parameter.

Remarks:

1. Any point $P(x,y,z)$ on a line L must satisfy the linear system (5.7).
2. There is exactly one value of t corresponding to each point on a line L.
3. The set of all points (x,y,z) which satisfies the linear system (5.7), is a line that passes through the point (x_0,y_0,z_0) and parallel to the vector with direction numbers $[v_1,v_2,v_3]$.

EXAMPLE 12 Find the parametric equations for the line that contains the point $(1,-1,-2)$ and it is parallel to the vector $3\hat{i} - 2\hat{j} + 5\hat{k}$. Also, find the points where the line passes through each coordinate plane.

Solution: The direction numbers are $[3,-2,5]$ and $x_0 = 1$, $y_0 = -1$ and $z_0 = -2$.

Then, the parametric form of a line is given by (5.7) as

$$x = 1 + 3t, \; y = -1 - 2t, \; z = -2 + 5t.$$

Now, the line will intersect the xy-plane, when $z = 0$

$\Rightarrow -2 + 5t = 0 \Rightarrow t = 2/5$.

$$\Rightarrow x = 1 + 3\left(\frac{2}{5}\right) = \frac{11}{5} \text{ and } y = -1 - 2\left(\frac{2}{5}\right) = -\frac{9}{5}.$$

Thus, the line will intersect xy-plane at the point $\left(\frac{11}{5}, -\frac{9}{5}, 0\right)$.

Similarly, the line will intersect the xz-plane at the point $\left(-\frac{1}{2}, 0, -\frac{9}{2}\right)$ and the yz-plane at the point $\left(0, -\frac{1}{3}, -\frac{11}{3}\right)$.

5.1.4 Systematic Form of a Line in $\mathbb{R}^3$

If the direction numbers of the vector $\vec{v} = v_1\hat{i} + v_2\hat{j} + v_3\hat{k}$ are non-zero, i.e., v_1, v_2 and v_3 all are non-zero, then we can solve the parametric equations (5.7) of the line L which is parallel to vector $\vec{v}$, for the value of t which gives the systematic form of a line L.

A line L in $\mathbb{R}^3$, containing a point (x_0,y_0,z_0) and parallel to the vector $\vec{v} = v_1\hat{i} + v_2\hat{j} + v_3\hat{k}$ such that the direction numbers v_1, v_2, v_3 all are non-zero. Then the point $P(x,y,z)$ is lie on the line L if and only if the coordinates of P satisfy the equation

$$\frac{x - x_0}{v_1} = \frac{y - y_0}{v_2} = \frac{z - z_0}{v_3}. \tag{5.8}$$

EXAMPLE 13 Find the systematic equations for line L through the points $P(1,-1,2)$ and $Q(2,1,3)$. Also, find the points of intersection with the coordinates plane.

Solution: Line L passes through the points P and Q, so it is parallel to the vector

$$\overrightarrow{PQ} = (2-1)\hat{i} + (1+1)\hat{j} + (3-2)\hat{k} = \hat{i} + 2\hat{j} + \hat{k}.$$

Thus, all the direction numbers are non-zero.
Then, the systematic equation of the line L is given by (5.8) as

$$\frac{x-1}{1} = \frac{y+1}{2} = \frac{z-2}{1} \tag{5.9}$$

Now, line L will intersect the *xz*-plane, when $y = 0$.
Then, by (5.8) we have

$$x - 1 = \frac{1}{2} \Rightarrow x = \frac{3}{2}$$

and
$$z - 2 = \frac{1}{2} \Rightarrow z = \frac{5}{2}.$$

Thus, line L will intersect the *xz*-plane at the point $\left(\frac{3}{2}, 0, \frac{5}{2}\right)$. Similarly, the line L will intersect the *xy*-plane at the point $(-1,-5,0)$ and *yz*-plane at the point $(0,-3,1)$.

Remark: In systematic form of a line L, if one or more direction numbers are zero, then the line is parallel to at least one coordinate plane.

If two lines in $\mathbb{R}^3$ are not parallel, i.e., their slopes are different, then they may not intersect with each other. Regarding, this we give the following definition.

Definition (Skew Lines): The two lines in $\mathbb{R}^3$ are said to be *skew*, if they have different direction numbers and they will not intersect with each other.

Remark: From the above discussion, we conclude that the line in a space may intersect with each other, be parallel or skew.

EXAMPLE 14 Find that the following lines are parallel, intersect or skew. If they are intersecting each other, then also find the point of intersection.

(a) $L_1 : \frac{x-4}{2} = \frac{y-6}{-3} = \frac{z+2}{5}$ and $L_2 : \frac{x}{4} = \frac{y+2}{-6} = \frac{z-3}{5}$

(b) $L_1 : \frac{x-3}{2} = \frac{y-1}{-1} = \frac{z-4}{1}$ and $L_2 : \frac{x+2}{3} = \frac{y-3}{-1} = \frac{z-2}{1}$

(c) $L_1 : \frac{x-1}{2} = \frac{y+1}{1} = \frac{z-2}{4}$ and $L_2 : \frac{x+2}{4} = \frac{y}{-3} = \frac{z+1}{1}$

Solution:

(a) Line L_1 has direction numbers $[2,-3,5]$ and line L_2 has direction numbers $[4,-6,10]$.
Since, $\langle 2,-3,5\rangle = 2\langle 4,-6,10\rangle$, then the equation $\langle 2,-3,5\rangle = t\langle 4,-6,10\rangle$ has a solution for t. Thus, lines L_1 and L_2 are parallel.

(b). The direction numbers of lines L_1 and L_2 are $[2,-1,1]$ and $[3,-1,1]$, respectively. There does not exist any real number t such that $\langle 2,-1,1\rangle = t\langle 3,-1,1\rangle$. Thus, the lines L_1 and L_2 are not parallel.
The parametric equations of the lines are:

$L_1 : x = 2t+3,\ \ y = 1-t,\ z = 4+t$ and $L_2 : x = 3s-2,\ y = 3-s,\ z = s+2$.

Let $P(x,y,z)$ be the point of intersection of L_1 and L_2, then we have

$$2t+3 = 3s-2 \Rightarrow 2t-3s = -5$$

$$1-t = 3-s \Rightarrow t-s = -2$$

$$4+t = s+2 \Rightarrow t-s = -2.$$

On solving, the first two equations, we get $t=-1$ and $s=1$.
The value of t and s also satisfy the third equation. So that the lines L_1 and L_2 are intersecting each other.
Now, $x = 2(-1)+3 = 1,\ y = 1-(-1)$ and $z = 4+(-1) = 3$.
Therefore, the point of intersection is $P(1,2,3)$.

(c) The direction numbers of lines L_1 and L_2 are $[2,1,4]$ and $[4,-3,1]$, respectively. There does not exist any real number t such that $\langle 2,1,4\rangle = t\langle 4,-3,1\rangle$. Thus, lines L_1 and L_2 are not parallel.
The parametric equations of the lines are

$L_1 : x = 1+2t,\ \ y = -1+t,\ z = 2+4t$ and $L_2 : x = -2+4s,\ y = -3s,\ z = -1+s$.

Let $P(x,y,z)$ be the point of intersection of L_1 and L_2, then we have

$$1+2t = -2+4s \Rightarrow 2t-4s = -3$$

$$-1+t = -3s \Rightarrow t+3s = 1$$

$$2+4t = -1+s \Rightarrow 4t-s = -3.$$

On solving the first two equations, we get $t = -\frac{1}{2}$ and $s = \frac{1}{2}$.

The values of t and s do not satisfy the third equation. So, that the lines L_1 and L_2 are not intersecting with each other.
Therefore, the lines L_1 and L_2 are skew.

Note:

1. We have used two parameters t and s for parametric form of two lines. These two lines may intersect at different times. For instance: in Example 14(c), the value of the parameter $t=-\frac{1}{2}$ and $s=\frac{1}{2}$ instead of $t=s$.
2. In Example 14(b), we have evaluated the intersection point by putting the value of t in parametric form of line L_1. We can also put the value of s in parametric form of line L_2 and it will give the same intersection point.

5.2 POLAR COORDINATES

A coordinate system represents the specified location of a point in the plane by means of coordinates relative to two perpendicular coordinate axes. Usually, we use Cartesian coordinate system, in which the location of a point is directed by distance from two perpendicular axes. In this section, we will study a new kind of coordinate system which was introduced by Newton, called the polar coordinate system.

5.2.1 Polar Coordinate Systems

This system is used to describe the path of a moving particle which has angular direction and distance from the fixed point.

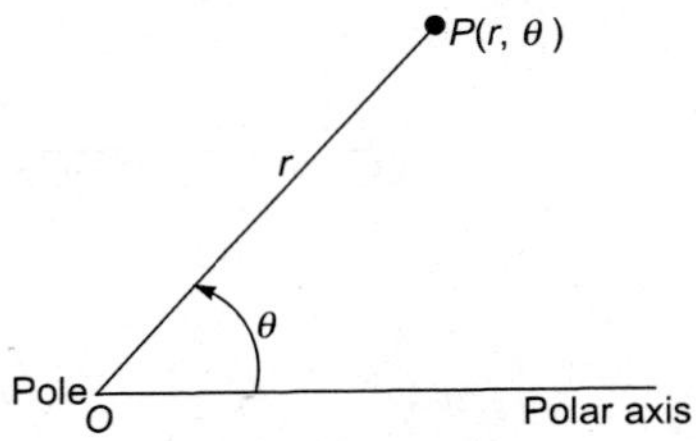

Figure 5.14

Definition: A polar coordinate system in a plane consists of a fixed point O, called the *pole* (or origin) and a ray originating from the pole O, called a *polar axis*.

In such a coordinate system, each point P in the plane can be associated with a pair of polar coordinates (r,θ), where r is the distance from the pole O to P and θ is angle (usually measured in radians) between the polar axis and the line OP (as shown in Figure 5.14).

The number r is called the *radial coordinate* of P and the number θ is called an *angular coordinate* (or polar angle) of P. The angle θ is positive when it is measured in the counterclockwise direction and negative when it is measured in clockwise direction.

If P is the pole, i.e., P is on the point O, then $r=0$ and so, $(0,\theta)$ are polar coordinates for any value of θ.

Now, since r is the distance from the point $P(r,\theta)$ to the pole. Then, r is non-negative. Also, the polar coordinate $(-r,\theta)$ can be defined as that the points $(-r,\theta)$ and (r,θ) lie on the same straight line which passes through the pole O and having the same distance $|r|$ (as shown in Figure 5.15) but

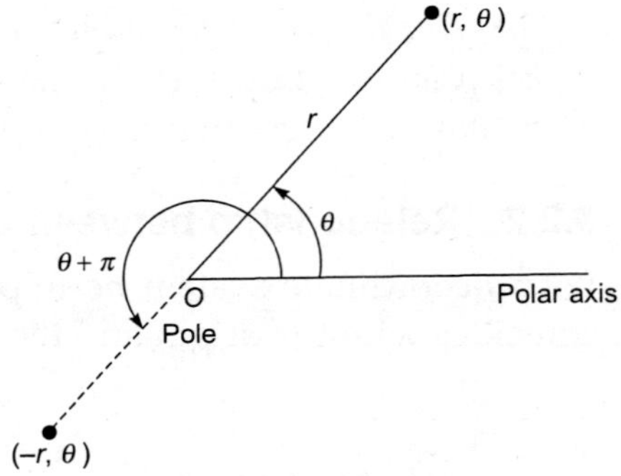

Figure 5.15

on the opposite side of O. It is known as the extension of polar coordinates through an angle π on the opposite side of the pole. Therefore, in general, $(-r,\theta)$ and $(r,\theta+\pi)$ represent the same polar coordinates.

EXAMPLE 15 Plot the points whose polar coordinates are given by:

(a) $\left(5,\dfrac{\pi}{3}\right)$ (b) $\left(6,\dfrac{2\pi}{3}\right)$

(c) $(3,3\pi)$ (d) $\left(2,\dfrac{5\pi}{4}\right)$

(e) $\left(3,-\dfrac{2\pi}{3}\right)$ (f) $(-2,3\pi/4)$

Solution:

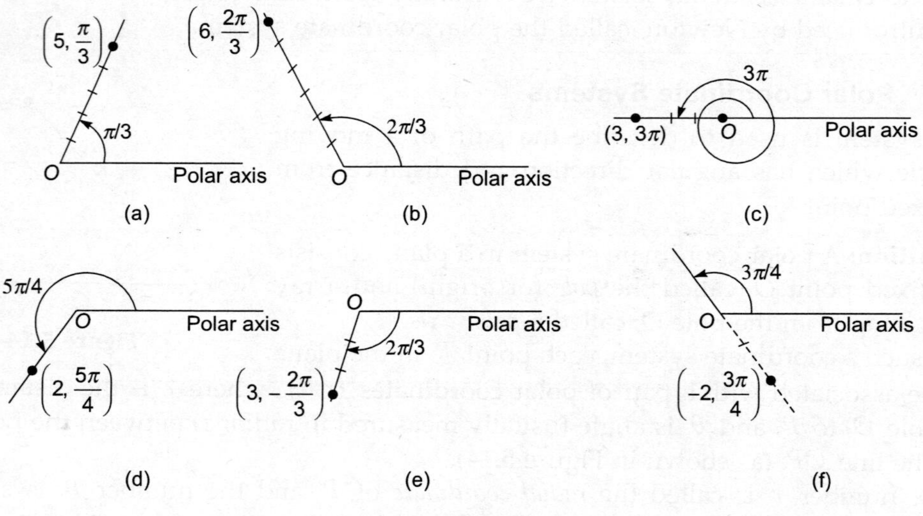

Figure 5.16

"In general, a complete counterclockwise rotation is given by an angle 2π, so that the point P has polar coordinates (r,θ), then $(r,\theta+2n\pi)$ and $(r,\theta-2n\pi)$ are also polar coordinates of P for any non-negative integer n, i.e., a point has infinitely many pairs of polar coordinates."

5.2.2 Relationship between Cartesian and Polar Coordinates

Let *xy*-coordinate system be imposed on the polar coordinates system in which pole coincides with the origin and the polar axis coincides with positive *x*-axis (Figure 5.17).

Let P be a point which has Cartesian coordinates (x,y) and polar coordinates (r,θ). Then, from the figure, we have

$$\cos\theta = \frac{x}{r} \text{ and } \sin\theta = \frac{y}{r}.$$

Therefore, the relation between Cartesian coordinates and polar coordinates are given by the equations

$$x = r\cos\theta \text{ and } y = r\sin\theta. \qquad (5.10)$$

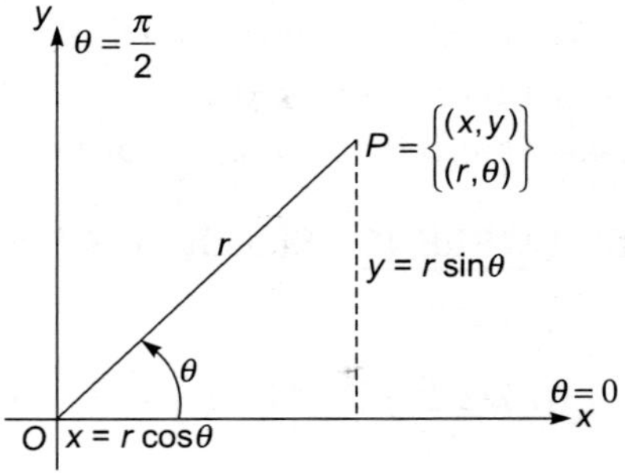

Figure 5.17

Therefore, if r and θ are known, then we can find the value of x and y. However, if x and y are known, then also we can find the value of r and θ by relation

$$x^2 + y^2 = r^2 \text{ and } \tan\theta = \frac{y}{x}. \qquad (5.11)$$

EXAMPLE 16 Find the Cartesian coordinates of the point whose polar coordinates are $\left(6, \frac{\pi}{3}\right)$.

Solution: Since $r = 6$ and $\theta = \frac{\pi}{3}$, then by equation (5.10), we have

$$x = 6\cos\frac{\pi}{3} = 6.\left(\frac{1}{2}\right) = 3$$

and
$$y = 6\sin\frac{\pi}{3} = 6.\left(\frac{\sqrt{3}}{2}\right) = 3\sqrt{3}.$$

Thus, $(3, 3\sqrt{3})$ are the required Cartesian coordinates of the point P.

EXAMPLE 17 Find the polar coordinates of the point P whose Cartesian coordinates are $(2,-2)$.

Solution: Since, $x = 2$ and $y = -2$. Then, by equation (5.11), we have

$$r^2 = x^2 + y^2 = (2)^2 + (-2)^2 = 8$$

$\Rightarrow r = \sqrt{8} = 2\sqrt{2}$

and $\tan\theta = -\frac{2}{2} = -1$

$\Rightarrow \theta = -\frac{\pi}{4} \text{ or } \frac{7\pi}{4}.$

Thus, point P has two polar coordinates $\left(2\sqrt{2}, -\frac{\pi}{4}\right)$ and $\left(2\sqrt{2}, \frac{7\pi}{4}\right)$.

5.2.3 Graphs of Polar Curves

Let the given equation of the polar curves be $r = f(\theta)$ or $f(r,\theta) = 0$.

A point in polar coordinates system has infinitely many different pairs of polar coordinates. Thus, the graph of the polar curve $r = f(\theta)$ consists all the points with at least one pair of polar coordinates (r,θ) and satisfy the given equation, $r = f(\theta)$.

EXAMPLE 18 Sketch the graph of the curve

(a) $r = 2$

(b) $\theta = \dfrac{\pi}{4}$

Solution:

(a) Since $r = 2$, the polar coordinates are $(2,\theta)$ which represent the distance of 2 units from the pole, for all value of θ. Thus, the graph is a circle with radius 2 and centered at pole (Figure 5.18(a)).

(b) Since $\theta = \dfrac{\pi}{4}$, the polar coordinates are $\left(r, \dfrac{\pi}{4}\right)$, for all values of r. The point $\left(r, \dfrac{\pi}{4}\right)$ lies on the line which makes an angle $\dfrac{\pi}{4}$ with polar axis. For $r \geq 0$, the point lies on the first quadrant and for $r < 0$, the point lies on the third quadrant. Since, there is no restriction on r, the graph is the entire line (Figure 5.18(b)).

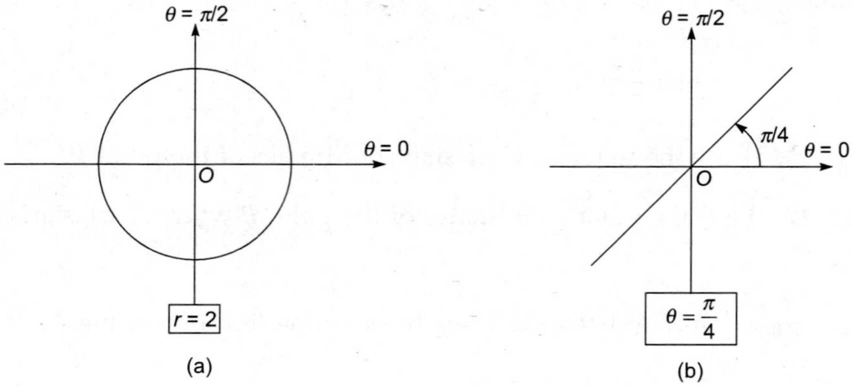

Figure 5.18

EXAMPLE 19 Sketch the graph of the curve $r = \theta$, $\theta \geq 0$.

Solution: Since the equation of the curve is $r = \theta$, so as θ increases, r also increases. Thus, the graphs of the given curve are spirals starting from the pole as θ increases in multiples of $\pi / 2$ (Figure 5.19).

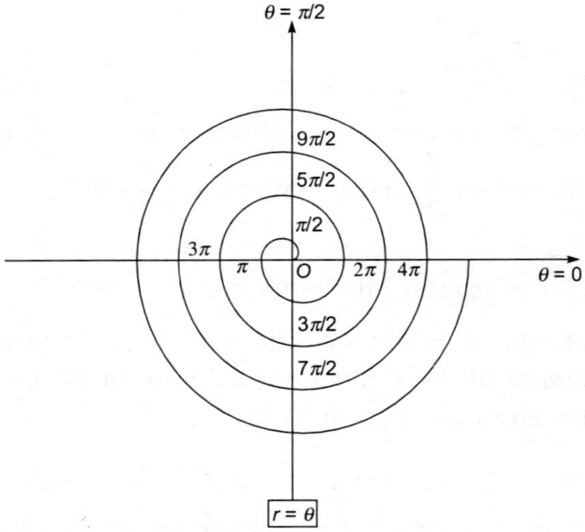

Figure 5.19

EXAMPLE 20 Sketch the graph of the curve $r = \cos\theta$ and also find the equation of the curve in Cartesian coordinates.

Solution: First of all, we find the values of r corresponding to different values of θ at the increment of $\frac{\pi}{6}$.

θ	0	$\frac{\pi}{6}$	$\frac{\pi}{3}$	$\frac{\pi}{2}$	$\frac{2\pi}{3}$	$\frac{5\pi}{6}$	π	$\frac{7\pi}{6}$	$\frac{4\pi}{3}$	$\frac{3\pi}{2}$	$\frac{5\pi}{3}$	$\frac{11\pi}{6}$	2π
r	1	$\frac{\sqrt{3}}{2}$	$\frac{1}{2}$	0	$-\frac{1}{2}$	$-\frac{\sqrt{3}}{2}$	-1	$\frac{-\sqrt{3}}{2}$	$\frac{-1}{2}$	0	$\frac{1}{2}$	$\frac{\sqrt{3}}{2}$	1

Plot the points from the table and while joining these points, we observe that the graph of the curve is a circle (Figure 5.20). We plot the points from $\theta = 0$ to $\theta = \pi$ as the θ beyond π yields the same points again.

Now, to convert the given curve in Cartesian form, we use the equations (5.10) and (5.11).

$$x = r\cos\theta \Rightarrow \cos\theta = \frac{x}{r}$$

and the given equation is $r = \cos\theta$

$$\Rightarrow r = \frac{x}{r} \Rightarrow r^2 = x$$

$$\Rightarrow x = r^2 = x^2 + y^2$$

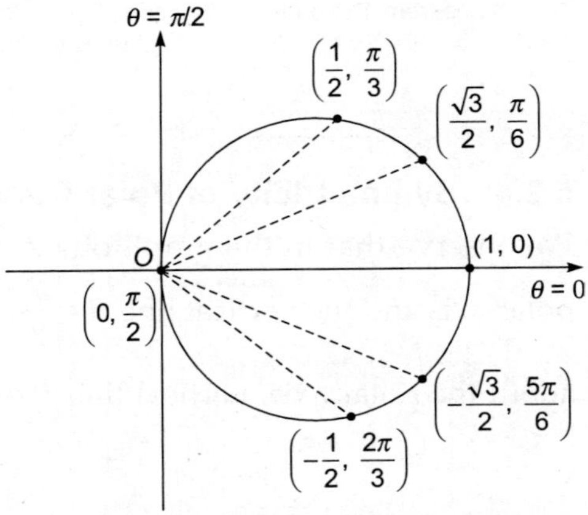

Figure 5.20

$$\Rightarrow \left(x-\frac{1}{2}\right)^2+y^2=\frac{1}{4}.$$

Therefore, equation of the curve in Cartesian form is $\left(x-\frac{1}{2}\right)^2+y^2=\frac{1}{4}$, which represents a circle with radius $\frac{1}{2}$ and is centred at $\left(\frac{1}{2},0\right)$.

EXAMPLE 21 Sketch the graph of the polar curve $r=\cos 2\theta$.

Solution: We sketch the curve $r=\cos 2\theta$, for $0\le\theta\le 2\pi$ and the variation in r corresponding to increase in θ is mentioned below of all the figures. The curve is called *four-petal rose* (as shown in Figure 5.21).

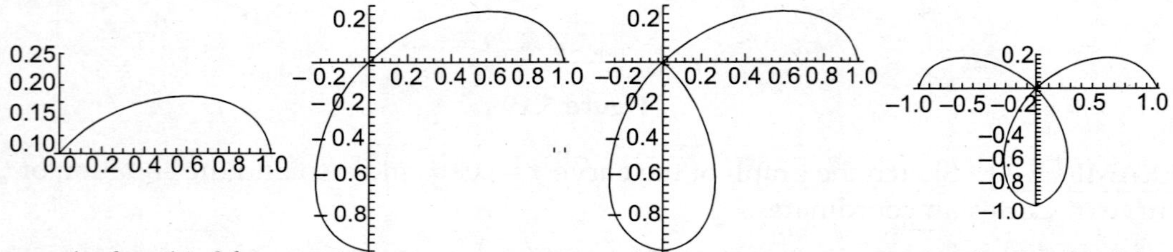

r varies from 1 to 0 for $0\le\theta\le\pi/4$ | *r* varies from 0 to – 1 for $\pi/4\le\theta\le\pi/2$ | *r* varies from – 1 to 0 for $\pi/2\le\theta\le 3\pi/4$ | *r* varies from 0 to 1 for $3\pi/4\le\theta\le\pi$

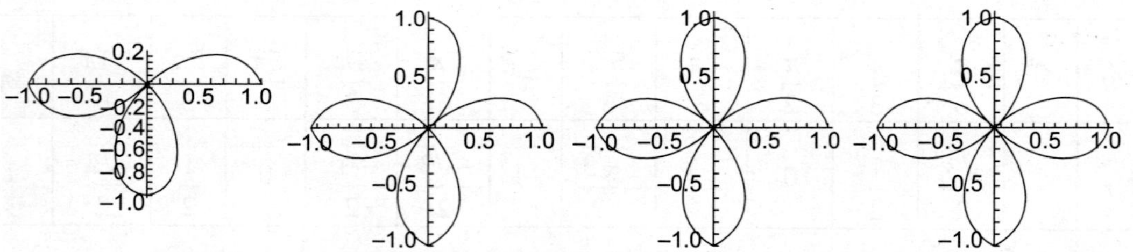

r varies from 1 to 0 for $\pi\le\theta\le 5\pi/4$ | *r* varies from 1 to –1 for $5\pi/4\le\theta\le 3\pi/2$ | *r* varies from –1 to –0 for $3\pi/2\le\theta\le v7\pi/4$ | *r* varies from 0 to 1 for $7\pi/4\le\theta\le 2\pi$

Figure 5.21

5.2.4 Symmetricity of Polar Curves

We observe that in the graph of $r=\cos\theta$ (as in Figure 5.19), it is symmetric about the polar axis and the vertical line $\theta=\frac{\pi}{2}$. So, for sketching the polar curve, the symmetricity about the polar axis, vertical line $\theta=\frac{\pi}{2}$ and the pole are very helpful.

The three rules of symmetricity are the following:

1. A polar curve is symmetric about the polar axis, if the polar equation of the curve remains same when θ is replaced by $-\theta$ (as in Figure 5.22(a)).
2. A polar curve is symmetric about the vertical line $\theta = \dfrac{\pi}{2}$, if the polar equation of the curve remains same when θ is replaced by $\pi - \theta$ (as in Figure 5.22(b)).
3. A polar curve is symmetric about the pole, if the polar equation of the curve remains same when θ by $\pi + \theta$ or r by $-r$ (as in Figure 5.22(c)).

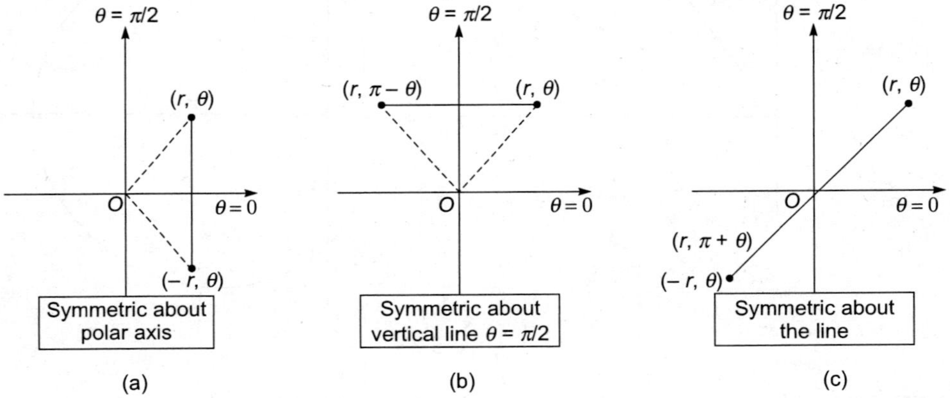

Figure 5.22

EXAMPLE 22 Sketch the graph of the polar curve $r = a(1+\cos\theta)$, $a > 0$ in polar coordinates.

Solution: To sketch the curve $r = a(1+\cos\theta)(a \geq 0)$, the following steps are involved:

1. Replace θ by $-\theta$ in the given equation of the curve and we observe that the given equation of the curve remains unchanged, i.e., the curve is symmetric about the polar axis.
   So, if we obtain the upper half of the curve, then we can also obtain the lower half by reflection about the polar axis.
2. Now, we find the values of r corresponding to different values of θ which are as:
   (a) At $\theta = 0$,then $r = 2a$.
   (b) As θ increases from 0 to $\pi/6$, then r decreases from $2a$ to $1.86a$.
   (c) As θ increases from $\pi/6$ to $\pi/3$, then r decreases from $1.86a$ to $3a/2$.
   (d) As θ increases from $\pi/3$ to $\pi/2$, then r decreases from $3a/2$ to a.
   (e) As θ increases from $\pi/2$ to $2\pi/3$, then r decreases from a to $0.13a$.
   (f) As θ increases from $2\pi/3$ to π, then r decreases from $0.13a$ to 0.

3. Joining these above points, we obtain the upper half of the curve.
4. The rest of the part can be obtained by reflecting the upper half of the curve (or we let θ increase from π to 2π.
5. The required curve is heart shaped and it is known as *Cardioid* (as shown in Figure 5.23).

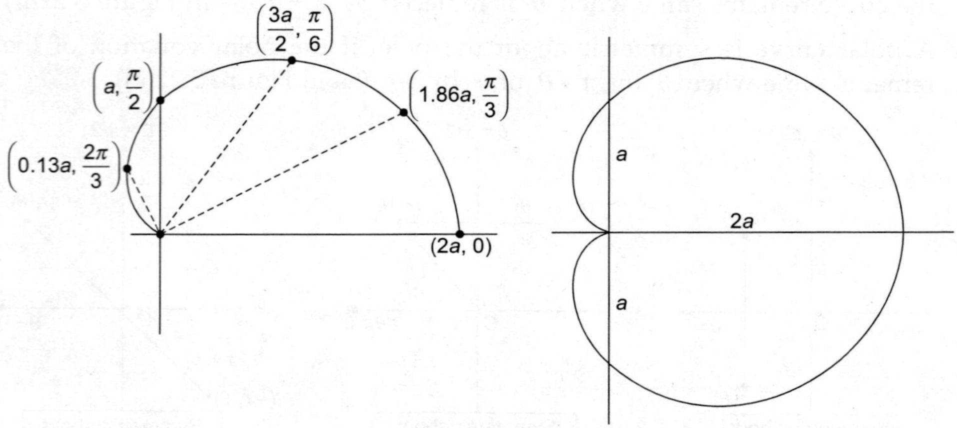

Figure 5.23

EXAMPLE 23 Sketch the graph of the polar curve $r = 1 + \sin\theta$ in polar coordinates.

Solution: To sketch the curve $r = (1 + \sin\theta)$, the following steps are involved:

1. Replace θ by $\pi - \theta$ in the given equation of the curve and we observe that the given equation of the curve remains unchanged, i.e., the curve is symmetric about the vertical line $\theta = \pi / 2$.
2. Now, we find the values of r corresponding to different values of θ which are:
   (a) At $\theta = 0$, then $r = 1$.
   (b) As θ increases from 0 to $\pi / 2$, r increases from 1 to 2.
   (c) As θ increases from $\pi / 2$ to π, r decreases from 2 to 1.
   (d) As θ increases from π to $3\pi / 2$, r decreases from 1 to 0.
   (e) As θ increases from $3\pi / 2$ to 2π, r increases from 0 to 1.
3. Joining these above points, we sketch the given curve.
4. The required curve is heart shaped and it is also known as *Cardioid* (as shown in Figure 5.24). We plot the points from $\theta = 0$ to $\theta = 2\pi$ as θ beyond 2π yields the same points again.

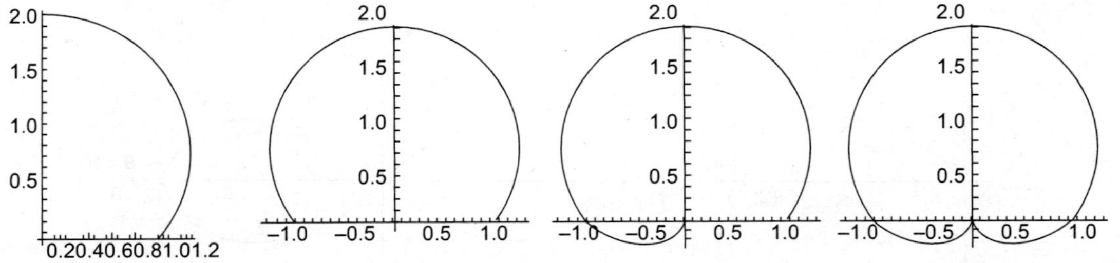

$0 \le \theta \le \pi/2$ $\quad$ $0 \le \theta \le \pi$ $\quad$ $0 \le \theta \le 3\pi/2$ $\quad$ $0 \le \theta \le 2\pi$

Figure 5.24

EXAMPLE 24 Sketch the graph of the polar curve $r^2 = 4\cos 2\theta$ in polar coordinates.

Solution: Since the given equation of curve does not express r as function of θ. So, on solving for r, we obtain two functions as:

$$r = -2\sqrt{\cos 2\theta} \text{ and } r = 2\sqrt{\cos 2\theta}.$$

Thus, to graph the given curve $r^2 = 4\cos 2\theta$, we will graph the two functions separately and then combine their graphs.

First, we will graph the curve $r = 2\sqrt{\cos 2\theta}$.

To sketch the curve $r = 2\sqrt{\cos 2\theta}$, the following steps are involved:

1. Replace θ by $-\theta$ or if we replace θ by $\pi - \theta$ in the given equation of the curve, we observe that the given equation of the curve remains unchanged, i.e., the curve is symmetric about both polar axis and the vertical line $\theta = \pi/2$. So, if we obtain the graph of the curve in the first quadrant, then we can also obtain the curve in the second quadrant by reflecting that portion about the vertical line $\theta = \pi/2$ and in the third and fourth quadrants by reflecting those two portions about the polar axis. In this way the entire graph can be obtained.
2. Now, we find the values of r corresponding to different values of θ which are as:
   (a) At $\theta = 0$, then $r = 2$.
   (b) As θ increases from 0 to $\pi/4$, then r decreases from 2 to 0.
   (c) As θ increases from $\pi/4$ to $\pi/2$, then there are no points on the polar graph.
3. Joining these above points, we sketch the given curve (as shown in Figure 5.25(a)).
4. Then, complete the graph by reflecting that portion as discussed above.
5. The required curve is propeller shaped and it is also known as *lemniscate* (as shown in Figure 5.25(b)).

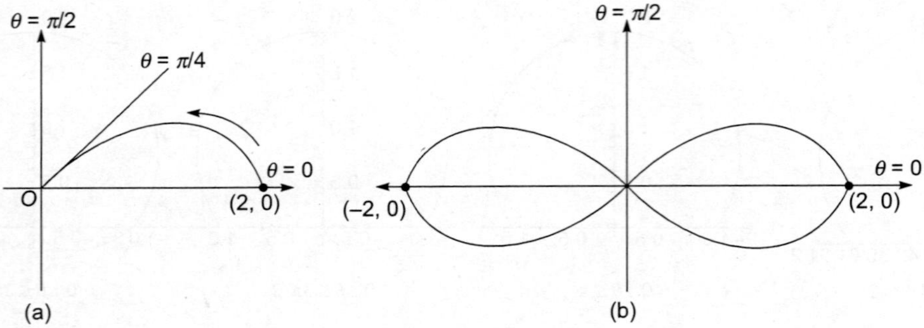

Figure 5.25

Note that there are gaps in the graph when $\frac{\pi}{4}<\theta<\frac{3\pi}{4}$ and $\frac{5\pi}{4}<\theta<\frac{7\pi}{4}$, since $\cos 2\theta$ is negative over those intervals.

5.2.5 Families of Lines and Rays through the Pole

Let a point P have polar coordinates (r,θ_0) in a polar coordinate system. Then, $\theta=\theta_0$ is the fixed angle with the polar axis for all values of r and so, the point (r,θ_0), for all value of r lies on the line that makes an angle $\theta=\theta_0$ with the polar axis and passes through the pole.

Conversely, all the points which lie on the line that makes an angle $\theta=\theta_0$ and passes through the pole, has a pair of polar coordinates (r,θ_0).

Therefore, we conclude that the equation $\theta=\theta_0$ represents the line that makes an angle θ_0 with the polar axis which passes through the pole and contains all the points that have a pair of polar coordinates (r,θ_0) (as shown in Figure 5.26(a)).

If r is restricted to be non-negative (as in Figure 5.26(b)), then the graph of the equation $\theta=\theta_0$ represents a ray that originates from the pole and makes an angle θ_0 with the polar axis.

Therefore, as θ_0 varies, the equation $\theta=\theta_0$ yields either a family of lines through the pole or a family of rays through the pole, depending on the restriction on r.

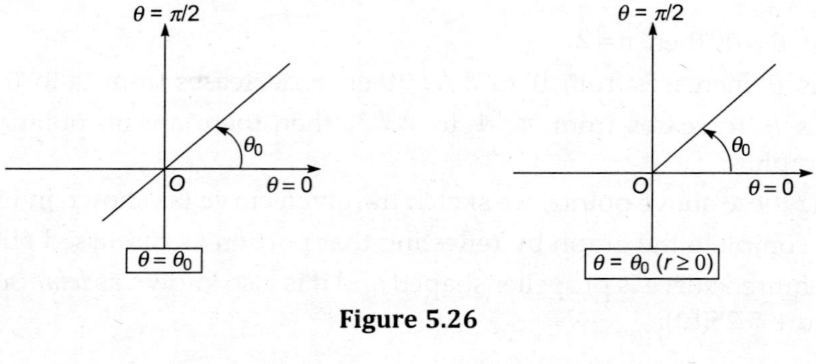

Figure 5.26

5.2.6 Families of Circles

We will consider three families of circles

$$r = a \tag{5.12}$$

$$r = a\cos\theta \tag{5.13}$$

$$r = a\sin\theta. \tag{5.14}$$

The equation $r = a$ represents a circle of radius a centred at pole. Thus, as a varies, the equation (5.12) yields a family of circles centred at the pole (Figure 5.27(a)).

The equation $r = a\cos\theta$ represents a circle of radius $\frac{a}{2}$, centred on the polar axis and the vertical line $\theta = \frac{\pi}{2}$ is the tangent to the circle at the pole (Figure 5.27(b)). Thus, as a varies, the equation (5.13) yields a family of circles centred at the polar axis and the vertical line $\theta = \frac{\pi}{2}$ is tangent to the circles at the pole (Figure 5.27(c)).

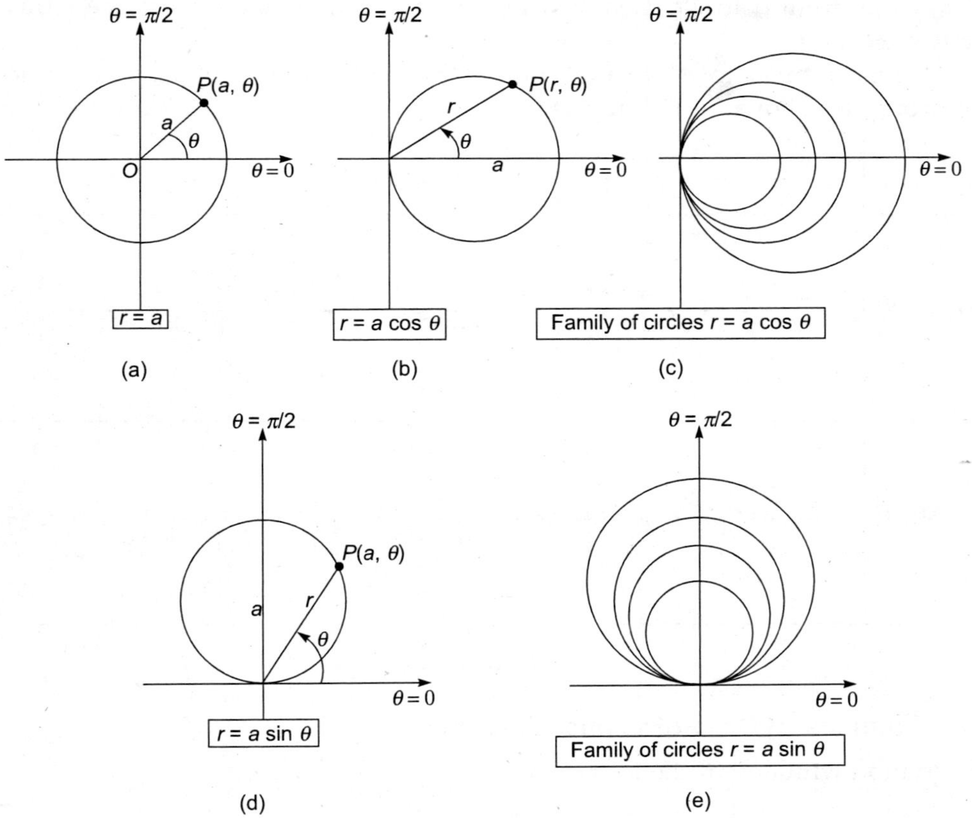

Figure 5.27

The equation $r = a\sin\theta$ represents a circle of radius $\frac{a}{2}$, centred on the vertical line $\theta = \frac{\pi}{2}$ and the polar axis is the tangent to the circle at the pole (Figure 5.27(d)). Thus, as a varies, equation (5.14) yields a family of circles centred at the vertical line $\theta = \frac{\pi}{2}$ and the polar axis is tangent to the circles at the pole (Figure 5.27(e)).

5.2.7 Families of Rose Curves

The equations of the form

$$r = a\sin n\theta \tag{5.15}$$

$$r = a\cos n\theta, \tag{5.16}$$

where $a > 0$ and n is a positive integer, in polar coordinates system, represents families of flower shaped curves called *roses*:

if n is odd, then the rose contains n equally spaced leaves of radius a;

if n is even, then the rose contains $2n$ equally spaced leaves of radius a;

if θ varies from 0 to 2π, then the rose with an even number of leaves are traced out exactly once; and

If θ varies from 0 to π, then the rose with an odd number of leaves are traced out exactly once (as shown in Figure 5.28).

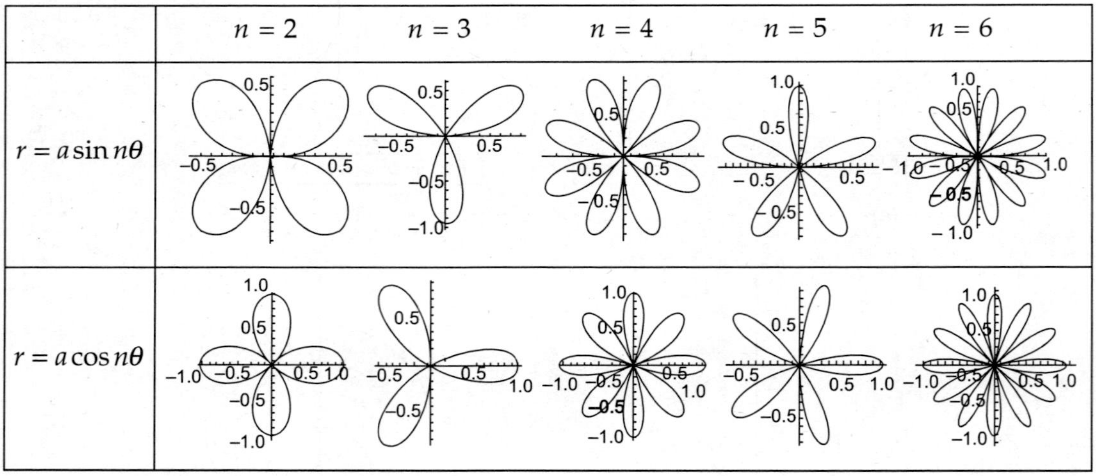

Figure 5.28

5.2.8 Families of Cardioids and Limacons

The equation with any of the four forms

$$r = a \pm b\sin\theta \tag{5.17}$$

$$r = a \pm b\cos\theta, \tag{5.18}$$

where $a>0$ and $b>0$, represents polar curves are called *limacons*. The shapes of limacons depend on the ratio of $\frac{a}{b}$. There are four possible shapes of limacon, which are as follows:

1. Limacon is of the shape "*Limacon with inner loop*", if $\frac{a}{b}<1$.
2. Limacon is of the shape "*Cardioid*", if $\frac{a}{b}=1$.
3. Limacon is of the shape "*Dimpled Limacon*", if $1<\frac{a}{b}<2$.
4. Limacon is of the shape "*Convex Limacon*", if $\frac{a}{b}\geq 2$.

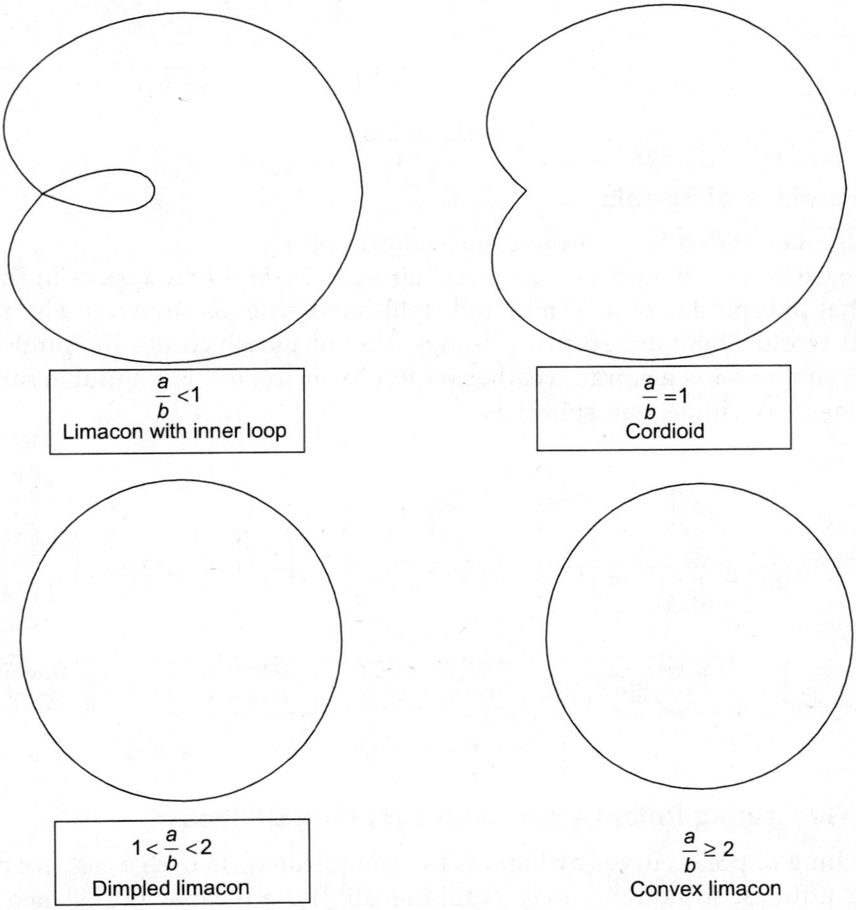

Figure 5.29

EXAMPLE 25 Sketch all the graphs of the limacons $r = a + \cos\theta$ for the range of a from 0.5 to 2.5 in steps of 0.5.

Solution: From Figure 5.30, we observe that the limacons evolve from the loop type to the convex type. As a starts increasing from 0.5, the loops get smaller and smaller till the cardioid reaches at $a = 1$ and then further, as a increases again from 1 to 2.5, the limacons evolve through the dimpled limacon into convex limacon.

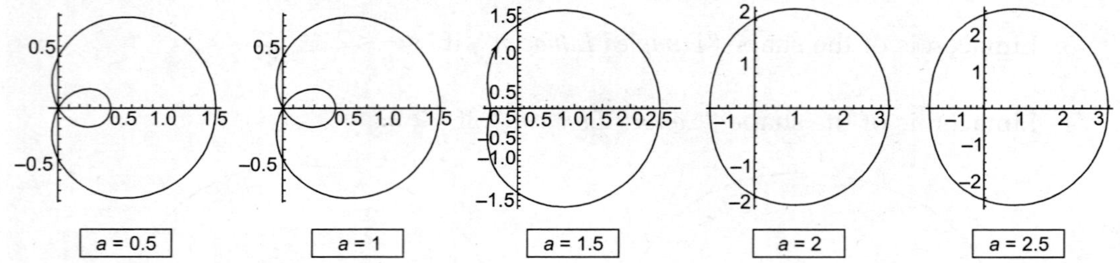

Figure 5.30

5.2.9 Families of Spirals

A *spiral* is a curve that loops around the central point.

The restriction on the polar angle and signs of constant that appear in the equation decide that the spiral is of left-hand and right-hand type (as shown in Figure 5.31).

In real world, there are so many things around us which are in spiral shape. For example: sunflower is a spiral, weather pattern over Iceland is a spiral, a sailor's loops rope forms an Archimedean spiral, etc.

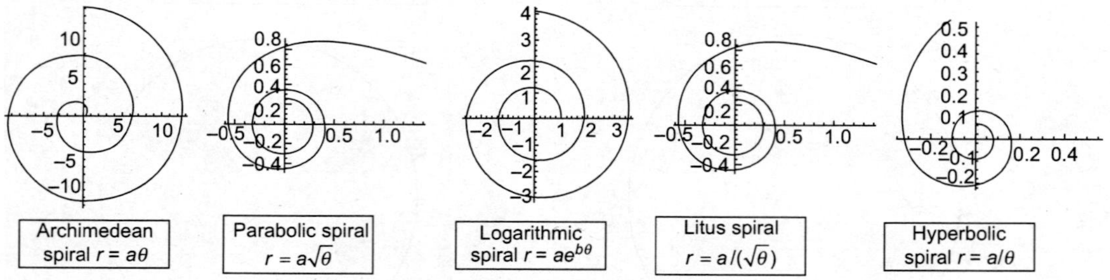

Figure 5.31

5.2.10 Generating Polar Curves with Graphing Utilities

The sketching of polar curves by hand is too complicated, so in this case we can use the graphing utilities. Although, many graphing utilities are capable to sketch the graph of the polar curves directly, but some are not. However, if a graphing utility is capable of sketching the parametric polar equations, then it can be used to sketch the graph of polar curves $r = f(\theta)$.

To find the parametric polar equations for the polar curves $r = f(\theta)$, substituting the value of r in (5.10), we get

$$x = f(\theta)\cos\theta \text{ and } y = f(\theta)\sin\theta . \tag{5.19}$$

These coordinates are known as the pair of parametric equations for the polar curves $r = f(\theta)$, where θ behaves as the parameter.

EXAMPLE 26 Express the polar equation $r = 10 + \sin\left(\frac{8\theta}{5}\right)$ parametrically and generate the polar graph from the parametric equations using a graphing utility.

Solution: Since $x = r\cos\theta$ and $y = r\sin\theta$

Put the value of r from the given equation of the curve, we get

$$x = \left(10 + \sin\left(\frac{8\theta}{5}\right)\right)\cos\theta \text{ and } y = \left(10 + \sin\left(\frac{8\theta}{5}\right)\right)\sin\theta .$$

Now, we need to determine the domain of θ, so that the entire graph can be sketched.

Let n be the smallest number of complete rotations such that after that the curve starts to repeat itself.

Thus, $10 + \sin\left(\frac{8}{5}(\theta + 2n\pi)\right) = 10 + \sin\left(\frac{8\theta}{5} + \frac{16n\pi}{5}\right) = 10 + \sin\frac{8\theta}{5}$

$$\Rightarrow \qquad \sin\left(\frac{8\theta}{5} + \frac{16n\pi}{5}\right) = \sin\frac{8\theta}{5}$$

So, we require that $\frac{16n\pi}{5}$ is an even multiple of π and the smallest n for which this occurs is $n = 5$.

The entire graph will be traced for $0 \le \theta \le 10\pi$ by the parametric equations (as shown in Figure 5.32).

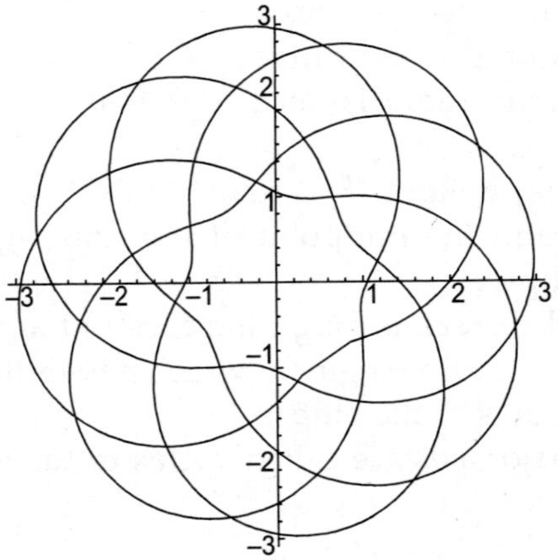

Figure 5.32

$$x = \left(10 + \sin\left(\frac{8\theta}{5}\right)\right)\cos\theta \text{ and } y = \left(10 + \sin\left(\frac{8\theta}{5}\right)\right)\sin\theta$$

5.3 CONIC SECTIONS

In this section, we will study the basic geometric properties of parabolas, ellipses and hyperbolas. These curves have wide range of applications in such fields as planetary motion, design of telescope, antenna, etc.

The circles, ellipses, parabolas and hyperbolas are known as *conic sections or conics* because they are obtained by the intersection of double-napped circular cones with the plane.

If the plane passes through the vertex of a double-napped circular cone, then the intersection is a point, a pair of intersecting lines or a single line. These curves are known as *degenerate conic sections.*

Definition (Parabola): A *parabola* is the set of all points in the plane that are equidistant from a fixed line and a fixed point that do not lie on the line.

The fixed line is known as directrix of the parabola and a fixed point is called the focus. A parabola is symmetric to the line which passes through the focus and intersect the directrix at right angle. This line is called the *axis or the axis of symmetry* of the parabola. A vertex of the parabola is a point where the axis intersects the parabola. The vertex is half way between the directrix and the focus of the parabola (Figure 5.33).

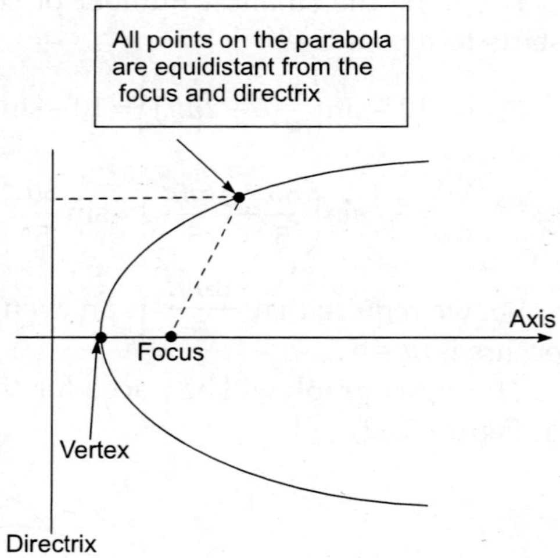

Figure 5.33

Definition (Ellipse): An *ellipse* is a set of all points in the plane, the sum of whose distances from two fixed points is a positive constant and greater than the distance between the fixed points.

These two fixed points F_1 and F_2 are called foci of the ellipse and the midpoint of the line segment joining F_1 and F_2 is called the centre of the ellipse.

The line segment which passes through the foci and across the ellipse is called the major axis of the ellipse and the line segment which passes through the centre and across the ellipse is called minor axis of the ellipse.

The endpoints of the major axis are called vertex of the ellipse (Figure 5.34).

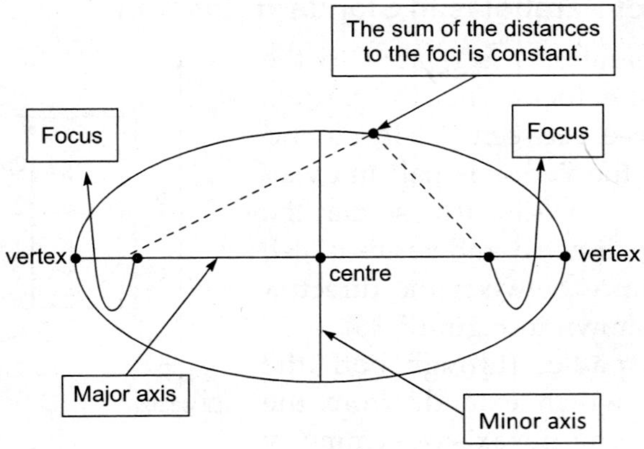

Figure 5.34

Definition (Hyperbola): A *hyperbola* is a set of all points in the plane, the difference of whose distances from two fixed points F_1 and F_2 is a positive constant and less than the distance between the fixed points.

The two fixed points F_1 and F_2 are called foci of the hyperbola. The line which passes through the foci across the two branches of hyperbola is called the focal axis and the midpoint of the line segment joining the foci is called the centre of the hyperbola.

A line which passes through the centre and perpendicular to the focal axis is called conjugate axis of the hyperbola.

The hyperbola intersects the focal axis at two points is called vertex of the hyperbola.

A pair of lines is associated with every hyperbola is called the asymptotes of the hyperbola. These lines intersect at the centre of the hyperbola and have the property that a point P moves along the hyperbola away from the centre such that the perpendicular distance between the point and the asymptotes tends to zero (Figure 5.35).

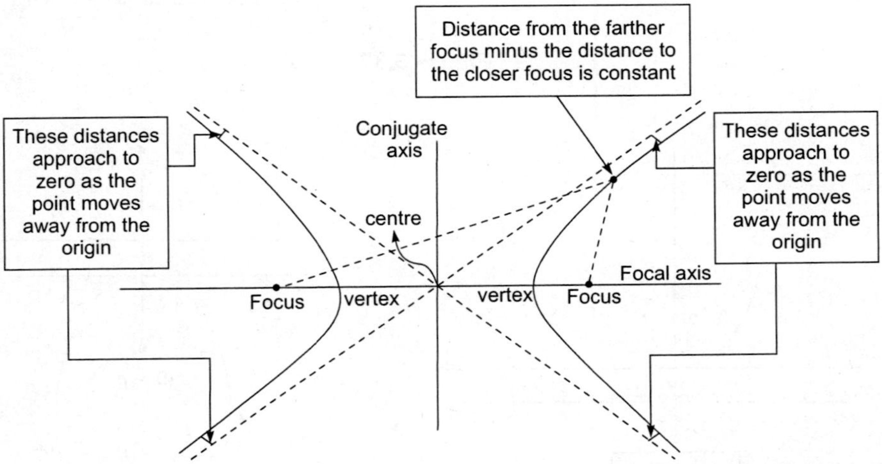

Figure 5.35

5.3.1 Equations of Parabolas in Standard Position

The study of the parabolas is related to the distance between the focus and the vertex. Let p be the distance between the focus and the parabola. Since, the vertex is half midway between the focus and the directrix, so that the distance between the directrix and vertex is also p and so, the distance between the directrix and focus is $2p$ (as shown in Figure 5.36).

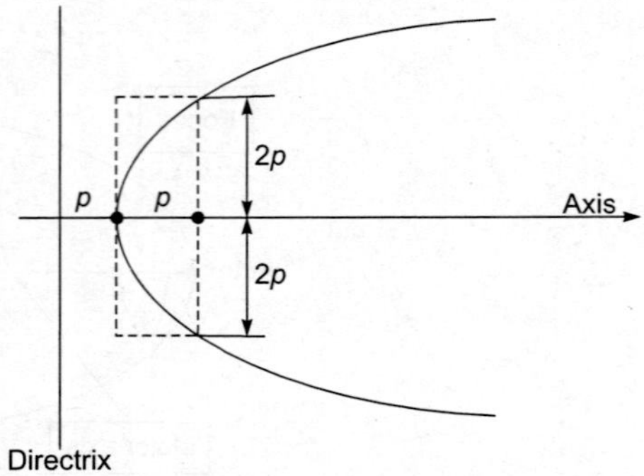

Figure 5.36

So, the parabola passes through both the corners of the box which extends from the vertex to the focus along the axis of symmetry and extends $2p$ units above and below the axis of symmetry.

If the vertex of the parabola is at the origin and the axis of symmetry is either along the x-axis or y-axis, then an equation of the parabola is in standard form, known as *standard equations* of a parabola and the position of the parabola is also known as *standard positions* of a parabola (as shown in Figures 5.37 and 5.38).

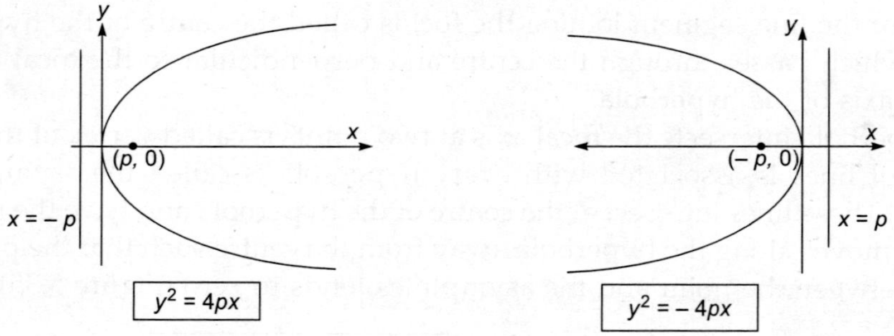

Figure 5.37

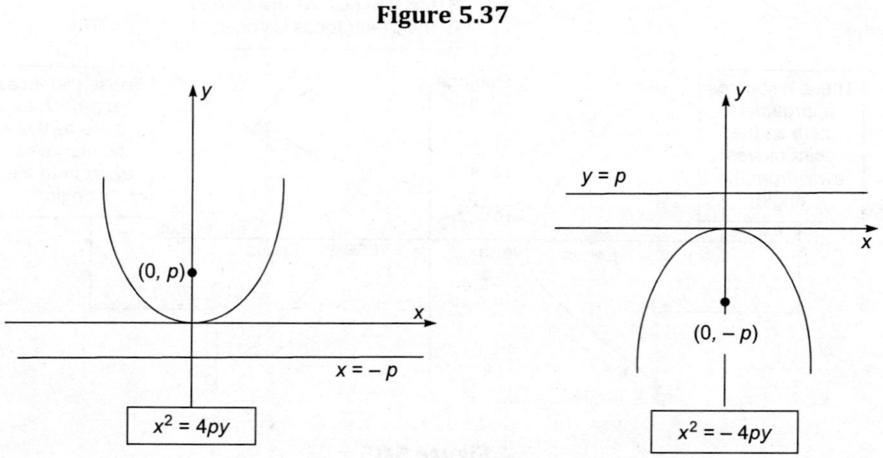

Figure 5.38

Now, we will derive the standard equation of a parabola. Let the parabola be symmetric along the x-axis, vertex at the origin with focus $F(p,0)$ and having directrix $x=-p$. Let $P(x,y)$ be any point on the parabola such that PQ is perpendicular on the directrix $x=-p$ at $Q(-p,y)$ (as shown in Figure 5.39).

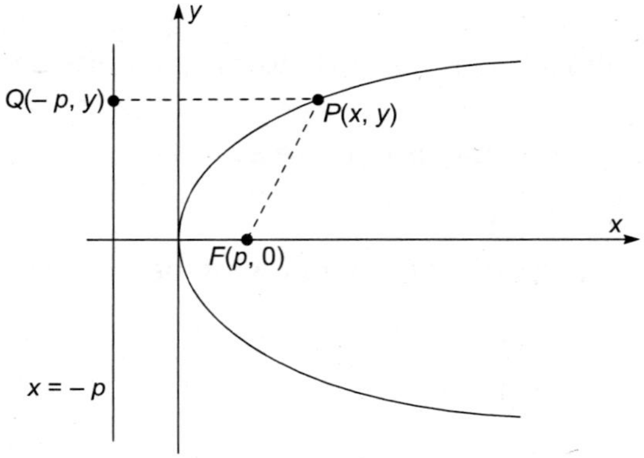

Figure 5.39

By the definition of parabolas, P is equidistant from the focus and directrix, therefore

$$PF = PQ.$$

$$\Rightarrow \qquad \sqrt{(p-x)^2+y^2}=\sqrt{(x+p)^2}$$

Squaring on both sides, we get

$$(p-x)^2+y^2=(x+p)^2$$

$$\Rightarrow \qquad y^2 = 4px. \tag{5.20}$$

Hence, Equation (5.20) is the standard equation of the parabola.

Similarly, the other standard equations which are given in Figures 5.37 and 5.38 can be derived.

General Strategy for Sketching Parabolas

The following four steps are the basic steps for sketching the parabolas from their standard equations:

1. Determine the axis of symmetry, whether it is along the x-axis or y-axis. If the standard equation of a parabola has a y^2 term, then the axis of symmetry is along the x-axis and has a x^2 term, then the axis of symmetry is along the y-axis.
2. Determine which way the parabola opens. Let the axis of symmetry be along the x-axis So, if the coefficient of x is positive in the standard equation of a parabola, then the parabola opens to the right and if the coefficient of x is negative in the standard equation of a parabola, then the parabola opens to the left.

Similarly, let the axis of symmetry be along the x-axis. So, if the coefficient of y is positive in the standard equation of a parabola, then the parabola opens up and if the coefficient of y is negative in the standard equation of a parabola, then the parabola opens down.

3. Determine the value of p from the standard equation of a parabola and then draw a box extending p units from the origin along the axis of symmetry in direction in which the parabola opens and extending $2p$ units on each side of the axis of symmetry.
4. Sketch the parabola from the origin on each side of the axis of symmetry such that its vertex is at the origin and it passes through the corners of the box.

EXAMPLE 27 Sketch the graphs of the following parabolas:

(a) $y^2 = 12x$ (b) $y^2 + 8x = 0$

(c) $x^2 = 16y$ (d) $x^2 + 4y = 0$

Solution:

(a) The given equation of parabola is $y^2 = 12x$. It contains y^2 terms, so it is symmetric along the x-axis and the coefficient of x is positive, so the parabola open towards right.

On comparing with standard equation of the parabola $y^2 = 4px$, we get $p = 3$. Thus, draw a box extending $p = 3$ units right from the origin and extending $2p = 6$ units above and below the axis of symmetry, then sketch the parabola such that it passes through the corners of the box.

The focus of the given parabola is $p = 3$ units from the vertex along the axis of symmetry in the direction where the parabola opens and so, the coordinates of focus is $(3,0)$ (as shown in Figure (5.40(a)).

(b) The given equation of parabola is $y^2 + 8x = 0$, i.e., $y^2 = -8x$. It contains y^2 terms, so it is symmetric along the x-axis and the coefficient of x is negative, so the parabola open towards left.

On comparing with standard equation of the parabola $y^2 = -4px$, we get $p = 2$. Now, draw a box extending $p = 2$ units left from the origin and extending $2p = 4$ units above and below the axis of symmetry, then sketch the parabola such that it passes through the corners of the box.

The focus of the given parabola is $p = 2$ units from the vertex along the axis of symmetry in the direction where the parabola opens and so, the coordinates of focus are $(-2,0)$ (as shown in Figure (5.40(b)).

(c) The given equation of parabola is $x^2 = 16y$. It contains x^2 terms, so it is symmetric along the y-axis and the coefficient of y is positive, so the parabola opens up.

On comparing with standard equation of the parabola $x^2 = 4py$, we get $p = 4$. Thus, draw a box extending $p = 4$ units up from the origin and extending $2p = 8$ units left and right the axis of symmetry, then sketch the parabola such that it passes through the corners of the box.

The focus of the given parabola is $p = 4$ units from the vertex along the axis of symmetry in the direction where the parabola opens and so, the coordinates of focus is $(0,4)$ (as shown in Figure (5.40(c)).

(d) The given equation of parabola is $x^2 + 4y = 0$, i.e., $x^2 = -4y$. It contains x^2 terms, so it is symmetric along the y-axis and the coefficient of y is negative, so the parabola opens down.

On comparing with standard equation of the parabola $x^2 = -4py$, we get $p = 1$. Thus, draw a box extending $p = 1$ units down from the origin and extending $2p = 2$ units left and right the axis of symmetry, then sketch the parabola such that it passes through the corners of the box.

The focus of the given parabola is $p = 1$ units from the vertex along the axis of symmetry in the direction where the parabola opens and so, the coordinates of focus is $(0,-1)$ (as shown in Figure (5.40(d)).

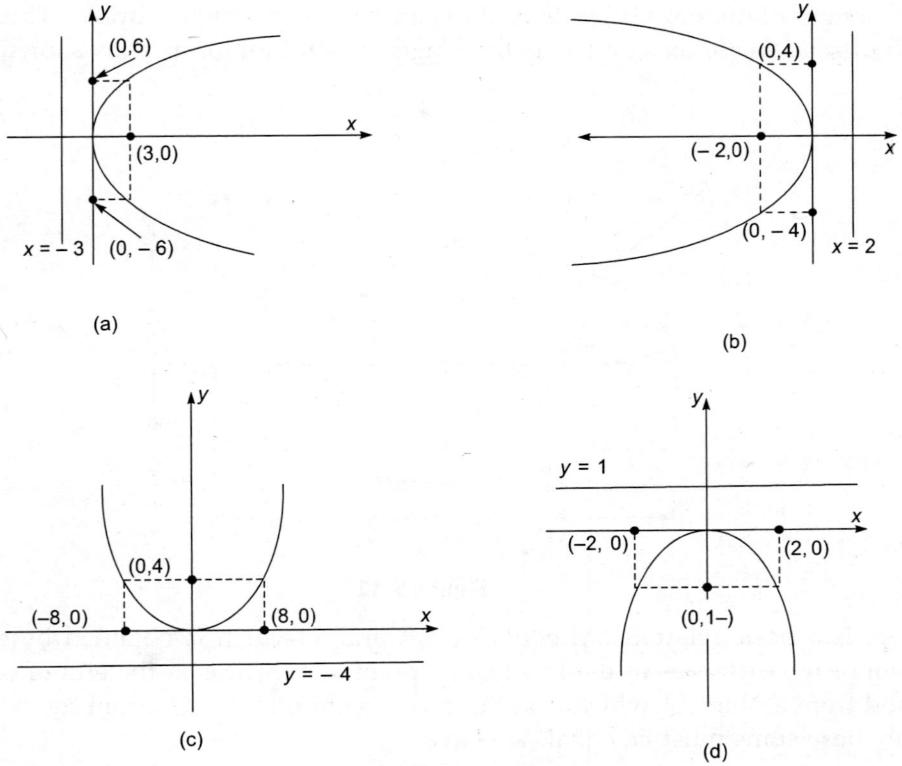

Figure 5.40

EXAMPLE 28 Find an equation of the parabola that is symmetric about the y-axis, has its vertex at the origin and passes through the point $(1,1)$.

Solution: Since the parabola is symmetric along the y-axis and having vertex at the origin, so the standard equations of the parabola are:

$$x^2 = 4py \text{ or } x^2 = -4py.$$

But, the parabola passes through $(1,1)$ that lies on the first quadrant. Thus, the parabola is opens up and so the standard equation of parabola is:

$$x^2 = 4py.$$

Since it passes through $(1,1)$, we have

$$1 = 4p \Rightarrow p = 1/4.$$

Therefore, the equation of the required parabola is

$$x^2 = y.$$

5.3.2 Equations of Ellipses in Standard Position

The study of the ellipse is related to the length of the major axis, length of the minor axis and distance between the foci. Traditionally, we denote the length of the major axis by $2a$, length of minor axis by $2b$ and distance between the foci by $2c$. Thus, a is the length of semi-major axis and b is the length of semi-minor axis (as shown in Figure 5.41).

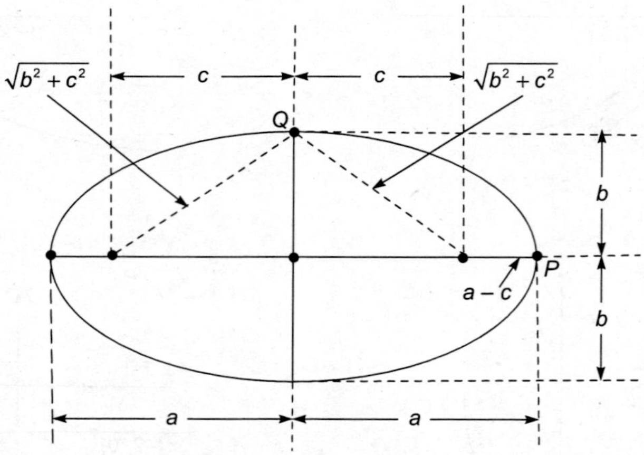

Figure 5.41

There is a basic relationship between a,b and c which is obtained by examining the sum of the distances to the foci from a point P which is at the end of semi-major axis and from a point Q which is at the end of semi-minor axis. From the definition of ellipse, these sums must be equal. We have

$$2\sqrt{b^2+c^2}=(a-c)+(a+c)$$

$$\Rightarrow \qquad a=\sqrt{b^2+c^2} \qquad (5.21)$$

or, equivalently

$$c=\sqrt{a^2-b^2}\,. \qquad (5.22)$$

Thus, by (5.21) the distance between the end of minor axis and the focus is a. Therefore, for all the points on the ellipse the sum of the distances to the foci is $2a$. It also follows that $a \geq b$ and equality holds when $c=0$.

Geometrically, this means that the length of the major axis of an ellipse is at least as large as the minor axis and if the foci of ellipse coincide, then the length of major axis is equal to length of minor axis. In the latter case, the ellipse is a circle.

If the centre of the ellipse is at the origin and the foci are on either x-axis or y-axis, then the equation of ellipse is in standard form, known as *standard equation* of an ellipse and the positions of an ellipse are known as *standard positions* of an ellipse (as shown in Figure 5.42).

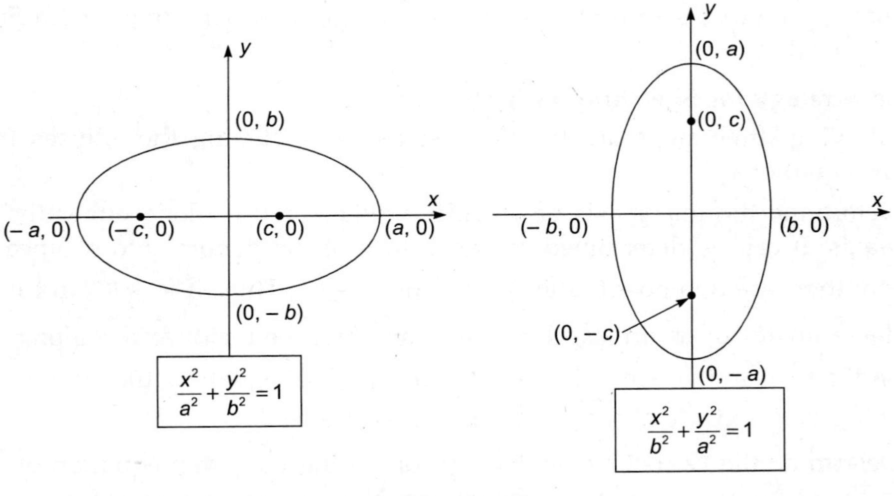

Figure 5.42

Now, we will derive the standard equation of an ellipse.

Let the ellipse be centred at the origin and foci $F_1(c,0)$ and $F_2(-c,0)$ on the x-axis. Let $P(x,y)$ be any point on the ellipse (Figure 5.43). Since the point $P(x, y)$ is on an ellipse, so the sum of the distances from $P(x, y)$ to the foci is $2a$,

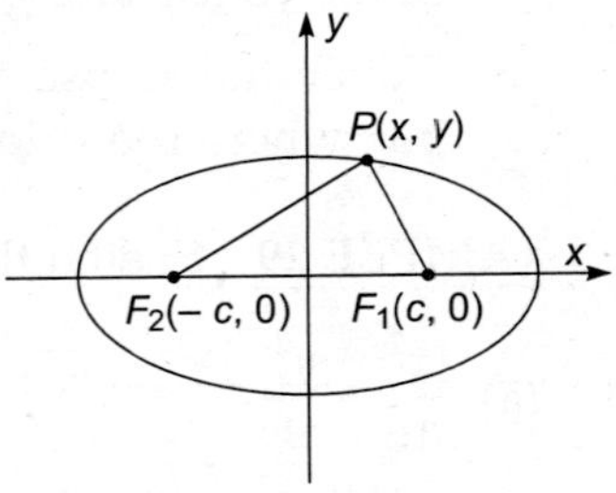

Figure 5.43

i.e.,

$$PF_1+PF_2=2a.$$

$$\Rightarrow \qquad \sqrt{(c-x)^2+y^2}+\sqrt{(c+x)^2+y^2}=2a$$

$$\Rightarrow \qquad \sqrt{(c-x)^2+y^2}=2a-\sqrt{(c+x)^2+y^2}\,.$$

Squaring on both sides, we get

$$(c-x)^2+y^2=4a^2+(c+x)^2+y^2-4a\sqrt{(c+x)^2+y^2}$$

$$\Rightarrow \sqrt{(c+x)^2+y^2}=a+\frac{cx}{a}.$$

Again, squaring on both sides, we get

$$(c+x)^2+y^2=a^2+\frac{c^2x^2}{a^2}+2cx$$

$$\Rightarrow \frac{x^2}{a^2}+\frac{y^2}{a^2-c^2}=1.$$

From (5.21), we have

$$\frac{x^2}{a^2}+\frac{y^2}{b^2}=1$$

which is the required standard equation of an ellipse.

Similarly, the other standard equations of an ellipse which are given in Figure 5.42 can be derived.

General Strategy for Sketching Ellipses

The following three steps are the basic steps for sketching the ellipses from their standard equations:

1. Determine the major axis of an ellipse, whether it is along either the x-axis or y-axis. It can be determined by the values of the denominators. Since $a \geq b$, so that there are two possibilities $a^2 > b^2$ or $a^2 = b^2$. Thus, if $a^2 > b^2$ and a^2 is in the denominator of x^2 in standard equation, then the major axis is along the x-axis and if a^2 is in the denominator of y^2 in standard equation, then the major axis is along the y-axis. If $a^2 = b^2$, then the ellipse is a circle.
2. Determine the value of a and b by comparing the given equation of an ellipse to the standard equations of the ellipse. Now, draw a box extending a units on both sides of the centre along the major axis and extending b units on both sides of the centre along the minor axis.
3. Sketch the ellipse so that its centre is at the origin and it touches each side of the box where the sides intersect with the coordinates axes x and y.

EXAMPLE 29 Sketch the graph of the following ellipses and show the foci:

(a) $\frac{x^2}{16}+\frac{y^2}{9}=1$

(b) $9x^2+y^2=9$

Solution:

(a) Since x^2 has the larger denominator, the major axis is along the x-axis.

Since $a \geq b \Rightarrow a^2 \geq b^2$, i.e., by comparing the given equation of an ellipse to the standard equation of the ellipse, we get $a^2 = 16$ and $b^2 = 9$.

So, that $a = 4$ and $b = 3$.

The foci lie c units on both sides of the centre along the major axis, where c is given by (5.22).

i.e., $$c = \sqrt{a^2 - b^2} = \sqrt{4^2 - 3^2} = \sqrt{7} \approx 2.6$$

Thus, the coordinates of the foci are $(\sqrt{7}, 0)$ and $(-\sqrt{7}, 0)$ which lie on the x-axis. Now, sketch the ellipse so that its centre is at the origin and it touches each side of the box where the sides intersect with the coordinates axes x and y (as shown in Figure 5.44(a)).

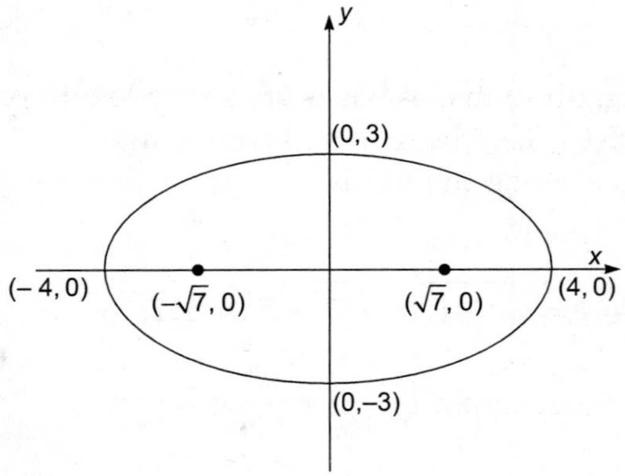

Figure 5.44(a)

(b) Equation of the given curve is

$$9x^2 + y^2 = 9.$$

Dividing by 9 on both sides, we get

$$\frac{x^2}{1} + \frac{y^2}{9} = 1.$$

Now, y^2 has the larger denominator, the major axis is along the y-axis.

Since $a \geq b \Rightarrow a^2 \geq b^2$, i.e., by comparing the given equation of an ellipse to the standard equation of the ellipse, we get $a^2 = 9$ and $b^2 = 1$. So that $a = 3$ and $b = 1$. Now, draw a box extending 3 units on both sides of the centre along the major axis and extending 1 unit on both sides of the centre along the minor axis and then, sketch the ellipse so that its centre is at the origin and it touches each side of the box where the sides intersect with the coordinates axes x and y.

The foci lie c units on both sides of the centre along the major axis, where c is given by (5.22),

i.e., $$c=\sqrt{a^2-b^2}=\sqrt{3^2-1^2}=\sqrt{8}\approx 2.8.$$

Thus, the coordinates of the foci are $\left(0,\sqrt{8}\right)$ and $\left(0,-\sqrt{8}\right)$ which lie on y-axis.

Now, sketch the ellipse so that its centre is at the origin and it touches each side of the box where the sides intersect the coordinates axes x and y (as shown in Figure 5.44(b)).

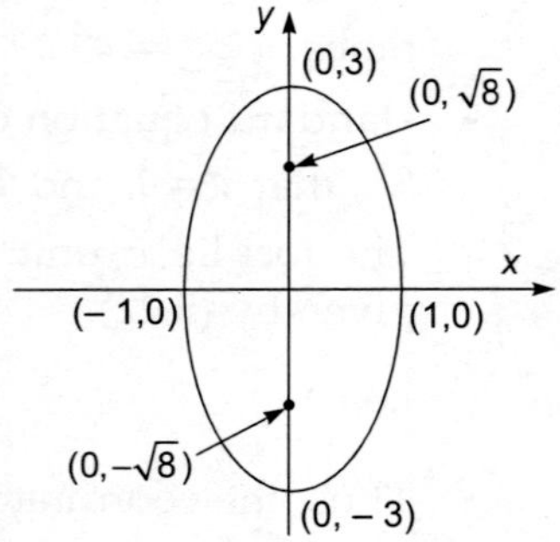

Figure 5.44(b)

EXAMPLE 30 Find the equation of an ellipse with length of major axis is 26 and foci are $(\pm 5,0)$.

Solution: Since the length of major axis is 26, so the length of semi-major axis is 13. Also, the foci are $(\pm 5,0)$, i.e., the foci are lie on x-axis.
Thus, the major axis is along the x-axis.

Now, $$a=13 \text{ and } c=5.$$

From (5.21), we have $b=\sqrt{a^2-c^2}=\sqrt{13^2-5^2}=\pm 12.$

Therefore, equation of an ellipse is $\dfrac{x^2}{169}+\dfrac{y^2}{144}=1.$

5.3.3 Equations of Hyperbolas in Standard Position

The study of hyperbola is related to the distance between the vertices by $2a$, the distance between the foci by $2c$ and the quantity $\sqrt{c^2-a^2}$ which is denoted by b.

From Figure 5.45, if V is the vertex of a hyperbola, the distance from V to the closer focus subtracted from the distance from V to the farther focus is given by

$$\left((c-a)+a+a\right)-(c-a)=2a.$$

Thus, for all the points on a hyperbola, the distance to the closer focus subtracted from the distance to the farther focus is $2a$.

The asymptotes to the hyperbola pass through the corners of a box which is made by extending b units on both sides of the centre along the conjugate axis and extending a units on both sides of the centre along the focal axis (as shown in Figure 5.44).

Thus, a is the length of semi-focal axis of the hyperbola and b is length of semi-conjugate axis.

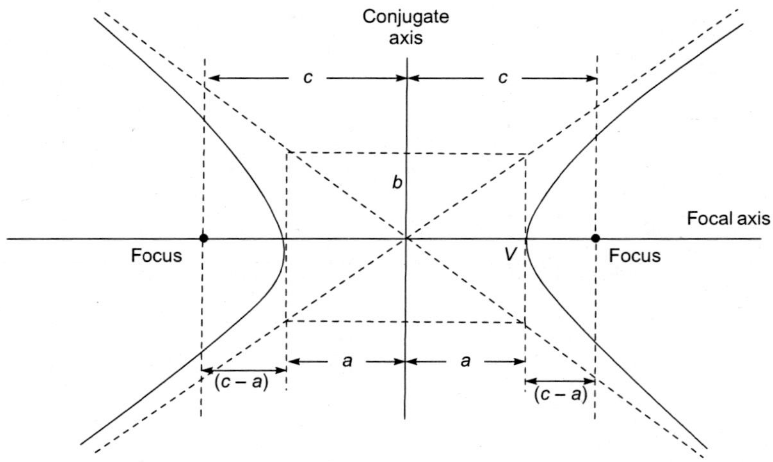

Figure 5.45

If the centre of the hyperbola is at the origin and the foci are on either x-axis or y-axis, then the equation of hyperbola is in standard form, known as *standard equation* of a hyperbola and the positions of a hyperbola are known as *standard positions* of a hyperbola (as shown in Figure 5.46 (a) and (b)).

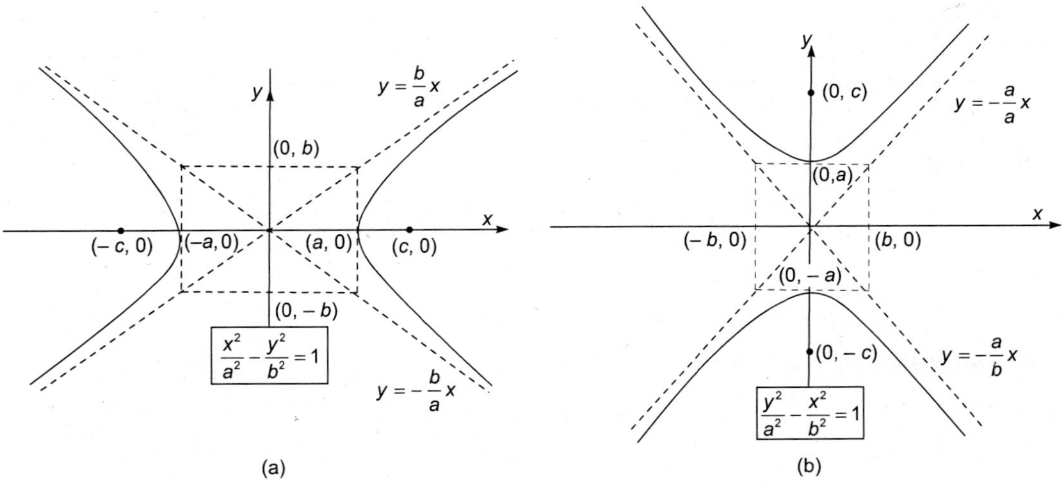

Figure 5.46

Now, we will derive the standard equation of a hyperbola.

Let the hyperbola be centred at the origin with foci $F_1(c,0)$ and $F_2(-c,0)$ are on the x-axis.

Let $P(x,y)$ be any point on the hyperbola (as shown in Figure 5.47)

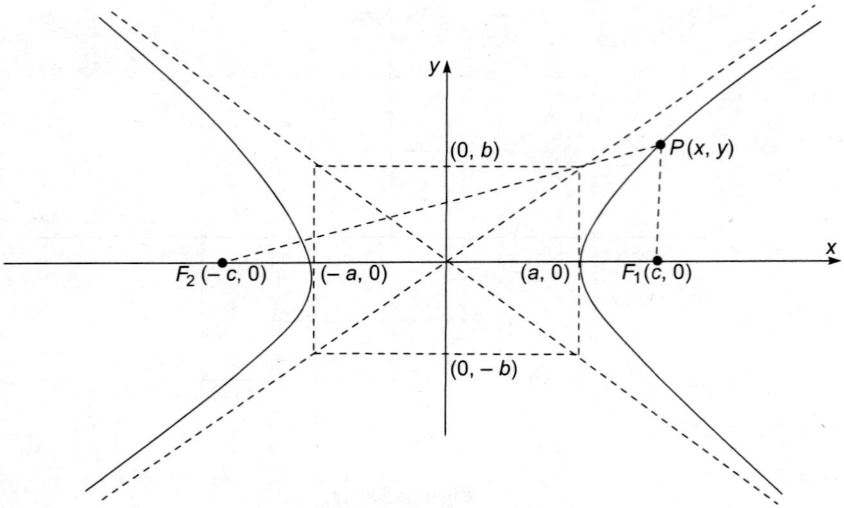

Figure 5.47

Since, for all the points on hyperbola, the distance to the closer focus subtracted from the distance to the farther focus is $2a$,

i.e., $$PF_2 - PF_1 = 2a$$

$$\Rightarrow \sqrt{(c+x)^2 + y^2} - \sqrt{(c-x)^2 + y^2} = 2a$$

$$\Rightarrow \sqrt{(c+x)^2 + y^2} = 2a + \sqrt{(c-x)^2 + y^2}.$$

Squaring on both sides, we get

$$(c+x)^2 + y^2 = 4a^2 + (c-x)^2 + y^2 + 4a\sqrt{(c-x)^2 + y^2}$$

$$\Rightarrow \quad \sqrt{(c-x)^2 + y^2} = \frac{cx}{a} - a.$$

Again, squaring on both sides, we get

$$(c-x)^2 + y^2 = \frac{c^2x^2}{a^2} + a^2 - 2cx$$

$$\Rightarrow \quad \frac{x^2}{a^2} + \frac{y^2}{a^2 - c^2} = 1$$

$$\Rightarrow \quad \frac{x^2}{a^2} - \frac{y^2}{b^2} = 1\left(as\, b = \sqrt{c^2 - a^2}\right)$$

which is the required standard equation of the hyperbola.

Similarly, the other standard equation of a hyperbola which is given in Figure 5.45(b), can be derived.

Now, we will derive the equations of the asymptotes to the hyperbola.

Let the equation of the hyperbola be

$$\frac{x^2}{a^2}-\frac{y^2}{b^2}=1$$

$$\Rightarrow \quad y^2=\frac{b^2}{a^2}\left(x^2-a^2\right)$$

$$\Rightarrow \quad y=\frac{b}{a}\sqrt{x^2-a^2} \text{ or } y=-\frac{b}{a}\sqrt{x^2-a^2}.$$

Thus, in the first quadrant, the vertical distance between the hyperbola $y=\frac{b}{a}\sqrt{x^2-a^2}$ and the line $y=\frac{b}{a}x$ (as in Figure 5.46(a)) is given as $\frac{b}{a}x-\frac{b}{a}\sqrt{x^2-a^2}$.

Now,
$$\lim_{x\to\infty}\frac{b}{a}x-\frac{b}{a}\sqrt{x^2-a^2}=\lim_{x\to\infty}\frac{b}{a}\left(x-\sqrt{x^2-a^2}\right)$$

$$=\lim_{x\to\infty}\frac{b}{a}\frac{\left(x-\sqrt{x^2-a^2}\right)\left(x+\sqrt{x^2-a^2}\right)}{\left(x+\sqrt{x^2-a^2}\right)}$$

$$=\lim_{x\to\infty}\frac{b}{a}\frac{a^2}{\left(x+\sqrt{x^2-a^2}\right)}$$

$$=0.$$

So, we obtained that the vertical distance between the hyperbola $y=\frac{b}{a}\sqrt{x^2-a^2}$ and the line $y=\frac{b}{a}x$ tends to zero as x tends to infinity.

Therefore, we conclude that the line $y=\frac{b}{a}x$ is asymptotes to the hyperbola $\frac{x^2}{a^2}-\frac{y^2}{b^2}=1$.

Similarly, other lines which are given in Figure 5.45(a) and (b), can be shown as the asymptotes to the hyperbolas.

The asymptotes of a hyperbola can be obtained by the following method.

The equation of the asymptotes of a hyperbola can be obtained by putting 0 for 1 on the right side of an equation of a hyperbola and then solve for $\boldsymbol{y}$ in terms of $\boldsymbol{x}$.

For instance:

Let the equation of a hyperbola be $\frac{\boldsymbol{x}^2}{\boldsymbol{a}^2}-\frac{\boldsymbol{y}^2}{\boldsymbol{b}^2}=1$.

Take $\frac{\boldsymbol{x}^2}{\boldsymbol{a}^2}-\frac{\boldsymbol{y}^2}{\boldsymbol{b}^2}=0\Rightarrow \boldsymbol{y}^2=\frac{\boldsymbol{b}^2}{\boldsymbol{a}^2}\boldsymbol{x}^2\Rightarrow \boldsymbol{y}=\pm\frac{\boldsymbol{b}}{\boldsymbol{a}}\boldsymbol{x}$, which is the required equation for the asymptotes.

General Strategy for Sketching Hyperbolas

The following four steps are the basic steps for sketching the hyperbolas from their standard equations:

1. Determine the focal axis of a hyperbola, whether it is along either the x-axis or y-axis. It can be determined by the negative sign in the equation of a hyperbola. If the negative sign precedes the y^2 term, then the focal axis is along the x-axis and if the negative sign precedes the x^2 term, then the focal axis is along the y-axis (as shown in Figure 5.46 (a) and (b)).
2. Determine the value of a and b by comparing the given equation of a hyperbola to the standard equations of the hyperbolas. Then, draw a box extending a units on both sides of the centre along the focal axis and extending b units on both sides of the centre along the conjugate axis.
3. Determine the equations for the asymptotes of a hyperbola and then draw the asymptotes along the diagonal of the box.
4. Sketch the hyperbola so that its centre is at the origin and it touches the two opposite outer sides of the box where they intersect the coordinates axes x and y.

EXAMPLE 31 Sketch the graph of the following hyperbolas and also show their foci, vertices and asymptotes.

(a) $\frac{x^2}{16}-\frac{y^2}{4}=1$

(b) $9y^2-4x^2=36$

Solution:

(a) Since the negative sign precedes the y^2 term, so the focal axis is along the x-axis. By comparing the given equation of hyperbola to the standard equation of hyperbola, we get $a^2=16$ and $b^2=4$. So, that $a=4$ and $b=2$.
The vertices of a hyperbola lie $a=4$ units on both sides of the centre along focal axis, thus the vertices of a hyperbola are $(\pm 4,0)$.
Now, draw a box extending 4 units on both sides of the centre along the focal axis and extending 2 units on both sides of the centre along the conjugate axis.
To find the equations of asymptotes, substitute 0 for 1 in the given equation of a hyperbola. This implies

$$\frac{x^2}{16}-\frac{y^2}{4}=0 \Rightarrow y=\pm 2x.$$

Thus, the equations for asymptotes to a hyperbola are $y=\pm 2x$.

The foci lie c units on both sides of the centre along the focal axis, where

$$b = \sqrt{c^2 - a^2} \Rightarrow c = \sqrt{a^2 + b^2} = \sqrt{4^2 + 2^2} = \sqrt{20} \approx 4.47$$

Thus, the coordinates of foci are $\left(\pm\sqrt{20}, 0\right)$.

Now, sketch the hyperbola so that its centre is at the origin and its vertices touch the two opposite outer sides of the box where the sides intersect with the coordinates axes x and y and asymptotes along the diagonal of the box (as shown in Figure 5.48(a)).

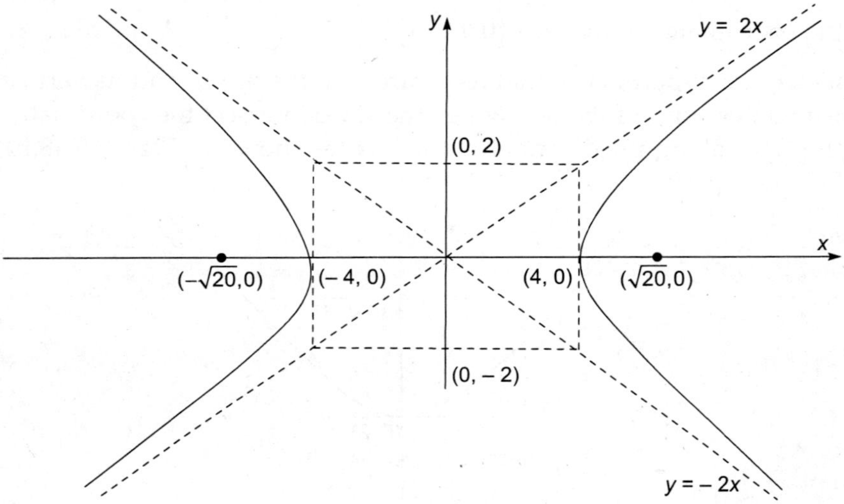

Figure 5.48(a)

(b) The given equation of curve is $9y^2 - 4x^2 = 36$

Dividing by 36 on both sides, we get

$$\frac{y^2}{4} - \frac{x^2}{9} = 1$$

Since the negative sign precedes the x^2 term, so the focal axis is along the y-axis. By comparing the given equation of hyperbola to the standard equation of hyperbola, we get $a^2 = 4$ and $b^2 = 9$. So $a = 2$ and $b = 3$.

The vertices of a hyperbola lie $a = 2$ units on both sides of the centre along focal axis, thus the vertices of a hyperbola are $(0, \pm 2)$.

Now, draw a box extending 2 units on both sides of the centre along the focal axis and extending 3 units on both sides of the centre along the conjugate axis.

To find equations of asymptotes, substitute 0 for 1 in the given equation of a hyperbola.

This implies

$$\frac{y^2}{4}-\frac{x^2}{9}=0 \Rightarrow y=\pm\frac{2}{3}x$$

Thus, the equations for asymptotes to a hyperbola are $y=\pm\frac{2}{3}x$.

The foci lie c units on both sides of the centre along the focal axis, where

$$b=\sqrt{c^2-a^2} \Rightarrow c=\sqrt{a^2+b^2}=\sqrt{2^2+3^2}=\sqrt{13}\approx 3.6$$

Thus, the coordinates of foci are $\left(0,\pm\sqrt{13}\right)$.

Now, sketch the hyperbola so that its centre is at the origin and its vertices touch the two opposite outer sides of the box where the sides intersect the coordinates axes x and y and asymptotes along the diagonal of the box (as shown in Figure 5.48(b)).

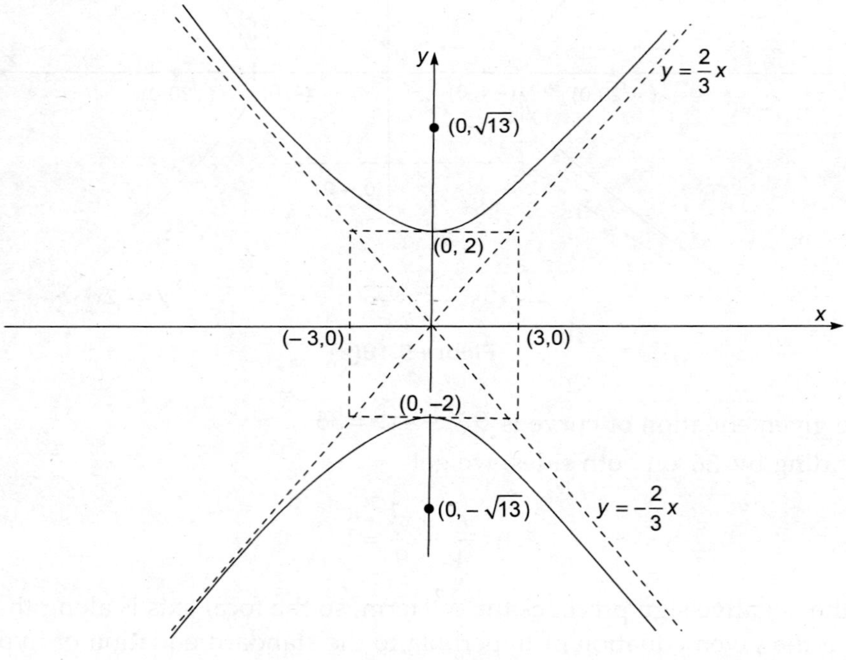

Figure 5.48(b)

EXAMPLE 32 Find an equation for a hyperbola that has the following conditions:

(a) vertices $(\pm 2, 0)$ and foci $(\pm 3, 0)$.

(b) vertices $(0, \pm 3)$ and asymptotes $y=\pm x$.

(c) asymptotes $y=\pm\frac{3}{2}x$ and $b=4$.

(d) asymptotes $y = \pm\frac{3}{4}x$ and $c = 5$.

Solution:

(a) Since the coordinates of foci are (±3, 0), i.e., the foci lie on x-axis. Then, the focal axis of a hyperbola is along the x-axis. Thus, the equation of a hyperbola is of the form $\frac{x^2}{a^2} - \frac{y^2}{b^2} = 1$.

Since vertices are (±2, 0) and foci are (±3, 0) then the value of $a = 2 \Rightarrow a^2 = 4$ and $c = 3$.

Also, we know that $b = \sqrt{c^2 - a^2} = \sqrt{3^2 - 2^2} = \sqrt{5} \Rightarrow b^2 = 5$.

The equation of asymptotes to a hyperbola is $y = \pm\frac{b}{a}x = \pm\frac{5}{2}x$.

Therefore, the equation of the hyperbola is $\frac{x^2}{4} - \frac{y^2}{5} = 1$.

(b) Since the vertices lie on the y-axis, the equation of the hyperbola is of the form $\frac{y^2}{a^2} - \frac{x^2}{b^2} = 1$ and the asymptotes are $y = \pm\frac{a}{b}x$.

Since the coordinates of the vertices are $(0, \pm 3)$, the value of $a = 3 \Rightarrow a^2 = 9$.

Since the equation of asymptotes are $y = \pm x = \pm\frac{3}{b}x$

$$\Rightarrow \quad 1 = \frac{a}{b} = \frac{3}{b}$$

$$\Rightarrow \quad b = 3.$$

Therefore, the equation of the hyperbola is $\frac{y^2}{9} - \frac{x^2}{9} = 1$.

(c) Let the equation of the hyperbola be $\frac{x^2}{a^2} - \frac{y^2}{b^2} = 1$.

Since the equation of the asymptotes are $y = \pm\frac{3}{2}x = \pm\frac{b}{a}x$ and $b = 4$, then

$$\Rightarrow \quad \frac{3}{2} = \frac{4}{a}$$

$$\Rightarrow \quad a = \frac{8}{3}.$$

Therefore, the equation of the hyperbola is $\frac{9x^2}{64} - \frac{y^2}{16} = 1$.

(d) Let the equation of the hyperbola be $\frac{x^2}{a^2}-\frac{y^2}{b^2}=1$

Since $c=5 \Rightarrow 5=\sqrt{a^2+b^2}$

$\Rightarrow \quad 25 = a^2 + b^2$ (5.23)

Since the equation of an asymptotes is $y=\pm\frac{3}{4}x \Rightarrow \frac{b}{a}=\frac{3}{4} \Rightarrow b=\frac{3a}{4}$. (5.24)
Substituting in (5.23), we get

$$25=a^2+\frac{9a^2}{16}$$

$$\Rightarrow \quad 25=\frac{25a^2}{16}$$

$$\Rightarrow \quad a^2 = 16$$

$$\Rightarrow \quad a = \pm 4.$$

By using (5.24), we have $b=\pm 3$.

Therefore, the equation of a hyperbola is $\frac{x^2}{16}-\frac{y^2}{9}=1$.

5.3.4 Translated Conics

When the vertex of the parabolas, centre of the ellipses and the hyperbolas are shifted to some other point from the origin, the conics are known as *translated conics*.

Let the vertex of the parabolas, centre of the ellipses and the hyperbolas be shifted to a point (h,k) from the origin, the equations of translated conics can be obtained by replacing x by $(x-h)$ and y by $(y-k)$ in their standard equations, which are given below:

Parabolas with vertex (h, k) and the axis of symmetry parallel to x-axis

$(y-k)^2 = 4p(x-h)$ (opens to right)

$(y-k)^2 = -4p(x-h)$ (opens to left)

Parabolas with vertex (h, k) and the axis of symmetry parallel to y-axis

$(x-h)^2 = 4p(y-k)$ (opens up)

$(x-h)^2 = -4p(y-k)$ (opens down)

Ellipses with centre (h, k) and the major axis parallel to x-axis

$$\frac{(x-h)^2}{a^2}+\frac{(y-k)^2}{b^2}=1\,(a \ge b)$$

Ellipses with centre (h, k) and the major axis parallel to y-axis

$$\frac{(x-h)^2}{b^2}+\frac{(y-k)^2}{a^2}=1\,(a\ge b)$$

Hyperbolas with centre (h, k) and the focal axis parallel to x-axis

$$\frac{(x-h)^2}{a^2}-\frac{(y-k)^2}{b^2}=1$$

Hyperbolas with centre (h, k) and the focal axis parallel to y-axis

$$\frac{(y-k)^2}{a^2}-\frac{(x-h)^2}{b^2}=1$$

EXAMPLE 33 Sketch the graph of the following conics:

(a) $(y-3)^2=6(x-2)$

(b) $x^2-4x+2y=1$

(c) $9(x-1)^2+16(y-3)^2=144$

(d) $x^2+9y^2+2x-18y+1=0$

(e) $\dfrac{(y+4)^2}{9}-\dfrac{(x-2)^2}{25}=1$

(f) $x^2-4y^2+2x+8y-7=0$

Solution:

(a) The given equation of the conic is $(y-3)^2=6(x-2)$.

It is of the form $(y-k)^2=4p(x-h)$, then $k=3, h=2$ and $p=\dfrac{3}{2}$.

So, the graph of the parabola with vertex $(2,3)$ opens towards right. Since $p=\dfrac{3}{2}$, i.e., the focus is $\dfrac{3}{2}$ units to the right of the vertex, which is located at the point $(7/2,3)$ and the directrix is $\dfrac{3}{2}$ units to the left of the vertex, which means that the equation of directrix is $x=1/2$. The graph of the parabola is shown in Figure 5.49(a).

(b) The given equation of the conic is $x^2-4x+2y=1$.

The equation involves x^2 term but not y^2, so take all the terms of x on one side, we get

$$x^2 - 4x = 1 - 2y$$
$$\Rightarrow (x-2)^2 - 4 = 1 - 2y$$
$$\Rightarrow (x-2)^2 = 5 - 2y$$
$$\Rightarrow (x-2)^2 = -2(y - 5/2)$$

It is of the form $(x-h)^2 = -4p(y-k)$, then $h = 2, k = 5/2$ and $p = \dfrac{1}{2}$.

So, the graph of the parabola with vertex $(2, 5/2)$, opens down and axis of symmetry is parallel to y-axis. Since $p = \dfrac{1}{2}$, i.e., the focus is $\dfrac{1}{2}$ unit below the vertex, which is located at the point $(2,2)$ and the directrix is $\dfrac{1}{2}$ unit above the vertex, which means that the equation of directrix is $y = 3$. The graph of the parabola is shown in Figure 5.49(b).

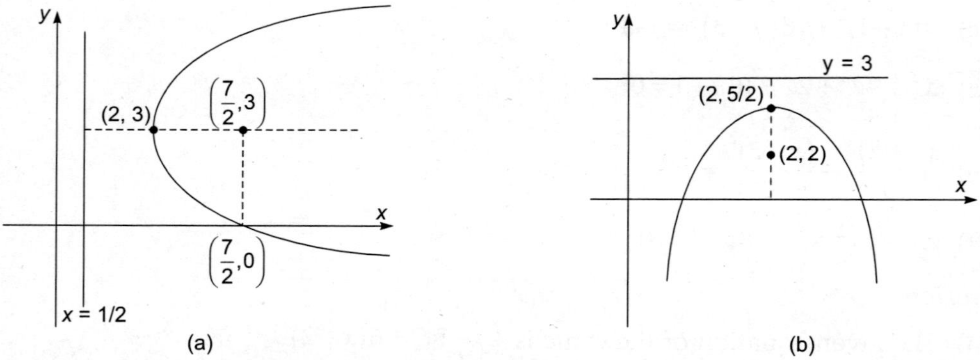

Figure 5.49

(c) The given equation of the conic is $9(x-1)^2 + 16(y-3)^2 = 144$.
Divide by 144 on both sides, we get

$$\frac{(x-1)^2}{16} + \frac{(y-3)^2}{9} = 144.$$

It is of the form $\dfrac{(x-h)^2}{a^2} + \dfrac{(y-k)^2}{b^2} = 1 \ (a \ge b)$, then $h = 1, k = 3, a^2 = 16, b^2 = 9$.

Thus, the graph of the equation is an ellipse with centre at $(1,3)$ and since the denominator of $(x-1)^2$ is larger, so the major axis is parallel to x-axis. Since $a = 4$, the major axis extends 4 units left to the centre and extends 4 units right to the centre, so its end points are $(-3,3)$ and $(5,3)$. Since $b=3$, the minor axis extends 3 units above the centre and extends 3 units below the centre, so its end points are $(1,6)$ and $(1,0)$.

Also, the foci lie c units left and right to the centre of the ellipse, where

$$c=\sqrt{a^2-b^2}=\sqrt{16-9}=\sqrt{7}.$$

Thus, the foci are at the points $\left(1-\sqrt{7},3\right)$ and $\left(1+\sqrt{7},3\right)$. The graph of the given ellipse is shown in Figure 5.49(c).

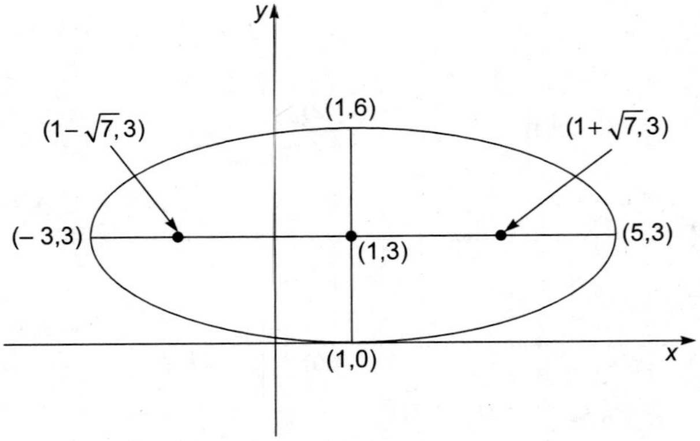

Figure 5.49(c)

(d) The given equation of the conic is $x^2+9y^2+2x-18y+1=0$.

Since the equation contains both x^2 and y^2 terms, so put all the terms of x and y on one side and the constant term on the other side of the equation, i.e.,

$(x^2+2x)+\left(9y^2-18y\right)=-1$

$\Rightarrow (x+1)^2+9(y-1)^2=9$

Dividing by 9 on both side, we get

$$\frac{(x+1)^2}{9}+\frac{(y-1)^2}{1}=1.$$

It is of the form $\frac{(x-h)^2}{a^2}+\frac{(y-k)^2}{b^2}=1\ (a\geq b)$, then $h=-1, k=1, a^2=9, b^2=1$.

Thus, the graph of the equation is an ellipse with centre at $(-1,1)$ and since the denominator of $(x+1)^2$ is larger, so the major axis is parallel to x-axis. Since $a=3$, the major axis extends 3 units left to the centre and extends 3 units right to the centre, so its end points are $(-4,1)$ and $(3,1)$. Since $b=1$, the minor axis extends 1 unit above the centre and extends 1 unit below the centre, so its end points are $(-1,2)$ and $(-1,0)$.

Also, the foci lie c units left and right to the centre of the ellipse, where

$$c=\sqrt{a^2-b^2}=\sqrt{9-1}=\sqrt{8}$$

Thus, the foci are at points $\left(-1-\sqrt{8},1\right)$ and $\left(-1+\sqrt{8},1\right)$. The graph of the given ellipse is shown in Figure 5.49(d).

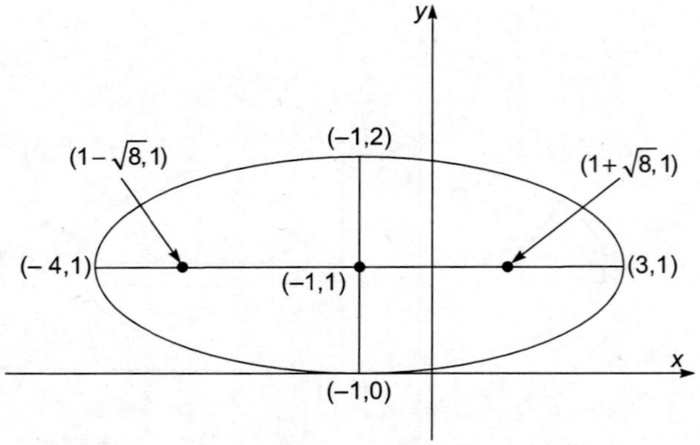

Figure 5.49(d)

(e) The given equation of the conic is $\dfrac{(y+4)^2}{9}-\dfrac{(x-2)^2}{25}=1$.

It is of the form $\dfrac{(y-k)^2}{a^2}-\dfrac{(x-h)^2}{b^2}=1$, then $k=-4, h=2$, $a^2=9$ and $b^2=25$.

Thus, the given equation represents hyperbola with centre $(2,-4)$ and focal axis is parallel to y-axis. Since $a=3$, the vertex is located 3 units above the centre and 3 units below the centre, i.e., the coordinates of the vertices are $(2,-1)$ and $(2,-7)$.

Since $c=\sqrt{a^2+b^2}=\sqrt{9+25}=\sqrt{34}\approx 5.83$.

Thus, the foci are located $\sqrt{34}$ units above and below the centre, so the coordinates of the foci are $\left(2,-4+\sqrt{34}\right)$ and $\left(2,-4-\sqrt{34}\right)$.

To find the equations for asymptotes,

Consider $\dfrac{(y+4)^2}{9}-\dfrac{(x-2)^2}{25}=0$

$$\Rightarrow \qquad y+4=\pm\frac{3}{5}(x-2)$$

$$\Rightarrow y=\frac{1}{5}(3x-26) \text{ and } y=-\frac{1}{5}(3x+14).$$

Therefore, the graph of the hyperbola is shown in Figure 5.49(e).

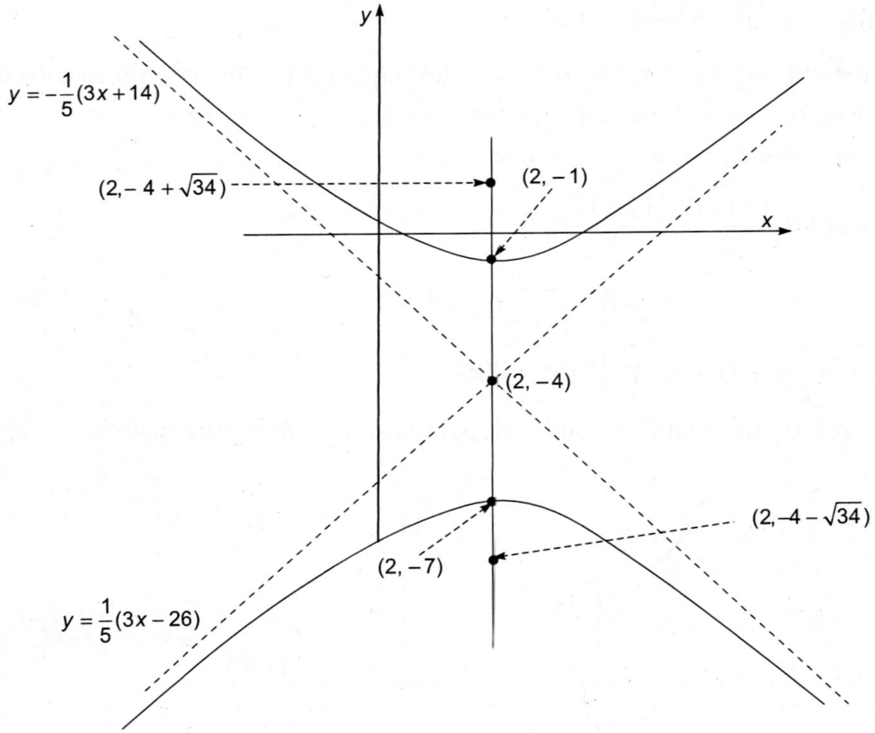

Figure 5.49(e)

(f) The given equation of the curve is $x^2-4y^2+2x+8y-7=0$.

This equation contains both the terms of x^2 and y^2, so put all the terms of x and y on one side and constant term on the other side of the equation, i.e.,

$$\left(x^2+2x\right)-4\left(y^2-2y\right)=7$$

$$\Rightarrow \qquad (x+1)^2-4(y-1)^2=4$$

$$\Rightarrow \qquad \frac{(x+1)^2}{4}-\frac{(y-1)^2}{1}=1.$$

It is of the from $\frac{(x-h)^2}{a^2}-\frac{(y-k)^2}{b^2}=1$, then $h=-1, k=1$, $a^2=4$ and $b^2=1$. Thus, the given equation represents hyperbola with centre $(-1,1)$ and focal axis is parallel to x-axis. Since $a=2$, the vertex is located 2 units left to the centre and 2 units right to the centre, i.e., the coordinates of the vertices are $(-3,1)$ and $(1, 1)$.

Since $c=\sqrt{a^2+b^2}=\sqrt{4+1}=\sqrt{5}$

Thus, the foci are located $\sqrt{5}$ units left and right to the centre, so the coordinates of the foci are $\left(-1-\sqrt{5},1\right)$ and $\left(-1+\sqrt{5},1\right)$.
To find the equations for asymptotes,

Consider $\frac{(x+1)^2}{4}-\frac{(y-1)^2}{1}=0.$

$$\Rightarrow \qquad y-1=\pm\frac{1}{2}(x+1)$$

$$\Rightarrow y=\frac{1}{2}(x+3) \quad \text{and} \quad y=-\frac{1}{2}(x-1).$$

Therefore, the graph of the hyperbola is shown in Figure 5.49(f).

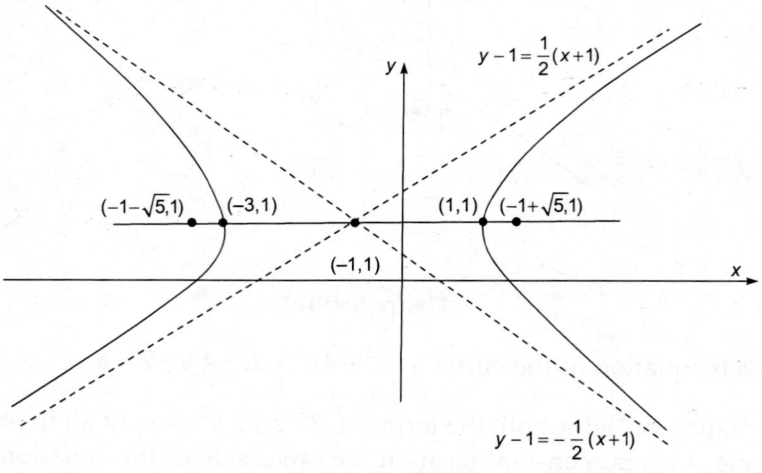

Figure 5.49(f)

EXAMPLE 34 Find the equation of the parabola with vertex $(1,1)$ and directrix $y=-2$.

Solution: Since the vertex is $(1,1)$ and the directrix is $y=-2$, then the axis of symmetry is parallel to y-axis and the parabola opens up.

So, the equation is $(x-h)^2=4p(y-k)$.

$$\Rightarrow \qquad (x-1)^2=4p(y-1)$$

Since the directrix $y=-2$, then the directrix is 3 units apart from the vertex and so, $p=3$. Then, we have

$$(x-1)^2=12(y-1)$$

which is the required equation of the parabola.

5.3.5 Reflection Properties of the Conics

Parabolas, ellipses and hyperbolas have specific type of reflection properties which make them useful in various applications.

The reflection properties of the conics are following.

Reflection Property of Parabolas

The tangent line at a point P on a parabola makes equal angles with the line through P parallel to the axis of symmetry and the line through P which passes through the focus (as shown in Figure 5.50(a)).

Reflection Property of Ellipses

The tangent line at a point P on an ellipse makes equal angles with the line joining the point P to the foci (as shown in Figure 5.50(b)).

Reflection Property of Hyperbolas

The tangent line at a point P on a hyperbola makes equal angles with the line joining the point P to the foci (as shown in Figure 5.50(c)).

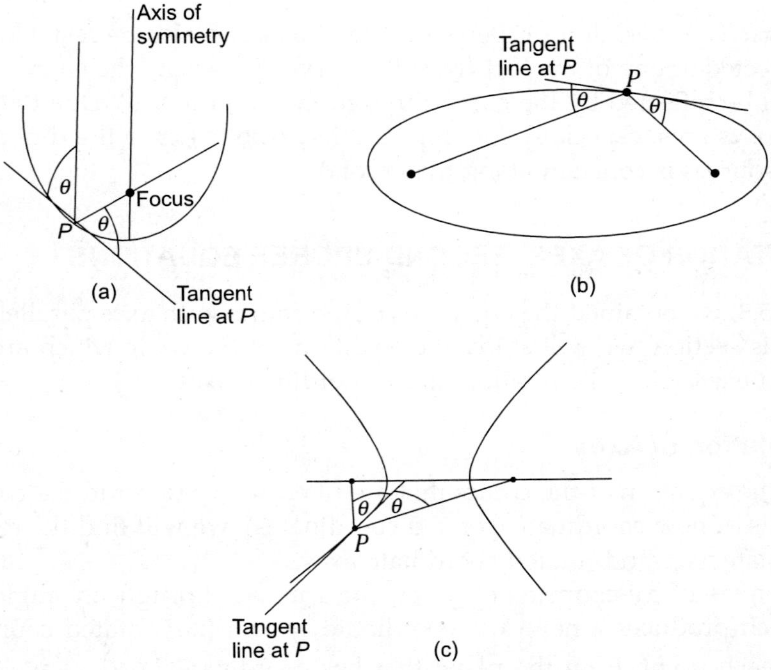

Figure 5.50

5.3.6 Application of the Conics

Parabolas: Fermat's principle states that light travels between two points along the path that requires the least time, as compared to other nearby path. From Fermat's principle, one may derive the law of reflection of light, that is, the light reflects off a surface at an angle equal to its angle of incidence, i.e., the angle of incidence is equal to angle of reflection. In particular, if a reflecting surface is generated by revolving a parabola about its axis of symmetry. Then by reflection property of parabolas, all the light rays entering parallel to axis will be reflected to the focus and conversely, if a source of light is coming out from the focus, then after reflection the light rays will be parallel to the axis. Therefore, the parabola has many important applications in real world such as the shape of automobile headlights are parabolic (the bulb in headlights placed at focus and light coming from bulb will be parallel beam of light rays, after reflection), mirrors in reflecting telescope (the parallel rays of light coming from the stars and planets will be reflected to an eyepiece which is located at the focus), in television, radio and radar signals which are based on parabolic shape of antenna, etc.

Ellipses: An ellipse is one of the most popular conics. Ellipses have many applications in our daily life such as whispering galleries are made in elliptical shape (In such rooms, the ceilings are made in elliptical shape, so that a sound made at one focus can be heard clearly at the other focus, i.e., any sound made at the focus will rebound and pass through the other focus, e.g., Statuary hall in the US capital building is elliptic, The Mormon Tabernacle in Salt Lake City has an elliptical ceiling.)

Hyperbolas: The reflection property of hyperbolas can also be restated as "a beam of light is directed at one of the foci (with the curve "between" the source and the focus), then it will be reflected by the curve through the other focus". Due to this, the mostly navigation systems, especially for ships are hyperbolic (since the difference in signals from two sources is constant along hyperbola).

5.4 ROTATION OF AXES: SECOND-DEGREE EQUATIONS

In section 5.3, we obtained the equations of the conics with axes parallel to coordinates axes. In this section, we will study the equations of the conic which are tilted relative to coordinate axes, i.e., the rotation of the coordinate axes.

5.4.1 Rotation of Axes

To obtain the equation of the conics that are tilted with respect to the coordinates axes, i.e., in terms of new coordinate (rotated coordinates), we will find the relation between the coordinate axes and rotated coordinate axes.

Let the axes of xy-coordinate system be rotated through an angle θ about the origin which produces a new $x'y'$-coordinate system (i.e., rotated coordinate system) such that each point P in the plane that has coordinate (x,y) as well as (x',y') (as shown in Figure 5.51(a)). Let r be the distance between the common origin of xy and $x'y'$-coordinate axes and a point P. Let the line through P to the origin make an angle α with the x'-axis (as shown in Figure 5.51(b)).

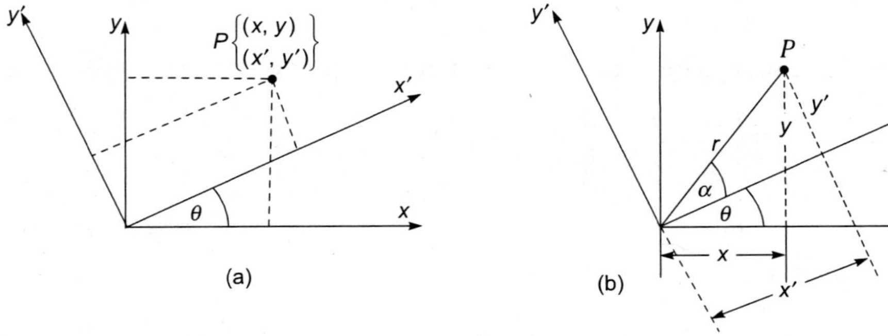

Figure 5.51

Now, from Figure 5.51(b)

$$x = r\cos(\theta+\alpha) \text{ and } y = r\sin(\theta+\alpha). \tag{5.25}$$

Also, $x' = r\cos\alpha$ and $y' = r\sin\alpha$. (5.26)

From (5.25), we have

$x = r(\cos\theta\cos\alpha - \sin\theta\sin\alpha)$ and $y = r(\sin\theta\cos\alpha + \cos\theta\sin\alpha)$

$\Rightarrow x = x'\cos\theta - y'\sin\theta$ and $y = x'\sin\theta + y'\cos\theta$ (by using (5.26)). (5.27)

Thus, we have the relationship between the coordinate axes and the rotated coordinate axes that are known as the *rotation equation.*

$$x = x'\cos\theta - y'\sin\theta$$
and
$$y = x'\sin\theta + y'\cos\theta$$

If these two equations solve for x' and y', we get

$$x' = x\cos\theta + y\sin\theta$$
and
$$y' = -x\sin\theta + y\cos\theta \tag{5.28}$$

EXAMPLE 35 Suppose that the xy-coordinate system is rotated through an angle of $\theta = 60°$, to produce new $x'y'$-coordinate axes. Find the equation of curve

$$\sqrt{3}xy + y^2 = 6$$

in $x'y'$-coordinate axes.

Solution: $\cos 60° = \frac{1}{2}$ and $\sin 60° = \frac{\sqrt{3}}{2}$

$\Rightarrow$ $$x = \frac{x'}{2} - \frac{\sqrt{3}}{2}y' \text{ and } y = \frac{\sqrt{3}}{2}x' + \frac{y'}{2}.$$

Substitute these values of x and y in the given equation of the curve, we get

$$\sqrt{3}\left(\frac{x'}{2} - \frac{\sqrt{3}}{2}y'\right)\left(\frac{\sqrt{3}}{2}x' + \frac{y'}{2}\right) + \left(\frac{\sqrt{3}}{2}x' + \frac{y'}{2}\right)^2 - 6 = 0$$

$\Rightarrow$ $$3x'^2 - y'^2 = 12$$

which is an equation of the hyperbola.

EXAMPLE 36 Find the new coordinates of the point $(2,4)$, if the coordinate axes are rotated through an angle of $\theta = 30°$.

Solution: By equation (5.28), we have

$$x' = x\cos 30° + y\sin 30° = \frac{2\sqrt{3}}{2} + \frac{4.1}{2} = \sqrt{3} + 2$$

and $$y' = -x\sin 30° + y\cos 30° = -2.\left(\frac{1}{2}\right) + 4.\left(\frac{\sqrt{3}}{2}\right) = -1 + 2\sqrt{3}.$$

Therefore, the new coordinates are $\left(\sqrt{3}+2, -1+2\sqrt{3}\right)$.

5.4.2 Second Degree Equation

In example 31(f), we have seen that the second degree equation of the form

$$Ax^2 + Cy^2 + Dx + Ey + F = 0 \tag{5.29}$$

represents the conics. So, the more general second degree equation is

$$Ax^2 + Bxy + Cy^2 + Dx + Ey + F = 0 \tag{5.30}$$

in which, if all A, B and C are not all zero, then it is called a bisecond-degree equation or quadratic equation in x and y.

If $B = 0$, then (5.30) reduces to (5.29) and the conics have their axes parallel to coordinate axes. But, if $B \neq 0$, then (5.30) has the cross product term Bxy and so, represents the conics that have their axes tilted relative to the coordinate axes.

So, if the cross product term Bxy can be eliminated from the equation (5.30) by the rotation of axes, then we say that the reducing form of the second degree equation that represents the conics has appropriate rotation of axes.

The following theorem enables us to determine an appropriate rotation of axes such that the cross product term Bxy can be eliminated:

Theorem

If the second degree equation is

$$Ax^2 + Bxy + Cy^2 + Dx + Ey + F = 0 \tag{5.31}$$

such that $B \neq 0$ and if $x'y'$-coordinate system is obtained by rotating the xy-axes through an angle θ satisfying

$$\cot 2\theta = \frac{A - C}{B}.$$

Then, in $x'y'$-coordinates, equation (5.31) will have the form

$$A'x'^2 + C'y'^2 + D'x' + E'y' + F' = 0.$$

EXAMPLE 37 Identify the curve $xy = -9$?

Solution: To eliminate the cross product term, we will rotate the coordinate axes. Comparing the given equation of the curve to the equation (5.31), we get

$$A = 0, B = 1, C = 0$$

Let the coordinate system be rotated through an angle θ, then θ must satisfy

$$\cot 2\theta = \frac{A - C}{B} = \frac{0 - 0}{1} = 0$$

$$\Rightarrow \qquad 2\theta = \frac{\pi}{2} \Rightarrow \theta = \frac{\pi}{4}.$$

From (5.27), we get

$$x = x'\cos\frac{\pi}{4} - y'\sin\frac{\pi}{4} = \frac{x'}{\sqrt{2}} - \frac{y'}{\sqrt{2}} \text{ and } y = x'\sin\frac{\pi}{4} + y'\cos\frac{\pi}{4} = \frac{x'}{\sqrt{2}} + \frac{y'}{\sqrt{2}}$$

Putting these values of x and y in the given equation of the curve $xy = -9$, we get

$$\left(\frac{x'}{\sqrt{2}} - \frac{y'}{\sqrt{2}}\right)\left(\frac{x'}{\sqrt{2}} + \frac{y'}{\sqrt{2}}\right) = -9.$$

$$\Rightarrow \qquad \frac{x'^2}{2} - \frac{y'^2}{2} = -9$$

$$\Rightarrow \qquad y'^2 - x'^2 = 18$$

which represents a hyperbola.

EXAMPLE 38 Identify the curve $9x^2 - 24xy + 16y^2 - 80x - 60y + 100 = 0$?

Solution: On comparing the given equation of curve with equation (5.31), we have

$$A = 9, B = -24, C = 16.$$

Then, $$\cot 2\theta = \frac{A-C}{B} = \frac{9-16}{-24} = \frac{7}{24}.$$

Since θ is to be chosen in the range $0<\theta<\pi/2$. Then by using right angled triangle relation (Hypotenuous^2 = Perpendicular^2 + Base^2), we get $\cos 2\theta = 7/25$ and

$$\cos\theta = \sqrt{\frac{1+\cos 2\theta}{2}} = \sqrt{\frac{\left(1+\frac{7}{25}\right)}{2}} = \frac{4}{5}$$

and $$\sin\theta = \sqrt{\frac{1-\cos 2\theta}{2}} = \sqrt{\frac{1-\frac{7}{25}}{2}} = \frac{3}{5}.$$

Then, from (5.27), we have

$$x = \frac{4}{5}x' - \frac{3}{5}y' \text{ and } y = \frac{3}{5}x' + \frac{4}{5}y'.$$

Putting these values of x and y in the given equation, we get

$$9\left(\frac{4}{5}x' - \frac{3}{5}y'\right)^2 - 24\left(\frac{4}{5}x' - \frac{3}{5}y'\right)\left(\frac{3}{5}x' + \frac{4}{5}y'\right) + 16\left(\frac{3}{5}x' + \frac{4}{5}y'\right)^2$$

$$-80\left(\frac{4}{5}x' - \frac{3}{5}y'\right) - 60\left(\frac{3}{5}x' + \frac{4}{5}y'\right) + 100 = 0$$

On simplification, we get

$$y'^2 = 4(x'-1)$$

which represents a parabola.

5.4.3 Discriminant

It is also possible to identify the graph of the second degree equation without rotating the coordinate axes. The following theorem enables us to determine the graph of the second degree equation without rotating the coordinate axes:

Theorem

Consider the second degree equation

$$Ax^2 + Bxy + Cy^2 + Dx + Ey + F = 0 \tag{5.32}$$

(a) If $B^2 - 4AC < 0$, then the second degree equation (5.32) represents an ellipse, a circle, a point or else has no graph.

(b) If $B^2 - 4AC > 0$, then the second degree equation (5.32) represents a hyperbola or a pair of intersecting lines.

(c) If $B^2 - 4AC = 0$, then the second degree equation (5.32) represents a parabola, a line, a pair of parallel lines or else has no graph.

The value of $B^2 - 4AC$ is known as the *discriminant* of the second degree equation.

The discriminant cannot be altered by any rotation of coordinate axes. Thus, the discriminant is said to be invariant under a rotation of coordinate axes.

EXAMPLE 39 By using the discriminant, identify the graph of the following conics:

(a) $x^2 - xy + y^2 - 2 = 0$

(b) $x^2 + 2\sqrt{3}xy + 3y^2 + 2\sqrt{3}x - 2y = 0$

(c) $2x^2 + 9xy + y^2 - 6x + y - 4 = 0$

Solution:

(a) On comparing with the second degree equation (5.32), we have

$$A = 1, B = -1, C = 1.$$

So, $B^2 - 4AC = 1 - 4 = -3 < 0$.

Thus, the given conic represents an ellipse.

(b) We have $A = 1, B = 2\sqrt{3}, C = 3$

$\Rightarrow B^2 - 4AC = 12 - 12 = 0$.

Thus, the given conic represents a parabola.

(c) We have $A = 2, B = 9, C = 1$

$\Rightarrow B^2 - 4AC = 81 - 8 = 73 > 0$.

Thus, the given conic represents a hyperbola.

Remark: In cases, where a second degree equation (5.32) represents a point, a line, a pair of parallel lines, a pair of intersecting lines or else has no graph. Then, the second degree equation (5.32) represents a *degenerate* conic section. So, if we allow for possible degeneracy, then every second degree equation (5.32) has a conic section as its graph.

Exercise

1. Sketch the trajectory of the particle whose parametric equations are

$$x = t - 3\sin t \text{ and } y = 4 - 3\cos t,\ t \in [0,10].$$

2. Find the graph of the following parametric equations and also, indicate the orientation on the graph:

(a) $x = e^{-t}\cos t$ and $y = e^{-t}\sin t$, for $t \geq 0$.

(b) $x = t^2 - 9$ and $y = \frac{1}{3}t$, for $-3 \le t \le 2$.

(c) $x = \sin \pi t$ and $y = \cos 2\pi t$, for $0 \le t \le \frac{1}{2}$.

3. Find the parametric equations of the following curves:

   (a) $y = 9x^2$

   (b) $r = 5\cos^3 \theta$ in polar coordinates

4. Find the parametric equation of a circle with radius 3 and centred at the origin, oriented counterclockwise.

5. Find the parametric equation of the ellipse $\frac{x^2}{9} + \frac{y^2}{4} = 1$, oriented counterclockwise.

6. Find the parametric equation of the parabola $y^2 = 4x + 9$, oriented from $(4,-5)$ to $(0,3)$.

7. Find the parametric equation of hyperbola $\frac{x^2}{16} - \frac{y^2}{9} = 1$.

8. Find the explicit relation between x and y by eliminating the parameter t and sketch the path described by the following parametric equations:

   (a) $x = -t$, $y = 3 - 2t$, $0 \le t \le 1$.

   (b) $x = t^3$, $y = t^2$, $t \ge 0$.

   (c) $x = 1 + \sin t$, $y = -2 + \cos t$, $0 \le t \le 2\pi$.

   (d) $x = 1 + \sin^2 t, y = -2 + \cos t$, $0 \le t \le \pi$.

   (e) $x = 4\tan 2t$, $y = 3\sec 2t, 0 \le t \le \pi$.

9. Find the parametric equations for the line that contains the point $(3,1,4)$ and is parallel to the vector $\boldsymbol{v} = -\boldsymbol{i} + \boldsymbol{j} - 2\boldsymbol{k}$. Find where this line passes through the coordinate planes and sketch the line.

10. Find symmetric equations for line L through the points $P(-1,3,7)$ and $Q(4,2,-1)$. Find the points of intersection with the coordinate planes and sketch the line.

11. Show that the following lines intersect, are parallel, skew. If, they intersect, give the point of intersection:

    (a) $x = 3 + 3t, y = 1 - 4t, z = -4 - 7t$ and $x = 2 + 3t, y = 5 - 4t, z = 3 - 7t$.

    (b) $L_1 : \frac{x-1}{2} = \frac{y+1}{1} = \frac{z-2}{4}$ and $L_2 : \frac{x+2}{4} = \frac{y}{-3} = \frac{z - \frac{1}{2}}{-1}$

(c) $L_1 : \frac{x+1}{2} = \frac{y-3}{-1} = \frac{z-2}{1}$ and $L_2 : \frac{x+1}{2} = \frac{y+1}{3} = \frac{z-3}{-4}$

12. Find the equation of the line L_1 that contains the point $P(-1,3,1)$ and orthogonal to the line $L_2 : x = 2-t, y = 1-2t, z = 5+t$.

13. Find the rectangular coordinates of the point P whose polar coordinates are the following:

(a) $(6, 2\pi/3)$ (b) $(4, \pi/6)$ (c) $\left(8, \frac{2\pi}{3}\right)$ (d) $\left(-6, -\frac{5\pi}{6}\right)$

(e) $(0, -\pi)$ (f) $\left(7, \frac{17\pi}{6}\right)$ (g) $(-5, 0)$.

14. Plot the following points in polar coordinates system:

(a) $(6, \pi/4)$ (b) $(1, -\pi/4)$ (c) $(-6, -\pi)$ (d) $(4, 7\pi/6)$

(e) $(-1, 9\pi/4)$.

15. Find the polar coordinates $(r, \theta)(r \geq 0, 0 \leq \theta \leq 2\pi)$ of the point P whose rectangular coordinates are the following:

(a) $\left(-2, -2\sqrt{3}\right)$ (b) $(-7, 0)$ (c) $\left(2\sqrt{3}, -2\right)$ (d) $(0, -2)$

(e) $(-8, -8)$ (f) $\left(-3, \sqrt{3}\right)$

16. Sketch the graph of the following curves in polar coordinates system:

(a) $r = 1$ (b) $\theta = -3\pi/4$ (c) $r = \sin\theta$ (d) $r = 4\cos\theta$

(e) $r = a(1-\cos\theta)(a \geq 0)$ (f) $r = 9\theta$

(g) $r^2 = 9\sin 3\theta$

17. Find the equation in polar coordinates for the following curves:

(a) $x = 2$ (b) $x^2 + y^2 = 16$

(c) $x^2 + y^2 + 6y = 0$ (d) $9xy = 4$

18. Sketch the graph of the following parabolas and also label the focus, vertex and directrix:

(a) $x^2 = 12y$ (b) $y^2 = 6x$

(c) $x^2 + 9y = 0$ (d) $y^2 + 10x = 0$

(e) $x^2 - 4y = 0$ (f) $(x+2)^2 = -(y+2)$

(g) $(y+1)^2 = -7(x+4)$ (h) $(x-1/2)^2 - 2(y-1) = 0$

(i) $x - y^2 + 4y - 2 = 0$ (j) $y^2 - 6y - 2x + 1 = 0$

19. Sketch the graph of the following ellipses and also label the foci, vertices and end of the major and minor axis:

(a) $\frac{x^2}{9}+\frac{y^2}{16}=1$ (b) $x^2+2y^2=4$

(c) $\frac{x^2}{4}+\frac{y^2}{25}=1$ (d) $4x^2+9y^2=36$

(e) $3(x+2)^2+4(y+1)^2=12$ (f) $(x+3)^2+4(y-5)^2=16$

(g) $4x^2+y^2+8x-10y+13=0$ (h) $9x^2+4y^2+18x-24y+9=0$

(i) $5x^2+9y^2-20x+54y+56=0$

20. Sketch the graph of the following hyperbolas and also label the focus, vertices and asymptotes:

(a) $\frac{x^2}{4}-\frac{y^2}{9}=1$ (b) $y^2-x^2=1$

(c) $\frac{y^2}{9}-\frac{x^2}{25}=1$ (d) $16x^2-25y^2=400$

(e) $4(x-2)^2-9(y-4)^2=36$ (f) $(y+3)^2-9(x+2)^2=36$

(g) $16(x+1)^2-8(y-3)^2=16$ (h) $16x^2-y^2-32x-6y=57$

(i) $4y^2-x^2+40y-4x+60=0$

21. Find the equation of the following parabolas that satisfies the given conditions:

(a) Vertex at $(0,0)$ and passes through $(5,2)$

(b) Vertex at $(0,0)$ and focuses at $(3,0)$

(c) Vertex at $(0,0)$ and directrix is $x=7$

(d) Focus at $(0,-3)$ and directrix is $y=3$

(e) Axis is $y=0$, passes through $(3,2)$ and $(2,-3)$

22. Find the equation of the following ellipses that satisfy the given conditions:

(a) Foci at $(0, \pm 2)$ and ends of major axis $(0, \pm 4)$

(b) Ends of major $(\pm 3, 0)$ and of minor axis $(0, \pm 2)$

(c) Length of minor axis 16 and foci at $(0, \pm 6)$

(d) Foci at $(\pm 1, 0)$ and $b=\sqrt{2}$

(e) Foci $(\pm 3, 0)$ and $a=4$.

23. Find the equation of the following hyperbolas that satisfies the given conditions:

    (a) Vertices $(0, \pm 8)$ and asymptotes $y = \pm\frac{4}{3}x$

    (b) Vertices $(\pm 1, 0)$ and asymptotes $y = \pm 2x$

    (c) Foci $(0, \pm 5)$ and asymptotes $y = \pm 2x$

    (d) Asymptotes are $y = 2x + 1$ and $y = -2x = 3$; passes through the origin.

24. Suppose that the axes of an xy-coordinate system are rotated through an angle of $\theta = 45°$ to obtain an $x'y'$-coordinate system. Find the equation of the curve $x^2 - xy + y^2 - 6 = 0$ in $x'y'$-coordinates.

25. Let be $x'y'$-coordinate system be obtained by rotating an xy-coordinate system through an angle of $\theta = 60°$. Find the $x'y'$-coordinates of the point whose xy-coordinates are $(-2,6)$.

26. Rotate the coordinate axes to eliminate the xy term and then identify the graph of the following equations:

    (a) $xy = 1$

    (b) $153x^2 - 192xy + 97y^2 - 30x - 40y - 200 = 0$

    (c) $x^2 + 4xy - 2y^2 - 6 = 0$

    (d) $x^2 + 2\sqrt{3}xy + 3y^2 + 2\sqrt{3}x - 2y = 0$

    (e) $52x^2 - 72xy + 73y^2 + 40x + 30y - 75 = 0$.

27. By using the discriminant, identify the graph of the following equations:

    (a) $8x^2 - 3xy + 5y^2 - 7x + 6 = 0$

    (b) $34x^2 - 24xy + 41y^2 - 25 = 0$

    (c) $x^2 + 4xy - 2y^2 - 6 = 0$

    (d) $6x^2 + 24xy - y^2 - 12x + 26y + 11 = 0$.

6 CHAPTER

Curvature

The shape of a plane curve C is described by the degree of bend or curve. For instance, a straight line has no bending while a circle has constant bending.

The *curvature* of a curve at a given point is a degree of rate of change of bending of curve at that point.

6.1 CURVATURE OF A CURVE AT A POINT

Let A be a fixed point on a curve. Let P and Q be two neighbouring points on a curve such that arc $AP = s$ and arc $AQ = s + \delta s$. Then, arc $PQ = \delta s$. Let the tangents be drawn at P and Q making angles ψ and $\psi + \delta\psi$, respectively with the positive direction of x-axis. Thus, $\delta\psi$ is an angle through which the tangent turns from P to Q.

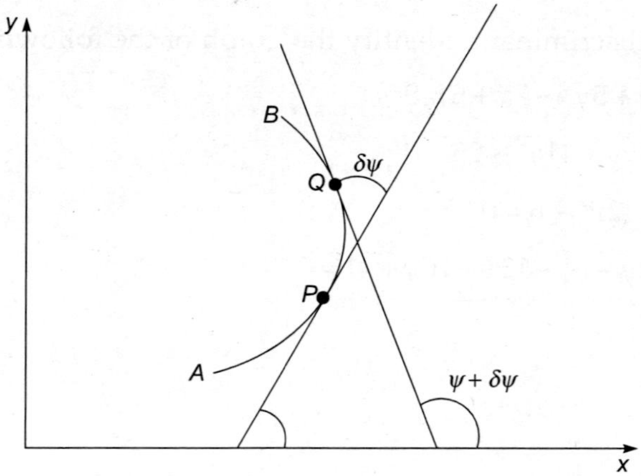

Figure 6.1

If δs is an arc length of PQ, then $\dfrac{\delta\psi}{\delta s}$ is an average bending of an arc PQ. Thus, the ratio of $\dfrac{\delta\psi}{\delta s}$ is called an average curvature of an arc PQ.

The limit of $\frac{\delta\psi}{\delta s}$ as Q approaches to P (i.e., $\delta s \to 0$) is defined to be the curvature of a curve at a point P. Thus, the curvature, k, of a curve at the point P is defined as

$$k = \lim_{\delta s \to 0} \frac{\delta\psi}{\delta s} = \frac{d\psi}{ds}, \tag{6.1}$$

where s is an arc length and ψ is an angle of the tangent at P makes with positive direction of x-axis.

Remark: It is clear that $\delta\psi$ is greater or less according as the whole bending of the curve from P to Q is greater or less.

6.2 CURVATURE OF A CIRCLE

Consider a circle with centre C and radius r. Let P and Q be two points on the circle such that arc $PQ = \delta s$. Let the tangents at P and Q make angles ψ and $\psi + \delta\psi$, respectively with the positive direction of x-axis. Then $\delta\psi$ is an angle through which the tangent turns from P to Q.

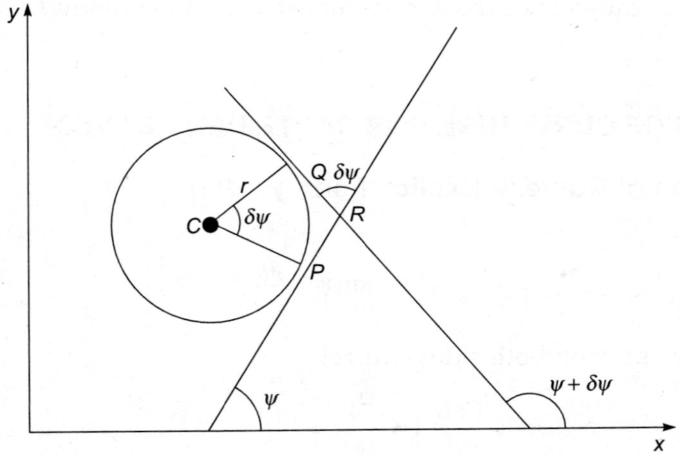

Figure 6.2

Now, $\delta s = r\delta\psi$ (as length of an arc $l = r\theta$ for a circle.)
By (6.1), we have

$$k = \lim_{\delta s \to 0} \frac{\delta\psi}{\delta s}$$

$$= \lim_{\delta\psi \to 0} \frac{\delta\psi}{r\delta\psi}.$$

$$\Rightarrow \quad k = \frac{1}{r}. \tag{6.2}$$

Thus, the curvature of a circle at any point is constant and it is equal to the reciprocal of radius of the circle.

6.3 RADIUS OF CURVATURE

The radius of curvature of a curve at any point P on a curve is defined as the radius of a circle C with centre O, which has the same tangent and the same curvature at P as the given curve, i.e., the radius of curvature = length of $OP = \rho$.

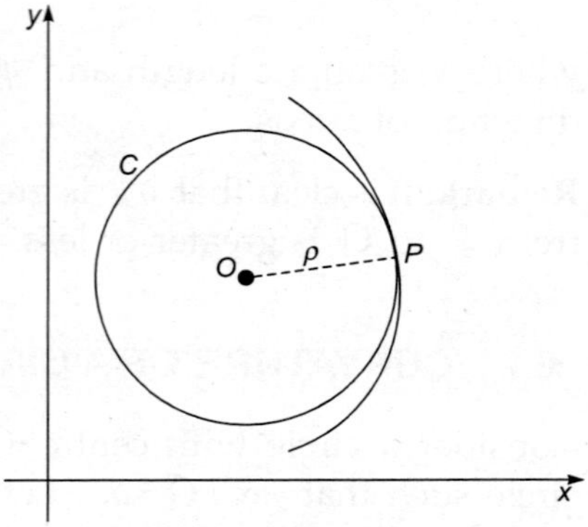

Figure 6.3

Now, by (6.2), we have

$$k = \frac{1}{\rho} \Rightarrow \rho = \frac{1}{k} = \frac{ds}{d\psi}.$$

Hence, the radius of curvature,

$$\rho = \frac{ds}{d\psi}.$$

The circle C is called the *circle of curvature* at P and the centre O is called the *centre of curvature* at P.

6.4 RADIUS OF CURVATURE FOR CARTESIAN CURVES

6.4.1 Equation of Curve in Explicit Form $y = f(x)$

We know that,

$$\tan\psi = \frac{dy}{dx}.$$

Differentiate w.r.t. x on both sides, we get

$$\sec^2\psi\left(\frac{d\psi}{dx}\right) = \frac{d^2y}{dx^2}.$$

$$\Rightarrow \qquad \frac{d^2y}{dx^2} = \sec^2\psi\left(\frac{d\psi}{ds}\right)\left(\frac{ds}{dx}\right)$$

$$\Rightarrow \qquad \frac{d^2y}{dx^2} = \sec^2\psi\left(\frac{1}{\rho}\right)\left(\frac{ds}{dx}\right).$$

Also, we know that

$$\frac{ds}{dx} = \sqrt{1+\left(\frac{dy}{dx}\right)^2}.$$

$$\Rightarrow \qquad \frac{d^2y}{dx^2} = \sec^2\psi\left(\frac{1}{\rho}\right)\sqrt{1+\left(\frac{dy}{dx}\right)^2}$$

$$= \sec^2\psi\left(\frac{1}{\rho}\right)\sqrt{1+\tan^2\psi}$$

$$\Rightarrow \qquad \rho = \frac{\left(1+\tan^2\psi\right)^{\frac{3}{2}}}{\frac{d^2y}{dx^2}}.$$

Therefore, the radius of curvature for Cartesian curves $y = f(x)$ is

$$\rho = \frac{\left(1+\left(\frac{dy}{dx}\right)^2\right)^{3/2}}{\frac{d^2y}{dx^2}}. \qquad (6.3)$$

Remarks

1. If $\frac{d^2y}{dx^2} > 0$, then by the above formula, the curvature k is positive and the given curve is concave upwards. Similarly, if $\frac{d^2y}{dx^2} < 0$, then the curvature k is negative and the given curve is concave downward.
2. The value of ρ has the same sign as that of $\frac{d^2y}{dx^2}$. But, since ρ is the radius of curvature, it is conventional to write the value of ρ with positive sign (i.e., in magnitude).
3. By definition, it is clear that the radius of curvature is independent on the choice of axes and depends only on the curve. Thus, if the tangent to the curve is parallel to y-axis (i.e., $\frac{dy}{dx} = 0$), then the formula of radius of curvature can be obtained by interchanging the axes of x and y. Therefore, the radius of curvature is also given by $\rho = \dfrac{\left(1+\left(\frac{dy}{dx}\right)^2\right)^{3/2}}{\frac{d^2y}{dx^2}}$.

6.4.2 Equation of Curve in Implicit Form $f(x, y) = 0$

The equation of given curve is $f(x,y)=0$.
Differentiate w.r.t. x and using chain rule, we get

$$\frac{\partial f}{\partial x}+\frac{\partial f}{\partial y}.\frac{dy}{dx}=0$$

$$\Rightarrow \qquad \frac{dy}{dx}=-\frac{f_x}{f_y}, \text{ provided } f_y \neq 0.$$

Again, differentiate w.r.t. x, we get

$$\frac{d^2y}{dx^2}=-\frac{\left(f_{xx}f_y^2-2f_{xy}f_xf_y+f_{yy}f_x^2\right)}{f_y^3}, \quad \text{provided } f_y \neq 0.$$

Since on putting the values of $\frac{dy}{dx}$ and $\frac{d^2y}{dx^2}$ in ρ given by $\rho=\frac{\left(1+\left(\frac{dy}{dx}\right)^2\right)^{3/2}}{\frac{d^2y}{dx^2}}$, we have

$$\rho=\frac{\left(1+\left(-\frac{f_x}{f_y}\right)^2\right)^{\frac{3}{2}}}{-\frac{\left(f_{xx}f_y^2-2f_{xy}f_xf_y+f_{yy}f_x^2\right)}{f_y^3}}.$$

Therefore, the radius of curvature for the Cartesian curves $f(x,y)=0$ is

$$\rho=-\frac{(f_x^2+f_y^2)^{3/2}}{f_{xx}f_y^2-2f_{xy}f_xf_y+f_{yy}f_x^2}.$$

6.4.3 Equation of Curve in Parametric Form

(I) If the equation of curve is in the form $x=f(t)$ and $y=g(t)$.

Then, $$\frac{dy}{dx}=\frac{dy}{dt}.\frac{dt}{dx}=\frac{dy/dt}{dx/dt}=\frac{g'(t)}{f'(t)}$$

and $$\frac{d^2y}{dx^2}=\frac{d}{dx}\left(\frac{dy}{dx}\right)=\frac{d}{dt}\left(\frac{dy}{dx}\right)\frac{dt}{dx}=\frac{\frac{d}{dt}\left(\frac{g'(t)}{f'(t)}\right)}{\frac{dx}{dt}}$$

$$= \frac{g''(t)f'(t) - f''(t)g'(t)}{(f'(t))^2} \times \frac{1}{f'(t)}$$

$$= \frac{g''(t)f'(t) - f''(t)g'(t)}{(f'(t))^3}.$$

Since on putting the values of $\frac{dy}{dx}$ and $\frac{d^2y}{dx^2}$ in ρ given by $\rho = \frac{\left(1+\left(\frac{dy}{dx}\right)^2\right)^{3/2}}{\frac{d^2y}{dx^2}}$, we have

$$\Rightarrow \qquad \rho = \frac{\left[1+\left(\frac{g'(t)}{f'(t)}\right)^2\right]^{3/2}}{\frac{g''(t)f'(t) - f''(t)g'(t)}{(f'(t))^3}}.$$

Therefore, the radius of curvature for parametric form $x = f(t)$ and $y = g(t)$ is

$$\rho = \frac{((f'(t))^2 + (g'(t))^2)^{3/2}}{g''(t)f'(t) - f''(t)g'(t)}$$

(II) If the equation of curve is in the form $x = f(s)$ and $y = g(s)$, where the parameter s is the arc length.

We know that $\cos\psi = \frac{dx}{ds}$ and $\sin\psi = \frac{dy}{ds}$.

Differentiate w.r.t. s, we get

$$\frac{d^2x}{ds^2} = -\sin\psi.\frac{d\psi}{ds} \text{ and } \frac{d^2y}{ds^2} = \cos\psi.\frac{d\psi}{ds}$$

$$\Rightarrow \qquad \frac{d^2x}{ds^2} = -\sin\psi.\frac{1}{\rho} \text{ and } \frac{d^2y}{ds^2} = \cos\psi.\frac{1}{\rho}$$

Squaring and adding, we get

$$\frac{1}{\rho^2} = \left(\frac{d^2x}{ds^2}\right)^2 + \left(\frac{d^2y}{ds^2}\right)^2.$$

EXAMPLE 1 Find the radius of curvature at the point (s,ψ) on the curve $s = a\log\tan\left(\frac{\pi}{4}+\frac{\psi}{2}\right)$, where a is any constant.

Solution: Since, $s = a\log\tan\left(\frac{\pi}{4}+\frac{\psi}{2}\right)$

Differentiate w.r.t. ψ, we get

$$\frac{ds}{d\psi} = a\sec^2\left(\frac{\pi}{4}+\frac{\psi}{2}\right).\frac{1}{2}.$$

Now, $$\rho = \frac{ds}{d\psi} = a\cdot\frac{1}{\tan\left(\frac{\pi}{4}+\frac{\psi}{2}\right)}\cdot\sec^2\left(\frac{\pi}{4}+\frac{\psi}{2}\right).\frac{1}{2}.$$

$$\Rightarrow \qquad \rho = a\cdot\frac{\cos\left(\frac{\pi}{4}+\frac{\psi}{2}\right)}{\sin\left(\frac{\pi}{4}+\frac{\psi}{2}\right)}\cdot\frac{1}{\cos^2\left(\frac{\pi}{4}+\frac{\psi}{2}\right)}\cdot\frac{1}{2}$$

$$= a\cdot\frac{1}{2\sin\left(\frac{\pi}{4}+\frac{\psi}{2}\right)\cos\left(\frac{\pi}{4}+\frac{\psi}{2}\right)} = a\cdot\frac{1}{\sin\left(\frac{\pi}{4}+\frac{\psi}{2}\right)}.$$

$$\Rightarrow \qquad \rho = a.\text{cosec}\left(\frac{\pi}{2}+\psi\right) = a\sec\psi.$$

EXAMPLE 2 Find the radius of curvature at a given point of the following curves:

(a) $\sqrt{x}+\sqrt{y}=\sqrt{a}$ at $\left(\frac{a}{4},\frac{a}{4}\right)$.

(b) $x^3+y^3=3axy$ at $\left(\frac{3a}{2},\frac{3a}{2}\right)$.

Solution:

(a) Equation of the curve is $\sqrt{x}+\sqrt{y}=\sqrt{a}$

Differentiate w.r.t. x, we get

$$\frac{1}{2}x^{-\frac{1}{2}}+\frac{1}{2}y^{-\frac{1}{2}}\left(\frac{dy}{dx}\right)=0$$

$$\Rightarrow \qquad \frac{dy}{dx} = -\sqrt{\frac{y}{x}}.$$

Again, differentiate w.r.t. x, we get

$$\frac{d^2y}{dx^2} = -\frac{x^{\frac{1}{2}}.\frac{1}{2}.y^{-\frac{1}{2}}.\frac{dy}{dx}-y^{\frac{1}{2}}.\frac{1}{2}.x^{-\frac{1}{2}}}{x}$$

$$= -\frac{\left(\sqrt{\frac{x}{y}}\left(-\sqrt{\frac{y}{x}}\right)-\sqrt{\frac{y}{x}}\right)}{2x} = \frac{\sqrt{x}+\sqrt{y}}{2x\sqrt{x}} = \frac{a}{2x\sqrt{x}}.$$

At $\left(\frac{a}{4},\frac{a}{4}\right)$, $\frac{dy}{dx} = -1$ and $\frac{d^2y}{dx^2} = \frac{4}{a}$.

Now, since $\rho = \dfrac{\left(1+\left(\frac{dy}{dx}\right)^2\right)^{3/2}}{\frac{d^2y}{dx^2}}$.

$$\Rightarrow \quad (\rho)_{\left(\frac{a}{4},\frac{a}{4}\right)} = \frac{(1+(-1)^2)^{3/2}}{4/a} = \frac{2^{3/2}}{2^2}\cdot a = \frac{a}{\sqrt{2}}.$$

(b) Equation of the curve is $x^3 + y^3 = 3axy$
Differentiate w.r.t. x, we get

$$3x^2 + 3y^2.\frac{dy}{dx} = 3a\left(y + x.\frac{dy}{dx}\right)$$

$$\Rightarrow \left(y^2 - ax\right)\frac{dy}{dx} = ay - x^2$$

$$\Rightarrow \quad \frac{dy}{dx} = \frac{ay-x^2}{y^2-ax}$$

Again, differentiate w.r.t. x, we get

$$\frac{d^2y}{dx^2} = \frac{\left(a\frac{dy}{dx} - 2x\right)(y^2-ax) - (ay-x^2)\left(2y\frac{dy}{dx} - a\right)}{(y^2-ax)^2}.$$

At $\left(\frac{3a}{2},\frac{3a}{2}\right)$, $\frac{dy}{dx} = -1$ and $\frac{d^2y}{dx^2} = -\frac{32}{3a}$.

So that, $(\rho)_{\left(\frac{3a}{2},\frac{3a}{2}\right)} = \dfrac{(1+1^2)^{3/2}}{-\frac{32}{3a}} = \dfrac{3a}{8\sqrt{2}}$.

EXAMPLE 3 Prove that, if ρ is the radius of curvature at any point P on the parabola $y^2 = 4ax$, then ρ^2 varies as $(SP)^3$, where S is the focus of the parabola.

Solution: Equation of the parabola is $y^2 = 4ax$.

In parametric form, equations of parabola are:

$$x = at^2 \text{ and } y = 2at,$$

i.e.,
$$f(t) = at^2 \text{ and } g(t) = 2at$$

Since
$$\rho = \frac{((f'(t))^2 + (g'(t))^2)^{3/2}}{g''(t)\, f'(t) - f''(t)\, g'(t)}.$$

$$\Rightarrow \quad \rho = \frac{((2at)^2 + (2a)^2)^{3/2}}{0 - (2a)(2a)} = \frac{(4a^2(1+t^2))^{3/2}}{-4a^2}$$

$$\Rightarrow \quad \rho = 2a(1+t^2)^{\frac{3}{2}} \text{ (in magnitude).}$$

Now, $SP =$ distance between focus $S(a,0)$ and a point $P\left(at^2, 2at\right)$

$$= \sqrt{\left(at^2 - a\right)^2 + (2at - o)^2}$$

$$= \sqrt{a^2\left(t^2 - 1\right)^2 + 4a^2t^2}$$

$$= a\left(t^2 + 1\right)$$

Thus, $\rho^2 = 4a^2\left(1+t^2\right)^3 = \frac{4}{a}(SP)^3$

Therefore, $\rho^2 \propto (SP)^3$.

EXAMPLE 4 The tangent at two points P and Q on the cycloid $x = a(\theta - \sin\theta)$, $y = a(1 - \cos\theta)$ are at right angles. Then, show that $\rho_1^2 + \rho_2^2 = 16a^2$, where ρ_1 and ρ_2 are the radii of curvature at points P and Q, respectively.

Solution: We have, $x = a(\theta - \sin\theta)$ and $y = a(1 - \cos\theta)$,

i.e.,
$$f(\theta) = a(\theta - \sin\theta) \text{ and } g(\theta) = a(1 - \cos\theta).$$

$$\Rightarrow \quad f'(\theta) = a(1 - \cos\theta) \text{ and } g'(\theta) = a\sin\theta$$

and
$$f''(\theta) = a\sin\theta \text{ and } g''(\theta) = a\cos\theta.$$

On putting the values of $f'(\theta)$, $g'(\theta)$, $f''(\theta)$ and $g''(\theta)$ in ρ, given by

$$\rho = \frac{\left\{(f'(\theta))^2 + (g'(\theta))^2\right\}^{3/2}}{g''(\theta)\, f'(\theta) - f''(\theta)\, g'(\theta)}, \text{ we have}$$

$$\rho = \frac{(a^2(1-\cos\theta)^2 + a^2\sin^2\theta)^{3/2}}{a^2\cos\theta(1-\cos\theta) - a^2\sin^2\theta} = \frac{\left(4a^2\sin^2\left(\frac{\theta}{2}\right)\right)^{3/2}}{-2a^2\sin^2\left(\frac{\theta}{2}\right)}.$$

$$\Rightarrow \qquad \rho = 4a\sin\left(\frac{\theta}{2}\right) \text{ (in magnitude).}$$

Let the points P and Q on the cycloid correspond to $\theta = \theta_1$ and $\theta = \theta_2$, respectively.

Then, the radii of curvature at P and Q are $\rho_1 = 4a\sin\left(\frac{\theta_1}{2}\right)$ and $\rho_2 = 4a\sin\left(\frac{\theta_2}{2}\right)$, respectively.

Now,
$$\frac{dy}{dx} = \frac{\frac{dy}{d\theta}}{\frac{dx}{d\theta}} = \frac{a\sin\theta}{1-\cos\theta} = \cot\left(\frac{\theta}{2}\right).$$

Since the tangents at P and Q are at right angles, the product of their slopes is equal to -1,

i.e.,
$$\cot\frac{\theta_1}{2}.\cot\frac{\theta_2}{2} = -1.$$

$$\Rightarrow \qquad \cot\frac{\theta_2}{2} = -\tan\frac{\theta_1}{2} = \cot\left(\frac{\pi}{2} + \frac{\theta_1}{2}\right)$$

$$\Rightarrow \qquad \frac{\theta_2}{2} = \frac{\pi}{2} + \frac{\theta_1}{2}$$

$$\Rightarrow \qquad \theta_2 = \pi + \theta_1.$$

Thus,
$$\rho_2 = 4a\sin\left(\frac{\theta_2}{2}\right) = 4a\sin\left(\frac{\pi}{2} + \frac{\theta_1}{2}\right) = 4a\cos\frac{\theta_1}{2}.$$

Hence,
$$\rho_1^2 + \rho_2^2 = \left(4a\sin\frac{\theta_1}{2}\right)^2 + \left(4a\cos\frac{\theta_1}{2}\right)^2 = 16a^2.$$

EXAMPLE 5 Show that the radius of curvature ρ at the point (x,y) on the ellipse $\frac{x^2}{a^2} + \frac{y^2}{b^2} = 1$ is $\frac{(ab)^2}{d^3}$, where d is the perpendicular distance from the centre upon the tangent at (x,y).

Solution: Parametric equations of the ellipse $\frac{x^2}{a^2} + \frac{y^2}{b^2} = 1$ are:

$x = a\cos\theta$ and $y = b\sin\theta$, where θ is the parameter,

i.e., $f(\theta) = a\cos\theta$ and $g(\theta) = b\sin\theta$.

$\Rightarrow$ $f'(\theta) = -a\sin\theta$ and $g'(\theta) = b\cos\theta$

$\Rightarrow$ $f''(\theta) = -a\cos\theta$ and $g''(\theta) = -b\sin\theta$.

On putting the values of $f'(\theta)$, $g'(\theta)$, $f''(\theta)$ and $g''(\theta)$ in ρ, given by

$$\rho = \frac{\left((f'(\theta))^2 + (g'(\theta))^2\right)^{3/2}}{g''(\theta)f'(\theta) - f''(\theta)g'(\theta)}, \text{ we have}$$

$$\rho = \frac{\left(a^2\sin^2\theta + b^2\cos^2\theta\right)^{3/2}}{ab(\sin^2\theta + \cos^2\theta)} = \frac{\left(a^2\sin^2\theta + b^2\cos^2\theta\right)^{3/2}}{ab}.$$

Equation of tangent at a point $(x, y) = (a\cos\theta, b\sin\theta)$ of the ellipse $\frac{x^2}{a^2} + \frac{y^2}{b^2} = 1$ is $\frac{x}{a}\cos\theta + \frac{y}{b}\sin\theta = 1$.

Since d is the length of the perpendicular from the centre $(0,0)$ to the above tangent,

$$d = \left|\frac{0 + 0 - 1}{\sqrt{\frac{\cos^2\theta}{a^2} + \frac{\sin^2\theta}{b^2}}}\right|.$$

{as perpendicular distance from the point (x, y) to the curve $Ax + By = C$ is $\left|\frac{Ax + By - C}{\sqrt{A^2 + B^2}}\right|$}

$$\Rightarrow \qquad d = \frac{ab}{\sqrt{a^2\sin^2\theta + b^2\cos^2\theta}}.$$

$$\Rightarrow \qquad d^3 = \frac{(ab)^3}{\left(a^2\sin^2\theta + b^2\cos^2\theta\right)^{3/2}}.$$

Now, $$\rho \times d^3 = \frac{\left(a^2\sin^2\theta + b^2\cos^2\theta\right)^{3/2}}{ab} \times \frac{(ab)^3}{\left(a^2\sin^2\theta + b^2\cos^2\theta\right)^{3/2}}.$$

Hence, $$\rho = \frac{(ab)^2}{d^3}.$$

EXAMPLE 6

(a) Find the radius of curvature for a curve with implicit equation $x^3+xy^2-6y^2=0$ at $(3,3)$.

(b) Find the radius of curvature at any point on the rectangular hyperbola $xy=c^2$.

Solution:

(a) Since the equation of curve is $x^3+xy^2-6y^2=0$,

i.e.,
$$f(x,y)=x^3+xy^2-6y^2.$$

This gives, $f_x=3x^2+y^2$, $f_y=2xy-12y$, $f_{xx}=6x$ and $f_{yy}=2x-12$.

On putting the values of $f'(\theta)$, $g'(\theta)$, $f''(\theta)$ and $g''(\theta)$ in ρ, given by

$$\rho=-\frac{\left((f_x)^2+(f_y)^2\right)^{3/2}}{f_{xx}f_y^2-2f_{xy}f_xf_y+f_{yy}f_x^2}\text{, we have}$$

$$\rho=-\frac{\left((36)^2+(-18)^2\right)^{3/2}}{18(-18)^2-2\times6\times36\times(-18)+(-6)\times36^2}=-5^{\frac{3}{2}}\text{ at (3, 3).}$$

Thus,
$$\rho=5^{\frac{3}{2}}\text{ (in magnitude).}$$

(b) Since the equation of the curve is $xy=c^2$

Differentiate w.r.t. x, we get

$$y+x\frac{dy}{dx}=0$$

$$\Rightarrow\qquad\frac{dy}{dx}=-\frac{y}{x}.$$

Again, differentiate w.r.t. x, we get

$$\frac{d^2y}{dx^2}=-\frac{1}{x}\frac{dy}{dx}+\frac{y}{x^2}=-\frac{1}{x}\left(\frac{-y}{x}\right)+\frac{y}{x^2}=\frac{2y}{x^2}.$$

On putting the values of $\frac{dy}{dx}$ and $\frac{d^2y}{dx^2}$ in ρ given by

$$\rho=\frac{\left(1+\left(\frac{dy}{dx}\right)^2\right)^{3/2}}{\frac{d^2y}{dx^2}}=\frac{\left(1+\left(\frac{-y}{x}\right)^2\right)^{3/2}}{\frac{2y}{x^2}}\text{, we have}$$

$$\rho=\frac{x^2+y^2}{2c^2}.$$

EXAMPLE 7 If ρ_1 and ρ_2 are the radii of curvature at the extremities of a focal chord of the parabola $y^2=4ax$, then prove that $\rho_1^{-\frac{2}{3}}+\rho_2^{-\frac{2}{3}}=l^{-\frac{2}{3}}$, where l is the semi-latus rectum of the parabola.

Solution: As in Example 3, the radius of curvature at any point (x,y) on the parabola, $\rho=2a\left(1+t^2\right)^{\frac{3}{2}}$.

Let $P\left(at_1^2,2at_1\right)$ and $Q\left(at_2^2,2at_2\right)$ be two extremities of a focal chord of the parabola $y^2=4ax$, then $t_1t_2=-1\Rightarrow t_2=-\frac{1}{t_1}$.

Then, the radius of curvature at P is $\rho_1=2a\left(1+t_1^2\right)^{\frac{3}{2}}$

and the radius of curvature at Q is $\rho_2=2a\left(1+t_2^2\right)^{\frac{3}{2}}$.

Now, $\rho_1^{-\frac{2}{3}}+\rho_2^{-\frac{2}{3}}=\left(2a\left(1+t_1^2\right)^{\frac{3}{2}}\right)^{-\frac{2}{3}}+\left(2a\left(1+t_2^2\right)^{\frac{3}{2}}\right)^{-\frac{2}{3}}$

$$=(2a)^{-\frac{2}{3}}\left[\left(1+t_1^2\right)^{-1}+\left(1+t_2^2\right)^{-1}\right]$$

$$=(2a)^{-\frac{2}{3}}\left[\left(1+t_1^2\right)^{-1}+\left(1+\frac{1}{t_1^2}\right)^{-1}\right]$$

$$=(2a)^{-\frac{2}{3}}\left[\left(1+t_1^2\right)^{-1}+\left(\frac{1+t_1^2}{t_1^2}\right)^{-1}\right]$$

$$=(2a)^{-\frac{2}{3}}\left(1+t_1^2\right)^{-1}\left(1+t_1^2\right)$$

$=(2a)^{-\frac{2}{3}}=l^{-\frac{2}{3}}$ (as the length of latus rectum of the parabola is $2a$).

EXAMPLE 8 Prove that, if ρ is the radius of curvature at a point P on the ellipse $\frac{x^2}{a^2}+\frac{y^2}{b^2}=1$, then $\rho=\frac{(CD)^3}{ab}$, where CD is the semi-diameter conjugate to CP of an ellipse.

Solution: The parametric equations of an ellipse are $x = a\cos\theta$ and $y = b\sin\theta$.

As in Example 5, the radius of curvature at $(x = a\cos\theta, y = b\sin\theta)$ on ellipse is

$$\rho == \frac{\left(a^2\sin^2\theta + b^2\cos^2\theta\right)^{3/2}}{ab}.$$

Let $P(a\cos\theta, b\sin\theta)$ be any point on the ellipse. Since CD is semi-diameter conjugate to CP, then the coordinates of D are $\left(a\cos\left(\frac{\pi}{2}+\theta\right), b\sin\left(\frac{\pi}{2}+\theta\right)\right) = (-a\sin\theta, b\cos\theta)$.

The length of $CD = \sqrt{(-a\sin\theta - 0)^2 + (b\cos\theta - 0)^2}$ (as $C(0,0)$ is the centre of the ellipse.)

$$= \sqrt{a^2\sin^2\theta + b^2\cos^2\theta}$$

$$\Rightarrow \qquad \rho = \frac{(CD)^3}{ab}.$$

EXAMPLE 9

(a) Find the radius of curvature at a point $(0,c)$ of the catenary $y = c\cos\frac{hx}{c}$.

(b) Show that the radius of curvature of catenary $y = c\cos\frac{hx}{c}$ is equal to the portion of the normal intercepted between the curve and x-axis.

(c) Show that the curvature of catenary $y = c\cos\frac{hx}{c}$ varies inversely as the square of the ordinate.

Solution:

(a) Since the equation of catenary is $y = c\cos\frac{hx}{c}$.
Differentiate w.r.t. x, we get

$$\frac{dy}{dx} = c\sinh\left(\frac{x}{c}\right)\times\frac{1}{c} = \sinh\left(\frac{x}{c}\right).$$

Again, differentiate w.r.t. x, we get

$$\frac{d^2y}{dx^2} = \cosh\left(\frac{x}{c}\right)\times\frac{1}{c}.$$

Now, the radius of curvature,

$$\rho = \frac{\left(1+\left(\frac{dy}{dx}\right)^2\right)^{3/2}}{\frac{d^2y}{dx^2}} = \frac{\left(1+\left(\sinh\left(\frac{x}{c}\right)\right)^2\right)^{3/2}}{\frac{1}{c}\left(\cosh\left(\frac{x}{c}\right)\right)^2} = c\left(\cosh\left(\frac{x}{c}\right)\right)^2. \qquad (6.4)$$

At (0, c), $\rho = \dfrac{(1+0^2)}{\frac{1}{c}} = c.$

(b) The portion of normal intercepted between the curve and x-axis

$$= y\sqrt{1+\left(\frac{dy}{dx}\right)^2} \text{ (as that is the length of the normal)}$$

$$= c\cosh\left(\frac{x}{c}\right)\sqrt{1+\left(\sinh\left(\frac{x}{c}\right)\right)^2}$$

$$= c\cosh\left(\frac{x}{c}\right)\sqrt{\left(\cosh\left(\frac{x}{c}\right)\right)^2}$$

$$= c\left(\cosh\left(\frac{x}{c}\right)\right)^2. \tag{6.5}$$

Then, by (6.4) and (6.5), $\rho =$ portion of normal intercepted between the curve and the x-axis.

(c) $\dfrac{1}{\rho} = \dfrac{1}{c\left(\cosh\left(\frac{x}{c}\right)\right)^2} = \dfrac{c^2}{cy^2} = \dfrac{c}{y^2}.$

$\Rightarrow$ the curvature varies inversely as the square of the ordinate.

EXAMPLE 10 If ρ is the radius of curvature at any point (x, y) on the curve $x^{\frac{2}{3}} + y^{\frac{2}{3}} = a^{\frac{2}{3}}$, then show that $\rho^3 = 27axy$.

Solution: The parametric equations of given curve $x^{\frac{2}{3}} + y^{\frac{2}{3}} = a^{\frac{2}{3}}$ are:

$$x = a\cos^3\theta \text{ and } y = a\sin^3\theta,$$

i.e., $f(\theta) = a\cos^3\theta$ and $g(\theta) = a\sin^3\theta$

$\Rightarrow f'(\theta) = -3a\cos^2\theta\sin\theta$ and $g'(\theta) = 3a\sin^2\theta\cos\theta$

$\Rightarrow f''(\theta) = -3a\cos\theta(\cos^2\theta - 2\sin^2\theta)$ and $g''(\theta) = 3a\sin\theta\left(2\cos^2\theta - \sin^2\theta\right).$

Now, $(f'(\theta))^2 + (g'(\theta))^2 = \left(-3a\cos^2\theta\sin\theta\right)^2 + \left(3a\sin^2\theta\cos\theta\right)^2 = 9a^2\sin^2\theta\cos^2\theta$

$$\text{and } g''(\theta)f'(\theta)-f''(\theta)g'(\theta)=(-3a\cos^2\theta\sin\theta)\left[3a\sin\theta\left(2\cos^2\theta-\sin^2\theta\right)\right]$$
$$-\left(3a\sin^2\theta\cos\theta\right)\left[-3a\cos\theta\left(\cos^2\theta-2\sin^2\theta\right)\right]$$
$$=9a^2\sin^2\theta\cos^2\theta\left[-2\cos^2\theta+\sin^2\theta+\cos^2\theta-2\sin^2\theta\right]$$
$$=-9a^2\sin^2\theta\cos^2\theta$$

Since, the radius of curvature, $\rho=\dfrac{\left((f'(\theta))^2+(g'(\theta))^2\right)^{3/2}}{g''(\theta)f'(\theta)-f''(\theta)g'(\theta)}=\dfrac{\left(9a^2\sin^2\theta\cos^2\theta\right)^{3/2}}{-9a^2\sin^2\theta\cos^2\theta}$

$\Rightarrow$ $\rho=3a\sin\theta\cos\theta$ (in magnitude)

$\Rightarrow$ $\rho^3=27a^3\sin^3\theta\cos^3\theta=27a\left(a\cos^3\theta\right)\left(a\sin^3\theta\right)$

$\Rightarrow$ $\rho^3=27axy.$

EXAMPLE 11 If ρ is the radius of curvature at any point on the parabola $y^2=4ax$, then show that ρ is equal to twice the part of the normal intercepted between the curve and the directrix.

Solution: As in Example 3, $\rho=2a\left(1+t^2\right)^{\frac{3}{2}}$

Let $P\left(at^2,2at\right)$ be any point on the parabola.

Then the equation of normal at P is $y+tx=2at+at^3$. It also meets the directrix which is at $x=-a$.

Let Q be that point on the directrix, then coordinates of Q are $\left(-a,3at+at^3\right)$.

Now, the length of $PQ=\sqrt{\left(-a-at^2\right)^2+\left(3at+at^3-2at\right)^2}$

$$=\sqrt{\left(1+t^2\right)^2a^2\left(1+t^2\right)}$$
$$=a\left(1+t^2\right)^{\frac{3}{2}}=\frac{\rho}{2}.$$

Hence, $\rho=2PQ.$

6.5 RADIUS OF CURVATURE FOR POLAR CURVES

6.5.1 Equation of Curve is in the Form $r=f(\theta)$

Let the equation of curve be in polar form as $r=f(\theta)$

Then, $\psi=\theta+\phi$, where ψ is an angle which the tangent at P makes with the positive direction of x-axis and ϕ is the angle between the tangent at P and the radius vector through P (as shown in Figure 6.4).

Now, the radius of curvature ρ is given by

$$\frac{1}{\rho} = \frac{d\psi}{ds} = \frac{d\theta}{ds} + \frac{d\phi}{ds} = \frac{d\theta}{ds} + \frac{d\phi}{d\theta} \cdot \frac{d\theta}{ds}$$

$$\Rightarrow \qquad \frac{1}{\rho} = \frac{d\theta}{ds}\left(1 + \frac{d\phi}{d\theta}\right) \tag{6.6}$$

Since $\tan\phi = r\dfrac{d\theta}{dr} = \dfrac{r}{\dfrac{dr}{d\theta}} = \dfrac{r}{r_1}$,

where $\quad r_1 = \dfrac{dr}{d\theta}$.

Differentiating w.r.t. θ, we get

$$\sec^2\phi \cdot \frac{d\phi}{d\theta} = \frac{r_1 . r_1 - r . r_2}{r_1^2},$$

where $$r_2 = \frac{d^2 r}{d\theta^2}.$$

$$\Rightarrow \qquad \left(1 + \tan^2\phi\right)\frac{d\phi}{d\theta} = \frac{r_1^2 - r \cdot r_2}{r_1^2}.$$

$$\Rightarrow \qquad \left(1 + \frac{r^2}{r_1^2}\right)\frac{d\phi}{d\theta} = \frac{r_1^2 - r \cdot r_2}{r_1^2}.$$

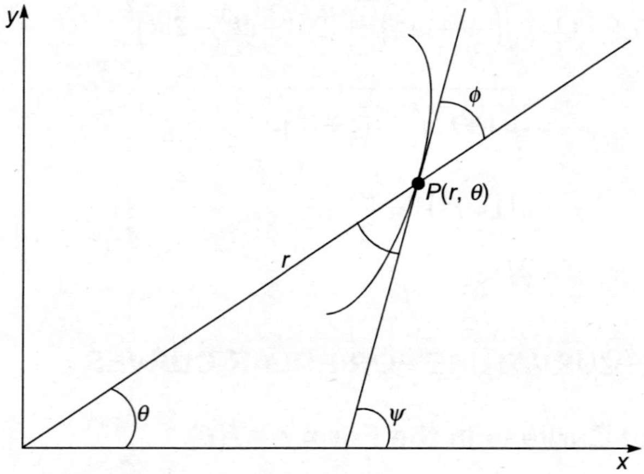

Figure 6.4

$$\Rightarrow \qquad \frac{d\phi}{d\theta} = \frac{r_1^2 - r \cdot r_2}{r_1^2\left(1 + \frac{r^2}{r_1^2}\right)} = \frac{r_1^2 - r \cdot r_2}{r_1^2 + r^2}. \tag{6.7}$$

Now, we know that $\sin\phi = r\dfrac{d\phi}{d\theta}$.

$$\Rightarrow \qquad \frac{d\theta}{ds} = \frac{\sin\phi}{r} = \frac{1}{r} \times \frac{r}{\sqrt{r^2 + r_1^2}} = \frac{1}{\sqrt{r^2 + r_1^2}} \tag{6.8}$$

Substituting (6.8) and (6.7) in (6.6), we get

$$\frac{1}{\rho} = \left(1 + \frac{r_1^2 - r \cdot r_2}{r_1^2 + r^2}\right) \times \frac{1}{\sqrt{r^2 + r_1^2}}.$$

$$\Rightarrow \qquad \frac{1}{\rho} = \frac{r^2 + 2r_1^2 - r \cdot r_2}{\left(r_1^2 + r^2\right)^{3/2}}.$$

Therefore, the radius of curvature for polar curves $r = f(\theta)$ is

$$\rho = \frac{\left(r^2 + r_1^2\right)^{3/2}}{r^2 + 2r_1^2 - r \cdot r_2}. \tag{6.9}$$

6.5.2 Equation of the Curve is in the Form $\frac{1}{r} = f(\theta)$.

Let the equation of the curve given be in the polar form $\dfrac{1}{r} = f(\theta)$,

i.e., we can also write it as $u = f(\theta)$, where $u = \dfrac{1}{r}$.

Now, since $u = \dfrac{1}{r} \Rightarrow r = \dfrac{1}{u}$.

Differentiate w.r.t. θ, we get

$$r_1 = -\frac{1}{u^2}\frac{du}{d\theta} = -\frac{u_1}{u^2}, \text{ where } u_1 = \frac{du}{d\theta}.$$

Again, differentiate w.r.t. θ, we get

$$r_2 = -\left(\frac{u_2 u^2 - u_1 \cdot 2u \cdot u_1}{u^4}\right) = \frac{2u \cdot u_1^2 - u^2 u_2}{u^4} = \frac{2u_1^2 - u \cdot u_2}{u^3}.$$

Put the value of r_1 and r_2 in the formula (6.9) for ρ, we get

$$\rho = \frac{\left(\dfrac{1}{u^2}+\dfrac{u_1^2}{u^4}\right)^{3/2}}{\dfrac{1}{u^2}+2\dfrac{u_1^2}{u^4}-\dfrac{2u_1^2-u\cdot u_2}{u^4}} = \frac{\dfrac{1}{u^6}\left(u^2+u_1^2\right)}{u^2+2u_1^2-2u_1^2+u\cdot u_2}.$$

Therefore, the radius of curvature for polar curves $\dfrac{1}{r}=f(\theta)$ is $\rho=\dfrac{\left(u^2+u_1^2\right)}{u^3\left(u+u_2\right)}$.

6.6 RADIUS OF CURVATURE FOR PEDAL CURVES

The equation of the pedal curves is given by $r=f(p)$, where $p=r\sin\phi$.

Now, since $p=r\sin\phi$, so

differentiate w.r.t. r, we get

$$\frac{dp}{dr}=\sin\phi+r\cos\phi\,.\frac{d\phi}{dr}$$

$$=r\frac{d\theta}{ds}+r\frac{dr}{ds}.\frac{d\phi}{dr}$$

$$=r.\left(\frac{d\theta}{ds}+\frac{d\phi}{ds}\right)=r.\frac{d}{ds}(\theta+\phi)=r.\frac{d\psi}{ds}.$$

$$\Rightarrow \qquad \frac{dp}{dr}=r\cdot\frac{1}{\rho}\quad\left(\text{as}\,\rho=\frac{ds}{d\psi}\right).$$

Therefore, the radius of curvature for pedal curves is $\rho=r\dfrac{dr}{dp}$.

6.7 RADIUS OF CURVATURE FOR POLAR TANGENTIAL CURVES

The equation of the polar tangential curves is given by $p=f(\psi)$, where ψ is the angle which the tangent at $P(x,y)$ makes with positive direction of x-axis and p is the perpendicular distance between the origin O and the tangent to the curve at P.

Let α be the angle between the perpendicular from the origin O on the tangent at P and the positive direction of x-axis.

i.e., $OT=p$ and $\angle TOM=\alpha$ (as shown in Figure 6.5), then $\psi=\alpha+\dfrac{\pi}{2}$.

The equation of tangent is

$$p = x\cos\alpha + y\sin\alpha$$

$$\Rightarrow \qquad p = x\cos\left(\psi - \frac{\pi}{2}\right) + y\sin\left(\psi - \frac{\pi}{2}\right)$$

$$\Rightarrow \qquad p = x\sin\psi - y\cos\psi.$$

Differentiate w.r.t. ψ, we get

$$\frac{dp}{d\psi} = x\cos\psi + y\sin\psi + \sin\psi\cdot\frac{dx}{d\psi} - \cos\psi\cdot\frac{dy}{d\psi}.$$

Now, $$\frac{dx}{d\psi} = \frac{dx}{ds}.\frac{ds}{d\psi} = \rho\cos\psi$$

$$\left(\text{as } \tan\psi = \frac{dy}{dx} = \frac{\frac{dy}{ds}}{\frac{dx}{ds}} \Rightarrow \cos\psi = \frac{dx}{ds}\right).$$

Similarly, $$\frac{dy}{d\psi} = \rho\sin\psi.$$

Thus, $$\frac{dp}{d\psi} = x\cos\psi + y\sin\psi + \sin\psi\cdot\rho\cdot\cos\psi - \cos\psi\cdot\rho\cdot\sin\psi = x\cos\psi + y\sin\psi.$$

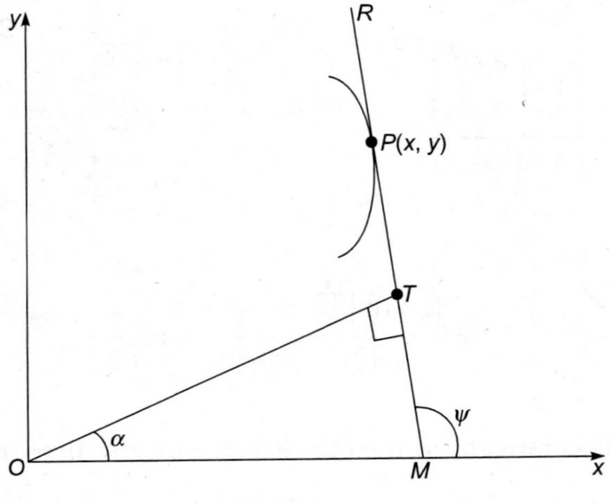

Figure 6.5

Again, differentiate w.r.t. ψ, we get

$$\frac{d^2p}{d\psi^2} = -x\sin\psi + \cos\psi\cdot\frac{dx}{d\psi} + y\cos\psi + \sin\psi\cdot\frac{dy}{d\psi}$$

$$= -(x\sin\psi - y\cos\psi) + \rho\left(\cos^2\psi + \sin^2\psi\right)$$

$$\Rightarrow \qquad \rho = \frac{d^2p}{d\psi^2} + (x\sin\psi - y\cos\psi).$$

Therefore, the radius of curvature for polar tangential curves is $\rho = p.+\dfrac{d^2p}{d\psi^2}$

6.8 RADIUS OF CURVATURE AT THE ORIGIN (BY NEWTON'S METHOD)

6.8.1 For Cartesian Curves $y = f(x)$.

Let a curve pass through the origin O such that the x-axis is tangent to the curve at origin (as shown in the Figure 6.6). Then, the value of $y_1 = \dfrac{dy}{dx}$ at the origin is zero, i.e., $y_1(0) = 0$.

Now, $\lim\limits_{\substack{x\to 0\\ y\to 0}} \dfrac{x^2}{2y}$ is an indeterminate form.

So that, by using L' Hospital's rule

$$\lim_{\substack{x\to 0\\ y\to 0}} \frac{x^2}{2y} = \lim_{\substack{x\to 0\\ y\to 0}} \frac{x}{y_1} = \lim_{\substack{x\to 0\\ y\to 0}} \frac{1}{y_2} = \frac{1}{y_2(0)}.$$

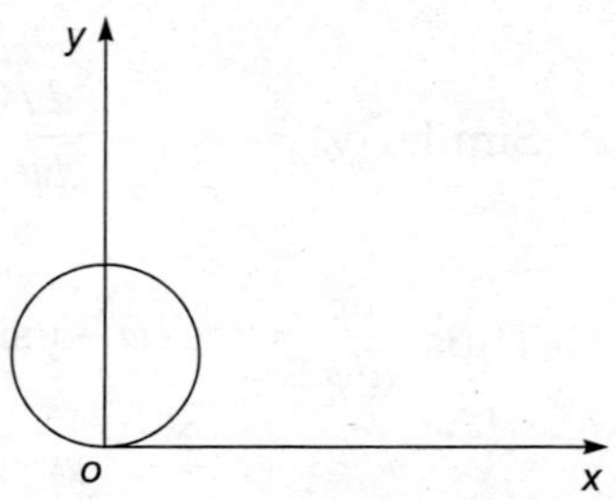

Figure 6.6

Since,
$$\rho = \frac{\left(1+\left(\dfrac{dy}{dx}\right)^2\right)^{3/2}}{\dfrac{d^2y}{dx^2}}$$

so at origin $(0,0)$

$$\rho = \frac{\left(1+0^2\right)^{3/2}}{y_2(0)} = \frac{1}{y_2(0)} = \lim_{\substack{x\to 0\\ y\to 0}} \frac{x^2}{2y}$$

Therefore, if x-axis is tangent to the curve at the origin, then $\rho(\text{at origin}) = \lim\limits_{\substack{x\to 0\\ y\to 0}} \dfrac{x^2}{2y}$.

Similarly, if y-axis is tangent to the curve at the origin, then $\rho(\text{at origin}) = \lim\limits_{\substack{x\to 0\\ y\to 0}} \dfrac{y^2}{2x}$.

These above formulae are known as Newton's formulae for finding the radius of curvature at the origin.

6.8.2 For Polar Curves $r = f(\theta)$

Let the equation of curve be $r = f(\theta)$ pass through the origin and the initial line $\theta = 0$ is tangent to the curve at origin.

$$\text{Now, } \rho = \lim_{\substack{x\to 0\\ y\to 0}} \frac{x^2}{2y} = \lim_{\theta\to 0} \frac{r^2\cos^2\theta}{2r\sin\theta} = \lim_{\theta\to 0} \frac{r}{2\theta}\cdot\frac{1}{\frac{\sin\theta}{\theta}}\cdot\cos^2\theta = \lim_{\theta\to 0}\frac{r}{2\theta}.$$

Therefore, the radius of curvature at the origin for the polar curve $r = f(\theta)$ is $\rho = \lim_{\theta\to 0} \frac{r}{2\theta}$.

EXAMPLE 12 Find the radius of curvature at any point of the cardioid $r = a(1+\cos\theta)$. Further, show that $\frac{\rho^2}{r}$ is constant.

Solution: Since, we have $r = a(1+\cos\theta)$,

differentiating w.r.t. θ, we have

$$r_1 = -a\sin\theta$$

Again, differentiate w.r.t. θ, we get

$$r_2 = -a\cos\theta.$$

$$\text{By (6.9), } \rho = \frac{\left(r^2 + r_1^2\right)^{3/2}}{r^2 + 2r_1^2 - r\cdot r_2} = \frac{\left(a^2(1+\cos\theta)^2 + a^2\sin^2\theta\right)^{3/2}}{a^2(1+\cos\theta)^2 + 2a^2\sin^2\theta + a^2\cos\theta(1+\cos\theta)}$$

$$= \frac{\left(2a^2(1+\cos\theta)\right)^{3/2}}{3a^2(1+\cos\theta)} = \frac{2^{3/2}}{3}\cdot a\cdot(1+\cos\theta)^{1/2}.$$

$$\Rightarrow \quad \rho = \frac{2}{3}\sqrt{2ar}.$$

Now, further $\rho^2 = \frac{8}{9}ar$

$$\Rightarrow \quad \frac{\rho^2}{r} = \frac{8}{9}a = \text{constant}.$$

EXAMPLE 13 Show that the radius of curvature at any point on the curve $p = \frac{r^{n+1}}{a^n}$, varies inversely as the $(n-1)^{th}$ power of the radius vector.

Solution: We have $p = \frac{r^{n+1}}{a^n}$.

$$\Rightarrow \qquad pa^n = r^{n+1}.$$

Differentiate w.r.t. p, we get

$$a^n = (n+1)r^n \cdot \frac{dr}{dp}.$$

$$\Rightarrow \qquad \frac{dr}{dp} = \frac{a^n}{(n+1)r^n}.$$

Since $\rho = r \cdot \frac{dr}{dp} = \frac{a^n}{(n+1)r^{n-1}}$, we have

$$\rho \propto \frac{1}{r^{n-1}}.$$

EXAMPLE 14 Show that the radius of curvature at any point on lamniscate $pa^2 = r^3$, varies inversely as the radius vector.

Solution: We have $pa^2 = r^3$

so differentiate w.r.t. p, we have

$$a^2 = 3r^2 \cdot \frac{dr}{dp}.$$

$$\Rightarrow \qquad \frac{dr}{dp} = \frac{a^2}{3r^2}.$$

Since $\rho = r \cdot \frac{dr}{dp} = r \cdot \frac{a^2}{3r^2} = \frac{a^2}{3} \cdot \frac{1}{r}$, we have

$$\rho \propto \frac{1}{r}.$$

EXAMPLE 15 If ρ_1 and ρ_2 be two radii of curvature at the extremities of any chord of the cardioid $r = a(1+\cos\theta)$ which passes through the pole of the cardioid. Show that $\rho_1^2 + \rho_2^2 = \frac{16}{9}a^2$.

Solution: Since $r = a(1+\cos\theta)$, then by Example 1, we have

$$\rho = \frac{2^{3/2}}{3} \cdot a \cdot (1+\cos\theta)^{1/2}.$$

$$\Rightarrow \qquad \rho = \frac{2^{3/2}}{3} \cdot a \cdot \left(2\cos^2\frac{\theta}{2}\right)^{1/2} = \frac{4}{3}a\cos\frac{\theta}{2}.$$

Let P and Q be two extremities of a chord of the cardioid which pass through the pole and have angles θ and $\pi + \theta$, respectively with positive direction of x-axis.

Then, the radius of curvature at P is $\rho_1 = \frac{4}{3} a \cos \frac{\theta}{2}$

and the radius of curvature at Q is $\rho_2 = \frac{4}{3} a \cos\left(\frac{\pi + \theta}{2}\right) = -\frac{4}{3} a \sin\left(\frac{\theta}{2}\right)$.

Now, $\rho_1^2 + \rho_2^2 = \frac{16}{9} a^2 \left(\cos^2 \frac{\theta}{2} + \sin^2 \frac{\theta}{2}\right) = \frac{16}{9} a^2$.

EXAMPLE 16 Find the radius of curvature of the curve $r = a \sin(n\theta)$ at the origin.

Solution: We have $r = a \sin(n\theta)$

and $$\rho(\text{at origin}) = \lim_{\theta \to 0} \frac{r}{2\theta} = \lim_{\theta \to 0} \frac{a \sin(n\theta)}{2\theta} \left(\frac{0}{0} \text{ form}\right)$$

$$= \lim_{\theta \to 0} \frac{na \cos(n\theta)}{2}$$

$$= \frac{na}{2}.$$

EXAMPLE 17 Find the radius of curvature of the curve $x^3 + y^3 - 2x^2 + 6y = 0$ at the origin.

Solution: Equate the lowest degree term from the equation of given curve to zero,

i.e., $$6y = 0 \Rightarrow y = 0.$$

Thus, the x-axis is tangent to the curve at the origin.

Now, we have $x^3 + y^3 - 2x^2 + 6y = 0$

Divide by $2y$ on both sides, we get

$$\frac{x^3}{2y} + \frac{y^3}{2y} - \frac{2x^2}{2y} + \frac{6y}{2y} = 0$$

$$\Rightarrow \qquad \frac{x^2}{2y} \cdot x + \frac{y^2}{2} - 2 \cdot \frac{x^2}{2y} + 3 = 0.$$

Taking limits as $x \to 0, y \to 0$ on both sides, we get

$$\rho.0 + 0 - 2.\rho + 3 = 0$$

$$\Rightarrow \qquad \rho = \frac{3}{2}.$$

EXAMPLE 18 Find the radius of curvature at origin for the cycloid $x = a(\theta + \sin\theta)$ and $y = a(1 - cos\theta)$ by Newton's method.

Solution: Let the initial line be tangent to the curve at the origin, then

$$\rho = \lim_{\substack{x\to 0\\ y\to 0}} \frac{x^2}{2y} = \lim_{\theta\to o} \frac{a^2(\theta+\sin\theta)^2}{2a(1-\cos\theta)} \left(\frac{0}{0}\text{ form}\right)$$

$$= \lim_{\theta\to 0} \frac{a^2 2(\theta+\sin\theta)(1+\cos\theta)}{2a\sin\theta} \left(\frac{0}{0}\text{ form}\right)$$

$$= \lim_{\theta\to 0} \frac{2a^2\left[(1+\cos\theta)^2 + (\theta+\sin\theta)(-\sin\theta)\right]}{2a\cos\theta}$$

$$= 4a.$$

EXAMPLE 19 Find the radius of curvature at the point on the curve $r = a(1+\cos\theta)$ where the tangent line is parallel to initial line.

Solution: We have, $r = a(1+\cos\theta)$, then by Example 1, we have

$$\rho = \frac{2^{3/2}}{3} \cdot a \cdot (1+\cos\theta)^{1/2}.$$

Since the tangent line is parallel to initial line, $\psi = 0$.

$$\Rightarrow \quad \theta + \phi = 0$$

$$\Rightarrow \quad \phi = -\theta$$

$$\Rightarrow \quad \tan\phi = -\tan\theta$$

$$\Rightarrow \quad r \cdot \frac{d\theta}{dr} = -\tan\theta$$

$$\Rightarrow \quad a(1+\cos\theta) \times \frac{1}{-a\sin\theta} = -\frac{\sin\theta}{\cos\theta}$$

$$\Rightarrow \quad \cos\theta + \cos^2\theta = \sin^2\theta$$

$$\Rightarrow \quad 2\cos^2\theta + \cos\theta - 1 = 0$$

$$\Rightarrow \quad \cos\theta = -1, -\frac{1}{2}.$$

If $\cos\theta = -1$, then $\rho = 0$ which is not possible.

$$\therefore \quad \cos\theta = \frac{1}{2} \Rightarrow \rho = \frac{2^{3/2}}{3} a\left(1 + \frac{1}{2}\right)^{1/2} = \frac{2a}{\sqrt{3}}.$$

EXAMPLE 20 Find the radius of curvature at the point $(2,0)$ on the curve $y^2 = (x-1)(x-2)$ by using Newton's formula.

Solution: On shifting the origin to the point $(2,0)$, the new coordinate axes are

$$x = X + 2 \quad \text{and} \quad y = Y.$$

Then, the given equation of curve becomes

$$Y^2 = (X+1)X.$$

i.e.,

$$Y^2 = X^2 + X, \tag{6.10}$$

Equate the lowest degree term to zero, we get

$$X = 0,$$

i.e., Y-axis is tangent to the curve in (6.10) at the origin.

So that,

$$\rho(at(2,0)) = \lim_{\substack{X\to 0\\ Y\to 0}} \frac{Y^2}{2X} = \lim_{\substack{X\to 0\\ Y\to 0}} \frac{X(X+1)}{2X} = \frac{1}{2}.$$

EXAMPLE 21 Show that the radius of curvature of the parametric curve $x = t - \frac{1}{3}t^3,\ y = t^2$ at the origin is $\frac{1}{2}$.

Solution: Let the x-axis be a tangent to the curve at the origin.

Thus,

$$\rho(\text{at origin}) = \lim_{\substack{x\to 0\\ y\to 0}} \frac{x^2}{2y}$$

$$= \lim_{t\to 0} \frac{\left(t - \frac{1}{3}t^3\right)^2}{2t^2} \left(\frac{0}{0}\text{ form}\right)$$

$$= \lim_{t\to 0} \frac{2\left(t - \frac{1}{3}t^3\right)\left(1 - t^2\right)}{4t} \left(\frac{0}{0}\text{ form}\right)$$

$$= \lim_{t\to 0} \frac{1}{2}\left[\left(1-t^2\right)^2 - \left(t - \frac{1}{3}t^3\right)(-2t)\right]$$

$$= \frac{1}{2}.$$

6.9 CENTRE AND CIRCLE OF CURVATURE

We have already defined centre and circle of curvature in Sections 6.2 and 6.3 of the present chapter. In this section we will discuss the coordinates of centre of curvature and the equation of circle of curvature.

Let xy be the coordinate axes with origin O'. Let $O(X,Y)$ be the centre of curvature corresponding to any point $P(x,y)$ on the curve $y = f(x)$, i.e., $OP = \rho$ (as shown in Figure 6.7).

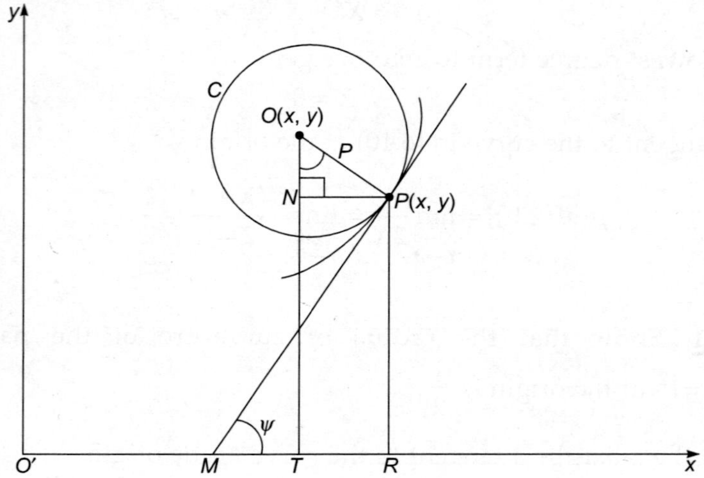

Figure 6.7

Let PM be the tangent to the curve at the point P and meets the x-axis at M and makes an angle ψ with positive direction of x-axis.

Let us draw a perpendicular PR and OT on x-axis from P and O, respectively. Also, draw a perpendicular PN form P on OT.

Now, $\angle PON + \angle PNO + \angle OPN = 180°$

$\Rightarrow \angle PON = 90° - \angle OPN$

$= 90° - (90° - \angle NPM)$

$= \angle NPM$

$= \angle PMT$ (as alternating angles)

$\Rightarrow \angle PON = \psi.$

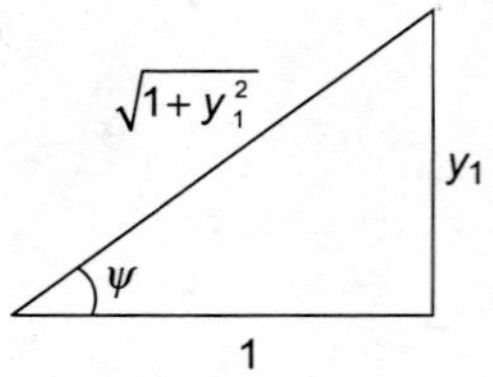

Figure 6.8

Now, $X = O'T = O'R - TR$

$= O'R - NP = x - \rho \sin\psi$

and $Y = OT = ON + NT = y + \rho\cos\psi.$

We know that, $\tan\psi = \dfrac{dy}{dx} = y_1.$

$$\Rightarrow \sin\psi = \frac{y_1}{\sqrt{1+y_1^2}} \text{ and } \cos\psi = \frac{1}{\sqrt{1+y_1^2}} \text{ (as shown in Figure 6.8).}$$

Since $X = x - \rho\sin\psi$ and $Y = y + \rho\cos\psi$,

$$X = x - \frac{\rho \cdot y_1}{\sqrt{1+y_1^2}} \text{ and } Y = y + \rho \cdot \frac{1}{\sqrt{1+y_1^2}}.$$

$$\Rightarrow X = x - \frac{\left(1+y_1^2\right)^{3/2}}{y_2} \cdot \frac{y_1}{\left(1+y_1^2\right)^{1/2}} \text{ and } Y = y + \frac{\left(1+y_1^2\right)^{3/2}}{y_2} \cdot \frac{1}{\left(1+y_1^2\right)^{1/2}}.$$

$$\left\{\text{as } \rho = \frac{\left(1+y_1^2\right)^{3/2}}{y_2}\right\}$$

$$\Rightarrow X = x - \frac{y_1\left(1+y_1^2\right)}{y_2} \text{ and } Y = y + \frac{\left(1+y_1^2\right)}{y_2}. \tag{6.11}$$

The (X, Y) coordinates given by (6.11) are the coordinates of centre of curvature. The equation of circle of curvature is given by

$$(x-X)^2 + (y-Y)^2 = \rho^2.$$

Remarks

1. If $y_2 > 0$, then the curve is concave at P. Then $X < x$ and $Y < y$.
2. If $y_2 < 0$, then the curve is convex at P.

6.10 EVOLUTE AND INVOLUTE

The locus of centre of curvature of a curve is called the *evolute* and the given curve is called *involute* of its evolute.

To Obtain the Evolute of a Curve

The following steps are involved to find the evolute of a curve.

1. Let (X, Y) be the centre of curvature of a curve, then

$$X = x - \frac{y_1\left(1+y_1^2\right)}{y_2} \text{ and } Y = y + \frac{\left(1+y_1^2\right)}{y_2}.$$

2. Eliminate the parameters x and y to obtain the relation between X and Y.
3. Generalize the relation of X and Y to obtain the required equation of evolute.

EXAMPLE 22 Find the coordinates of centre of curvature of the parabola $y^2 = 4ax$ at any point (x,y). Also, find the equation of evolute of the parabola.

Solution: The given equation of the parabola is $y^2 = 4ax$

Differentiate w.r.t. x, we get

$$2yy_1 = 4a \Rightarrow y_1 = \frac{4a}{2y} = \frac{4a}{2\times 2\sqrt{ax}} = a^{1/2}\, x^{-1/2}.$$

Again, differentiate w.r.t. x, we get

$$y_2 = a^{1/2}\left(\frac{-1}{2}\right)x^{-3/2}.$$

Let (X,Y) be the coordinates of centre of curvature of the parabola. Then,

$$X = x - \frac{y_1\left(1+y_1^2\right)}{y_2} \text{ and } Y = y + \frac{\left(1+y_1^2\right)}{y_2}.$$

$$\Rightarrow \quad X = x - \frac{a^{1/2}x^{-1/2}}{-\frac{1}{2}a^{1/2}x^{-3/2}}\left(1+\frac{a}{x}\right) \text{ and } Y = y + \frac{\left(1+\frac{a}{x}\right)}{-\frac{1}{2}a^{1/2}x^{-3/2}}.$$

$$\Rightarrow \quad X = x + 2x\left(1+\frac{a}{x}\right) \text{ and } Y = y - 2x\sqrt{\frac{x}{a}}\left(1+\frac{a}{x}\right).$$

$$\Rightarrow \quad X = 3x + 2a \text{ and } Y = 2\sqrt{ax} - 2x\sqrt{\frac{x}{a}} - 2\sqrt{ax}.$$

$$\Rightarrow \quad X = 3x + 2a \text{ and } Y = -2x\sqrt{\frac{x}{a}}.$$

Thus, $(X,Y) = \left(3x+2a,\ -2x\sqrt{\frac{x}{a}}\right)$ are the coordinates of centre of curvature of the parabola $y^2 = 4ax$.

Now, to obtain an equation of evolute, eliminate the parameters x and y from above X and Y, we get

$$x = \frac{X-2a}{3} \quad \Rightarrow \quad Y = -2\left(\frac{X-2a}{3}\right)\sqrt{\frac{X-2a}{3}}\cdot\frac{1}{\sqrt{a}}.$$

$$Y^2 = \frac{4}{9}\frac{(X-2a)^3}{3}\cdot\frac{1}{a}.$$

$$\Rightarrow \quad 27aY^2 = 4(X-2a)^3.$$

Hence, the equation of evolute of the parabola is $27aY^2 = 4(X-2a)^3$.

EXAMPLE 23 Find the equation of circle of curvature of the curve $y = x^3 + 2x^2 + x + 1$ at the point $(0,1)$.

Solution The equation of the given curve is

$$y = x^3 + 2x^2 + x + 1.$$

Differentiate w.r.t. x, we get

$$y_1 = 3x^2 + 4x + 1.$$

Again, differentiate w.r.t. x, we get

$$y_2 = 6x + 4.$$

At $(0,1)$, $\quad y_1 = 1$ and $y_2 = 4$.

Let (X,Y) be the coordinates of centre of curvature of the given curve, then

$$X = x - \frac{y_1\left(1+y_1^2\right)}{y_2} \text{ and } Y = y + \frac{\left(1+y_1^2\right)}{y_2}$$

$$\Rightarrow \quad X|_{(0,1)} = \frac{-1}{4}(1+1) = \frac{-1}{2} \text{ and } Y|_{(0,1)} = 1 + = \frac{(1+1)}{4} = \frac{3}{2}$$

and since

$$\rho = \frac{\left(1+y_1^2\right)^{3/2}}{y_2},$$

$$\rho|_{(0,1)} = \frac{(1+1)^{3/2}}{4} = \frac{1}{\sqrt{2}}.$$

Therefore, the equation of circle of curvature is $\left(x+\frac{1}{2}\right)^2 + \left(y-\frac{3}{2}\right)^2 = \frac{1}{2}$.

EXAMPLE 24 Find the coordinates of the centre of curvature at any point (x,y) on the ellipse $\frac{x^2}{a^2} + \frac{y^2}{b^2} = 1$. Also, find the evolute of the ellipse.

Solution: The equation of the ellipse is $\frac{x^2}{a^2} + \frac{y^2}{b^2} = 1$.

Differentiate w.r.t. x, we get

$$2\frac{x}{a^2} + 2\frac{y}{b^2}y_1 = 0$$

$$\Rightarrow \quad y_1 = \frac{-b^2}{a^2} \times \frac{x}{y}.$$

Again, differentiate w.r.t. x, we get

$$y_2 = \frac{-b^2}{a^2} \cdot \frac{1}{y}\left(1 - \frac{xy_1}{y}\right)$$

Let (X,Y) be the coordinates of the centre of curvature of the ellipse, then

$$X = x - \frac{y_1\left(1+y_1^2\right)}{y_2} \text{ and } Y = y + \frac{\left(1+y_1^2\right)}{y_2}.$$

$$\Rightarrow \quad X = \frac{x^3}{a^4}\left(a^2 - b^2\right) \text{ and } Y = \frac{y^3}{b^4}\left(b^2 - a^2\right).$$

Therefore, the centre of curvature of the ellipse is $(X,Y) = \left(\frac{x^3}{a^4}\left(a^2-b^2\right), \frac{y^3}{b^4}\left(b^2-a^2\right)\right)$.

Now, to obtain the evolute of the ellipse, eliminate the parameters x and y from X and Y, we get

$$aX = \frac{x^3}{a^3}\left(a^2 - b^2\right) \quad \text{and} \quad bY = \frac{y^3}{b^3}\left(b^2 - a^2\right).$$

On squaring and adding, we get

$$(aX)^{\frac{2}{3}} + (bY)^{\frac{2}{3}} = \left(a^2 - b^2\right)^{\frac{2}{3}}\left(\frac{x^2}{a^2} + \frac{y^2}{b^2}\right) = \left(a^2 - b^2\right)^{\frac{2}{3}}.$$

Hence, $(aX)^{\frac{2}{3}} + (bY)^{\frac{2}{3}} = \left(a^2 - b^2\right)^{\frac{2}{3}}$ is an evolute to the ellipse.

Exercise

1. (a) Show that the radius of curvature at the point (s,φ) on the curve $s = c\log\sec\varphi$ is given by $c\tan\varphi$.

   (b) Find the radius of curvature at the end of latus rectum of the curve $y^2 = 16x$.

2. Find the radius of curvature of the lemniscate $\left(x^2+y^2\right)^2 = a^2\left(x^2-y^2\right)$ at a point where the tangent is parallel to x-axis.

3. Let $\rho_1 \, and \, \rho_2$ be the radii of curvature at the points (x,y) on the curves $y = c^2$, $x^3 = 3c^2y$ which have same abscissa. Prove that the ratio of ρ_1 and ρ_2 varies as square root of ratio of the ordinates.

4. Let ρ be the radius of curvature at any point of the curve $x = a\cos^3\theta$, $y = a\sin^3\theta$. Prove that $\rho = 3d$ where d is the length of the perpendicular from the origin to the tangent.

5. Prove that the radius of curvature at any point of the curve $x = a\left(\cos t + \log \tan \frac{1}{2} t\right)$, $y = a \sin t$ is inversely proportional to the length of the normal intercepted between the point and the x-axis.
6. Show that the radius of curvature of the cycloid $x = a(\theta - \sin\theta), y = a(1 - \cos\theta)$ at any point is twice the length of the normal at that point.
7. Show that the radius of curvature for the curve $\sqrt{x/a} - \sqrt{y/b} = 1$ at the points where it touches the coordinate axes is given by $\frac{2b^2}{a}$.
8. Prove that the radius of curvature at $\theta = \frac{\pi}{6}$ for the curve $x = a\cos^4\theta,\ y = a\sin^4\theta$ is $\frac{16}{5\sqrt{10a}}$.
9. Prove that $\left(\rho_1^{\frac{2}{3}} + \rho_2^{\frac{2}{3}}\right)(ab)^{\frac{2}{3}} = a^2 + b^2$, where ρ_1 and ρ_2 are radii of curvature at the extremities of two conjugate semi-diameter of the ellipse $\frac{x^2}{a^2} + \frac{y^2}{b^2} = 1$.
10. Find the radius of curvature for the curve $x = a\cos\theta(1 + \sin\theta), y = a\sin\theta(1 + \cos\theta)$ at $\theta = \frac{\pi}{4}$.
11. Find the radius of curvature for the curve $r^2 = a^2\cos 2\theta$.
12. Prove that the radius of curvature at any point on the curve $\theta = \frac{\sqrt{r^2 - a^2}}{a} - \cos^{-1}\frac{a}{r}$ is given by $\rho = \sqrt{r^2 - a^2}$.
13. Show that the radius of curvature for the curve $r^n = a^n \cos n\theta$ is $\frac{a^n r^{-n+1}}{n+1}$.
14. Let ρ_1 and ρ_2 be the radii of curvature at the points in which the curves $r = a\theta$ and $r\theta = a$ intersect. Then, prove that ρ_1 and ρ_2 are in the ratio $1:3$.
15. Find the radius of curvature at origin for the following curves

    (a) $y^2 - 3xy - 4x^2 + x^3 + x^4 y + y^5 = 0$

    (b) $y^4 + x^3 + a\left(x^2 + y^2\right) - a^2 y = 0$.
16. Find the radius of curvature at the origin for the curve $y^2 = x^2 \frac{(a+x)}{(a-x)}$.
17. For the curve $x^{2/3} + y^{2/3} = a^{2/3}$, prove that the equation of the evolute is

$$(x+y)^{2/3} + (x-y)^{2/3} = 2a^{2/3}.$$

18. Find the coordinates of the centre of curvature for any point (x, y) on the hyperbola $\frac{x^2}{a^2} - \frac{y^2}{b^2} = 1$. Find its evolute also.

19. Find the circle of curvature for the curve $\sqrt{x} + \sqrt{y} = \sqrt{a}$, at the point $\left(\frac{a}{4}, \frac{a}{4}\right)$.

20. Show that for the curve $x^2 = 4ay$ the equation of the evolute is $4(y - 2a)^3 = 27ax^2$.

21. Show that the evolute of the cycloid $x = a(\theta - \sin\theta), y = a(\sin\theta - \theta\cos\theta)$ is another equal cycloid.

7 CHAPTER

Asymptotes

In this chapter, we will study the curves that can be extended to infinity. For instance, the parabola, hyperbola, etc. In such cases, it is interesting to investigate the variation of the point on the curve at which the tangent is drawn, moving away from the origin along the curve to infinity. The study of such type of variation of the point on the curve at which the tangent is drawn requires the concept of an asymptote.

Before giving the definition of asymptote, we will first discuss finite and infinite branches of a plane curve.

7.1 BRANCHES OF A CURVE

Let us consider an equation of circle $x^2 + y^2 = a^2$ which has centre $(0,0)$ and radius a.

Now, solving the equation for y in terms of x, we get

$$y = \pm\sqrt{a^2 - x^2}.$$

Thus, $y = \sqrt{a^2 - x^2}$ represents the upper half of the circle and $y = -\sqrt{a^2 - x^2}$ represents the lower half of the circle.

Therefore, these two above equations represent the two branches of a circle $x^2 + y^2 = a^2$. But these branches lie within the finite part of *xy*-plane bounded by $x = \pm a$ and $y = \pm a$.

These two branches of the circle are known as *finite branches* of a curve.

Now, consider the equation of parabola $y^2 = 4ax$. Its solution is $y = \pm 2\sqrt{ax}$. Then y tends to $\pm\infty$ as the point P on the parabola moves to infinity along the curve, i.e., both the branches of the parabola can be extended to infinity. These branches are known as *infinite branches* of curve.

7.2 ASYMPTOTES

Since the point $P(x,y)$ on an infinite branch of a curve tends to infinity along the curve, if either x or y or both tend to infinity. This yields the following definition of asymptotes:

Definition: A straight line at a finite distance from the origin is said to be an *asymptote* (or *rectilinear asymptote*) of an infinite branch of a curve, if the point $P(x,y)$ tends to infinity along the branch of the curve, the perpendicular distance of the point P from that straight line tends to zero.

Geometrically, line AB will be asymptote of the curve $y = f(x)$, if the perpendicular distance PQ (as shown in Figure 7.1) from the point P to the line AB tends to zero as a point $P(x,y)$ tends to infinity along the branch of the curve.

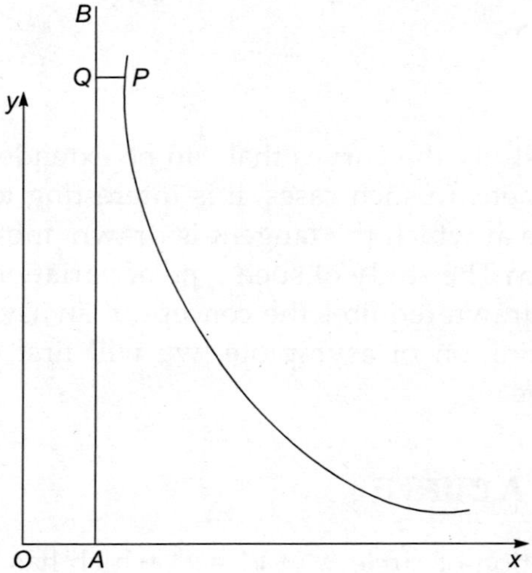

Figure 7.1

Illustrations

1. The line $x = -a$ is an asymptote to the curve $x(x^2 + y^2) = a(x^2 - y^2)$ (as shown in Figure 8.3)
2. The line $y = \pm 1$ parallel to x-axis is an asymptote of $y^2x^2 = x^2 + a^2$ (as shown in Figure 8.7)

7.2.1 Asymptotes of the Cartesian Curves

Let the equation of the curve be $f(x,y) = 0$.

The equation of line AB which is not parallel to y-axis is

$$y = mx + c, \tag{7.1}$$

where m is slope of the line. Let $P(x,y)$ be a point on the infinite branch of the curve.

As $P(x,y)$ tends to infinity along the line (7.1), then the abscissa x must tend to infinity. Thus, the perpendicular distance PQ from the point P to the line AB tends to zero,

i.e.,
$$\lim_{x \to \infty} PQ = 0$$

$$\Rightarrow \qquad \lim_{x\to\infty} \frac{y - mx - c}{\sqrt{1+m^2}} = 0.$$

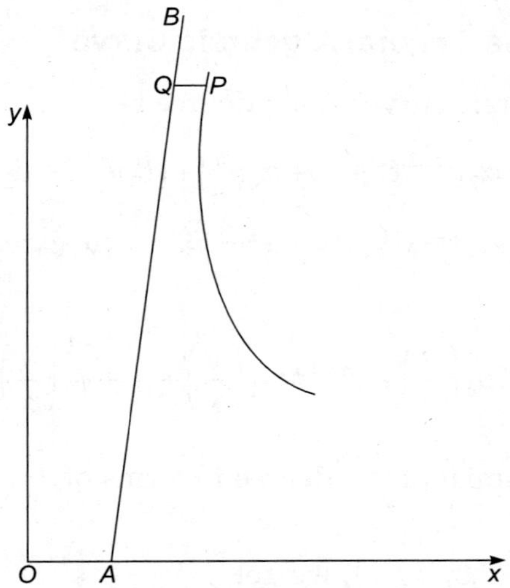

Figure 7.2

{as perpendicular distance from the point (x,y) to the curve $Ax+By=C$ is $\left|\frac{Ax+By-C}{\sqrt{A^2+B^2}}\right|$}

$$\Rightarrow \qquad \lim_{x\to\infty} (y - mx) = c \text{ (as } 1+m^2 \neq 0).$$

$$\Rightarrow \qquad \lim_{x\to\infty} \frac{y - mx}{x} = c \cdot \lim_{x\to\infty} \frac{1}{x} = 0.$$

$$\Rightarrow \qquad \lim_{x\to\infty} \left(\frac{y}{x} - m\right) = 0.$$

$$\Rightarrow \qquad \lim_{x\to\infty} \frac{y}{x} = m.$$

Thus, $m = \lim_{x\to\infty} \frac{y}{x}$ and $c = \lim_{x\to\infty} (y - mx)$, so that the line $y = mx + c$ is an asymptote of the given curve which is not parallel to y-axis.

To find the asymptote to the curve f(x, y) = 0 which is not parallel to y-axis, the following steps are involved:

1. Find $m = \lim_{x\to\infty} \frac{y}{x}$

2. Then, find $c=\lim_{x\to\infty}(y-mx)$

3. The line $y=mx+c$ will be the required asymptote to the curve $f(x,y)=0$.

7.2.2 Asymptote to the General Algebraic Curve

Let the equation of algebraic curve of nth degree be

$$\left(\alpha_o y^n+\alpha_1 y^{n-1}x+\alpha_2 y^{n-2}x^2+\ldots+\alpha_n x^n\right)+\left(\beta_1 y^{n-1}+\beta_2 y^{n-2}x+\ldots+\beta_n x^{n-1}\right)$$
$$+\left(\gamma_1 y^{n-2}+\gamma_2 y^{n-3}x+\ldots+\gamma_n x^{n-2}\right)+\ldots+\left(v_0 x+v_1 y\right)+\omega=0$$

It can be written as

$$x^n\phi_n\left(\frac{y}{x}\right)+x^{n-1}\phi_{n-1}\left(\frac{y}{x}\right)+\cdots+\phi_0\left(\frac{y}{x}\right)=0 \tag{7.2}$$

where $\phi_r\left(\frac{y}{x}\right)$ is a polynomial of rth degree in terms of $\frac{y}{x}$. (7.3)

Dividing the equation (7.3) by x^n, we get

$$\phi_n\left(\frac{y}{x}\right)+\frac{1}{x}\phi_{n-1}\left(\frac{y}{x}\right)+\frac{1}{x^2}\phi_{n-2}\left(\frac{y}{x}\right)+\cdots=0. \tag{7.4}$$

Let $y=mx+c$ be an asymptote of the curve which is not parallel to *y*-axis such that m and c are finite.

Taking limits as $x\to\infty$ in (7.4), we get

$$\phi_n(m)=0,\text{ where } m=\lim_{x\to\infty}\frac{y}{x}. \tag{7.5}$$

This equation enables us to find the value of m of the asymptote $y=mx+c$.

Now, differentiating (7.4) w.r.t. *x*, we get

$$\phi_n'\left(\frac{y}{x}\right)\left(\frac{\frac{dy}{dx}x-y}{x^2}\right)+\left(\frac{1}{x}\phi_{n-1}'\left(\frac{y}{x}\right)\right)\left(\frac{\frac{dy}{dx}x-y}{x^2}\right)-\frac{1}{x^2}\phi_{n-1}\left(\frac{y}{x}\right)$$
$$+\left(\frac{1}{x^2}\phi_{n-2}'\left(\frac{y}{x}\right)\right)\left(\frac{\frac{dy}{dx}x-y}{x^2}\right)-\frac{2}{x^3}\phi_{n-2}\left(\frac{y}{x}\right)+\ldots=0.$$

Multiply by x^2 and then taking limits as $x\to\infty$, we get

$$\phi_n'(m)(-c)-\phi_{n-1}(m)=0\ \{\text{by using } m=\lim_{x\to\infty}\frac{y}{x}\text{ and } c=\lim_{x\to\infty}(y-mx)\}. \tag{7.6}$$

But the value of c in (7.5) depends on the nature of the roots of an equation (7.5).

Following cases arise:

I. If an equation (7.5) has distinct roots, then $\phi_n'(m) \neq 0$.
Thus, the value of c is given by (7.6) as

$$c = -\frac{\phi_{n-1}(m)}{\phi_n'(m)} \tag{7.7}$$

Let $m_1, m_2, \ldots\ldots, m_n$ be distinct roots of $\phi_n(m) = 0$. Then the value of c corresponding to each value of m obtained by (7.7) are $c_1, c_2, \ldots\ldots c_n$, respectively.
Then, the required asymptotes are

$$y = m_1 x + c_1, y = m_2 x + c_2, \ldots\ldots\ldots\ldots y = m_n x + c_n .$$

II. If an equation (7.5) has some repeated roots, then $\phi_n'(m) = 0$, for some repeated values of m. Then, there are two possibilities:

(i) If $\phi_{n-1}(m) \neq 0$, then equation (7.6) does not determine any value of c. Therefore, no asymptotes are obtained for that repeated value of m.

(ii) If $\phi_{n-1}(m) = 0$, then the equation (7.6) reduces to

$$0.(-c) - 0 = 0.$$

Let μ be one of the repeated roots of the equation (7.5).

Let $y - \mu x = \eta$, i.e., $\frac{y}{x} = \mu + \frac{\eta}{x}$. Then by equation (7.2), we have

$$x^n \phi_n\left(\mu + \frac{\eta}{x}\right) + x^{n-1}\phi_{n-1}\left(\mu + \frac{\eta}{x}\right) + x^{n-2}\phi_{n-2}\left(\mu + \frac{\eta}{x}\right) + \ldots + x\phi_1\left(\mu + \frac{\eta}{x}\right) + \phi_0\left(\mu + \frac{\eta}{x}\right) = 0.$$

Expanding each term of expression, by using Taylor's theorem, we get

$$x^n\left[\phi_n(\mu) + \left(\frac{\eta}{x}\right)\phi_n'(\mu) + \frac{\left(\frac{\eta}{x}\right)^2}{2!}\phi_n''(\mu) + \ldots\right] + x^{n-1}\left[\phi_{n-1}(\mu) + \left(\frac{\eta}{x}\right)\phi_{n-1}'(\mu) + \frac{\left(\frac{\eta}{x}\right)^2}{2!}\phi_{n-1}''(\mu) + \ldots\right] + \ldots = 0.$$

$$\Rightarrow \quad x^n \phi_n(\mu) + x^{n-1}\left[\eta\phi_n'(\mu) + \phi_{n-1}(\mu)\right] + x^{n-2}\left[\frac{1}{2}\eta^2\phi_n''(\mu) + \eta\phi_{n-1}'(\mu) + \phi_{n-2}(\mu)\right] + \ldots = 0. \tag{7.8}$$

Dividing by x^{n-2} and taking limits as $x \to \infty$ in (7.8) (since $\phi_n'(\mu)=0$ and $\phi_{n-1}(\mu)=0$) we get

$$\lim_{x\to\infty}\left(\frac{1}{2}\eta^2\phi_n''(\mu)+\eta\phi_{n-1}'(\mu)+\phi_{n-2}(\mu)\right)=0.$$

$$\Rightarrow \qquad \frac{1}{2}c^2\phi_n''(\mu)+c\phi_{n-1}'(\mu)+\phi_{n-2}(\mu)=0 \ \left(\text{as } \lim_{x\to\infty}\eta=c\right).$$

So, if $\phi_n''(\mu)\neq 0$, then the above quadratic equation gives two value of c, say c_1 and c_2 corresponding to repeated values of m which is μ.

Then, the required asymptotes to the given curve are $y=\mu x+c_1$ and $y=\mu x+c_2$.

Similarly, if $\phi_n''(\mu)=\phi_{n-1}'(\mu)=\phi_{n-2}(\mu)=0$, then equation (7.8), we have

$$x^{n-3}\left[\frac{1}{3!}\eta^3\phi_n'''(\mu)+\frac{1}{2!}\eta^2\phi_{n-1}''(\mu)+\eta\phi_{n-2}'(\mu)+\phi_{n-3}(\mu)\right]$$
$$+x^{n-4}\left[\frac{1}{4!}\eta^4\phi_n^{iv}(\mu)+\ldots\right]+\ldots=0.$$

Dividing by x^{n-3} and taking limit as $x\to\infty$ in the above equation, we get

$$\frac{1}{6}c^3\phi_n'''(\mu)+\frac{1}{2}c^2\phi_{n-1}''(\mu)+c\phi_{n-2}'(\mu)+\phi_{n-3}(\mu)=0.$$

So, if $\phi_n'''(\mu)\neq 0$, then the above quadratic equation enables us to find three values of c, say c_1, c_2 and c_3 corresponding to three repeated values of m, which is μ.

Then, the required asymptotes to the given curve are $y=\mu x+c_1, y=\mu x+c_2$ and $y=\mu x+c_3$.

Again, if $\phi_n'''(\mu)=0$, then we can proceed in a similar way to obtain the value of c corresponding to the repeated values of m.

Remarks

1. The highest degree of $\phi_n(m)=0$ is n, i.e., equation (7.5) can have maximum n roots. Thus, the maximum number of asymptotes to the given curve cannot exceed n which are not parallel to y-axis.
2. We cannot find the asymptotes parallel to y-axis by the above discussed method as the line $y=mx+c$ which is considered to be an asymptote has a finite value of the slope m.

7.2.3 Asymptotes Parallel to Coordinate Axes

I. Asymptotes parallel to y-axis

Let the equation of algebraic curve be

$$y^n\phi_n(x)+y^{n-1}\phi_{n-1}(x)+y^{n-2}\phi_{n-2}(x)+\ldots+\phi_0(x)=0, \tag{7.9}$$

where $\phi_r(x)$, $r=0,1,\ldots,n$ is a polynomial in terms of x.

Dividing the equation (7.9) by y^n, we get

$$\phi_n(x)+\frac{1}{y}\phi_1(x)+\frac{1}{y^2}\phi_2(x)+\cdots+\frac{1}{y^n}\phi_0(x)=0. \tag{7.10}$$

Let $x=d$ be an asymptote of the given curve parallel to y-axis. Then, y alone tends to infinity as the point $P(x,y)$ tends to infinity along the curve,

i.e., $$\lim_{y\to\infty} x = d.$$

Taking limits as $y\to\infty$ in equation (7.10), we get

$$\phi_n(d)=0.$$

Thus, d is a root of the equation $\phi_n(x)=0$. So, if $d_1, d_2, \ldots d_n$ (say) are n roots of the equation $\phi_n(x)=0$, then the required asymptotes are $x=d_1, x=d_2, \ldots\ldots, x=d_n$.

Thus, we conclude that the asymptotes parallel to y-axis are obtained by equating the coefficient of highest power of y to zero from the given equation of curve.

Remark: If the coefficient of highest power of y in the given equation of curve is constant or contains any imaginary term, then there are no asymptotes parallel to y-axis.

II. Asymptotes Parallel to x-axis.

Proceeding in similar way as we have discussed in (I), the asymptotes parallel to x-axis are obtained by equating the coefficient of the highest power of x to zero from the given equation of curve.

Remark: If the coefficient of the highest power of x in the given equation of curve is constant or contains any imaginary term, then there are no asymptotes parallel to x-axis.

The asymptotes which are not parallel to coordinate axes, they are known as oblique asymptotes.

Working Rule: To find the asymptotes of the algebraic curve, we summarize all discussed concepts into the following steps:

1. The asymptotes parallel to coordinate axes:
   (a) Equate the coefficient of the highest power of y to zero from the given equation of curve to obtain the asymptotes parallel to y-axis. But, if the coefficient of the highest power of y is constant or contains any imaginary term, then there are no asymptotes parallel to y-axis.
   (b) Equate the coefficient of the highest power of x to zero from the given equation of curve to obtain the asymptotes parallel to x-axis. But, if the coefficient of the highest power of x is constant or contains any imaginary term, then there are no asymptotes parallel to x-axis.

2. The asymptotes which are not parallel to coordinate axes:
Let $y = mx + c$ be an asymptote which is not parallel to coordinate axes. To obtain the value of m and c, the following steps are involved.

(a) Find the polynomial $\phi_n(m)$, by putting $x = 1$ and $y = m$ in highest the degree term of the given equation of the curve.

(b) Solve $\phi_n(m) = 0$ for the values of m.
Let $m_1, m_2, \ldots\ldots, m_n$ be the real roots of $\phi_n(m) = 0$. (Imaginary are not to be considered.)

(c) Find $\phi_{n-1}(m)$, by putting $x = 1$ and $y = m$ in the next lowest degree terms of the given equation of the curve. Again, find $\phi_{n-2}(m)$, by putting $x = 1$ and $y = m$ in the next lowest degree terms of the given equation of the curve and so on.

(d) If $m_1, m_2, \ldots\ldots, m_n$ are distinct real roots of $\phi_n(m) = 0$, then the value of c corresponding to each value of m is given by

$$c = -\frac{\phi_{n-1}(m)}{\phi_n'(m)}.$$

Let $c_1, c_2, \ldots\ldots, c_n$ be the values of c corresponding to $m_1, m_2, \ldots\ldots, m_n$, respectively.
Then the required asymptotes of the given curve are:

$$y = m_1 x + c_1, y = m_2 x + c_2, \ldots\ldots\ldots\ldots\ldots\ldots\ldots\ldots y = m_n x + c_n.$$

(e) If some roots of $\phi_n(m) = 0$ are repeated. Then $\phi_n'(m) = 0$ for that repeated values of m.

(I) If $\phi_{n-1}(m) \neq 0$, then there are no asymptotes of the given curve for that repeated value of m.

(II) If $\phi_{n-1}(m) = 0$ for the repeated value of m.

(i) Let two roots of $\phi_n(m) = 0$ repeat. Then the values of c are found by the equation

$$\frac{1}{2}c^2\phi_n''(m) + c\phi_{n-1}'(m) + \phi_{n-2}(m) = 0.$$

This quadratic equation will give two values of c and therefore, there are two parallel asymptotes of the given curve corresponding to two repeated values of m.

(ii) Let three roots of $\phi_n(m) = 0$ repeat. Then $\phi_n''(m) = \phi_{n-1}'(m) = \phi_{n-2}(m) = 0$ for that repeated values of m. Then the value of c corresponding to repeated value of m is given by

$$\frac{1}{6}c^3\phi_n'''(m) + \frac{1}{2}c^2\phi_{n-1}''(m) + c\phi_{n-2}'(m) + \phi_{n-3}(m) = 0.$$

This cubic equation will give three values of c and therefore, there are three parallel asymptotes of the given curve corresponding to three repeated values of m.

Note: To find the asymptotes of the algebraic curve, we have to find the parallel asymptotes, if any, as well as the oblique asymptotes. The total number of asymptotes of the given curve (including parallel and the oblique asymptotes) can never exceed the highest degree of y in the given equation of curve.

EXAMPLE 1 Find the asymptotes parallel to either of the axes:

(a) $(x^3+a^3)y=bx^3$.

(b) $x^3+(x-a)y^2=0$.

(c) $x^2y^2=x^2b^2+y^2a^2$.

(d) $x^3+3xy^2+y^2+2x+y=0$.

(e) $x^2\left(y^2-a^2\right)=a^2y^2$.

Solution:

(a) The given equation of curve is $\left(x^3+a^3\right)y=bx^3$

i.e., $x^3y+a^3y=bx^3$

$\Rightarrow (y-b)x^3+a^3y=0.$

Equate the coefficient of the highest power of x, i.e., coefficient of x^3 to zero, we get

$$y-b=0 \Rightarrow y=b,$$

which is the asymptote parallel to x-axis.

Now, equate the coefficient of the highest power of y, i.e., coefficient of y to zero, we get

$$x^3+a^3=0 \Rightarrow x=-a,$$

which is the asymptote parallel to y-axis.

(b) The equation of curve is $x^3+(x-a)y^2=0$

Equate the coefficient of the highest power of y, i.e., coefficient of y^2 to zero, we get

$$x-a=0 \Rightarrow x=a,$$

which is the asymptote parallel to y-axis.

But the coefficient of the highest power of x, i.e., coefficient of x^3 is constant, so there is no asymptotes parallel to x-axis.

(c) The equation of curve is $x^2y^2=x^2b^2+y^2a^2$

$\Rightarrow x^2y^2-x^2b^2-y^2a^2=0.$

Equate the coefficient of the highest power of y, i.e., coefficient of y^2 to zero, we get

$$x^2 - a^2 = 0$$

$$\Rightarrow \quad x = \pm a$$

which are the asymptotes parallel to y-axis.

Again, equate the coefficient of highest power of x, i.e., coefficient of x^2 to zero, we get

$$y^2 - b^2 = 0$$

$$\Rightarrow \quad y = \pm b$$

which are the asymptotes parallel to x-axis.

These asymptotes of the curve form a rectangle of sides $x = \pm a$ and $y = \pm b$.

(d) The equation of curve is $x^3 + 3xy^2 + y^2 + 2x + y = 0$

$$\Rightarrow \quad x^3 + (3x+1)y^2 + 2x + y = 0.$$

Equate the coefficients of highest power of y, i.e., coefficient of y^2 to zero, we get

$$3x + 1 = 0 \Rightarrow x = -\frac{1}{3}$$

which is the asymptote parallel to y-axis.

But the coefficient of the highest power of x is constant. So, there is no asymptote parallel to x-axis.

(e) The equation of curve is $x^2(y^2 - a^2) = a^2y^2$

$$\Rightarrow \quad x^2y^2 - x^2a^2 - a^2y^2 = 0.$$

Equate the coefficient of the highest power of x, i.e., coefficient of x^2 to zero, we get

$$y^2 - a^2 = 0 \Rightarrow y = \pm a$$

which are the asymptotes parallel to x-axis.

Again, equate the coefficient of highest power of y, i.e., coefficient of y^2 to zero, we get

$$x^2 - a^2 = 0 \Rightarrow x = \pm a$$

which are the asymptotes parallel to x-axis.

These asymptotes of the curve form a square of sides $x = \pm a$ and $y = \pm a$.

EXAMPLE 2 Find the asymptotes of the following curves:

(a) $6x^2 - 2y^2 + x + (x+2)y + 1 = 0$.

(b) $xy^2 - x^3 = 2ay^2 - a^3$.

(c) $x^3 + (3x^2 - 4y^2)y - x + y + 3 = 0$.

(d) $x^4+y^4+2x^2y^2-2xy\left(x^2+y^2\right)-10x^2(x-y)+12y^2+2x+y=0$.

(e) $x^3+y^3+3xy(x+y)+2\left(x^2+y^2\right)+4xy-x-9y+2=0$.

Solution:

(a) The equation of the given curve is $6x^2-2y^2+xy+x+2y+1=0$.

Since the coefficients of x^2 and y^2, the highest degree terms in x and y are constants, thus the curve has no asymptotes parallel to the x-axis or y-axis.

Now, we find oblique asymptotes, by putting $x=1, y=m$ in the given equation of curve, we get

$$\phi_2(m)=6+m-2m^2 \quad \text{and} \quad \phi_1(m)=1+2m.$$

To find the value of m, let $\phi_2(m)=0$ which gives

$$6+m-2m^2=0$$

$$\Rightarrow \quad 2m^2-m-6=0$$

$$\Rightarrow (m-2)(2m+3)=0$$

$$\Rightarrow \quad m=2, -3/2.$$

Thus, the equation $\phi_2(m)=0$ has distinct roots, so that the value of c is given

$$c=\frac{-\phi_1(m)}{\phi_2'(m)}.$$

So, when $m=\frac{-3}{2}$, $\qquad c=\frac{-(1+2m)}{1-4m}=2/7$

and when $m=2$, $\qquad c=\frac{-(1+2m)}{1-4m}=5/7.$

Substituting the corresponding values of m and c in the equation $y=mx+c$, the required asymptotes are $y=\frac{-3}{2}x+\frac{2}{7}$ and $y=2x+\frac{5}{7}$,

i.e., $\qquad 14y+21x=4$ and $7y=14x+5$.

(b) The equation of the given curve is $y^2x-2ay^2-x^3+a^3=0$.

Since the coefficient of x^3, highest degree term in x is constant. Thus, there is no asymptote parallel to x-axis.

Now, equating the coefficient of y^2 to zero, we get

$$x-2a=0 \Rightarrow x=2a$$

which is an asymptote parallel to y-axis.

Now, we find oblique asymptotes, by putting $x=1, y=m$ in the given equation of curve, we get

$$\phi_3(m)=m^2-1 \text{ and } \phi_2(m)=-2am^2.$$

To find the value of m, let $\phi_3(m)=0$ which gives

$$m^2-1=0$$

$$\Rightarrow \qquad m = 1, -1.$$

Thus, the equation $\phi_3(m)=0$ has distinct roots, so the value of c is

$$c=\frac{-\phi_2(m)}{\phi_3'(m)}.$$

When $m=1$, $c=\dfrac{-\left(-2am^2\right)}{2m}=am=a$

and when $m=-1, c=\dfrac{-\left(-2am^2\right)}{2m}=am=-a.$

Substituting the corresponding values of m and c in equation $y=mx+c$, the required asymptotes are $x=2a$, $y=x+a$, $y=-x-a$.

(c) The equation of the given curve is $x^3+3x^2y-4y^3-x+y+3=0$.

Since the coefficients of x^3 and y^3, the highest degree terms in x and y are constants. Thus, the curve has no asymptotes parallel to the x-axis or y-axis.
Now, we find oblique asymptotes, by putting $x=1, y=m$ in the given equation of curve, we get

$$\phi_3(m)=1+3m-4m^3 \text{ and } \phi_1(m)=-1+m.$$

To find the value of m, let $\phi_3(m)=0$ which gives

$$4m^3-3m-1=0$$

$$\Rightarrow (m-1)\left(4m^2+4m+1\right)=0$$

$$\Rightarrow \qquad (m-1)(2m+1)^2=0$$

$$\Rightarrow \qquad m=1, \frac{-1}{2}, \frac{-1}{2}.$$

Thus, the equation $\phi_3(m)=0$ has two equal roots, so that the value of c for $m=-1/2$ is given by

$$\frac{c^2}{2}\phi_3''(m)+c\phi_2'(m)+\phi_1(m)=0.$$

Putting $m=\dfrac{-1}{2}$, we get

$$\frac{c^2}{2}\phi_3''\left(-\frac{1}{2}\right)+c\phi_2'\left(-\frac{1}{2}\right)+\phi_1\left(-\frac{1}{2}\right)=0$$

$$\Rightarrow \qquad \frac{c^2}{2}(12)+c.0-\frac{3}{2}=0$$

$$\Rightarrow \qquad 6c^2=\frac{3}{2}$$

$$\Rightarrow \qquad c=\pm\frac{1}{2}.$$

Since $m-1$ is distinct root of $\phi_3(m)=0$, so the value of c is given by

$$c=\frac{-\phi_2(m)}{\phi_3'(m)}.$$

Put $m=1$, we get

$c=0.$

Substituting the corresponding values of m and c in equation $y=mx+c$, the required asymptotes are $y=x$,$y=\frac{-1}{2}x+\frac{1}{2}$, $y=\frac{-1}{2}x-\frac{1}{2}$,

i.e., $y-x=0$, $2y+x-1=0$, $2y+x+1=0$.

(d) The equation of the given curve is

$$x^4+y^4+2x^2y^2-2x^3y-2xy^3-10x^3+10x^2y+12y^2+2x+y=0.$$

Since the coefficients of x^4 and y^4 (the highest degree terms in x and y) are constants. Thus, the curve has no asymptotes parallel to the x-axis or y-axis.

Now, we find oblique asymptotes, by putting $x=1$,$y=m$ in the given equation of the curve, we get

$$\phi_4(m)=1+m^4+2m^2-2m-2m^3,\ \phi_3(m)=-10+10m,\ \phi_2(m)=12m^2$$

and $\phi_1(m)=2+m$.

To find the value of m, let $\phi_4(m)=0$ which gives

$$1+m^4+2m^2-2m-2m^3=0$$

$$\Rightarrow \qquad \left(m^2+1\right)^2-2m\left(1+m^2\right)=0$$

$$\Rightarrow \qquad \left(1+m^2\right)\left(1+m^2-2m\right)=0$$

$$\Rightarrow \qquad \left(m^2+1\right)(m-1)^2=0.$$

The only real values of m are 1,1 which are equal.

Thus, the equation $\phi_4(m)=0$ has two real and equal roots, so the value of c for $m=1$ is given by

$$\frac{c^2}{2}\varphi_4''(m)+c\varphi_3'(m)+\varphi_2(m)=0$$

$$\Rightarrow \quad \frac{c^2}{2}\left(12m^2-12m+4\right)+10c+12m^2=0$$

$$\Rightarrow \quad \frac{c^2}{2}(12-12+4)+10c+12=0$$

$$\Rightarrow \quad c^2+5c+6=0$$

$$\Rightarrow \quad c=-2,-3.$$

Substituting the corresponding values of m and c in equation $y=mx+c$, the required asymptotes are $y=x-2, y=x-3$.

(e) The equation of the given curve is

$$x^3+y^3+3x^2y+3xy^2+2x^2+2y^2+4xy-x-9y+2=0$$

Since the coefficients of x^3 and y^3 are constants, the highest degree terms in x and y are constants. Thus, the curve has no asymptotes parallel to the x-axis or y-axis. Now, we find oblique asymptotes, by putting $x=1, y=m$ in the given equation of the curve, we get

$\phi_3(m)=1+m^3+3m+3m^2=(m+1)^3, \phi_2(m)=2+2m^2+4=2(1+m)^2$ and

$\phi_1(m)=-1-9m.$

To find the value of m, let $\phi_3(m)=0$ which gives

$$(m+1)^3=0$$

$\Rightarrow m=-1,-1,-1.$

Thus, the equation $\phi_3(m)=0$ has three equal roots, so the value of c for $m=-1$ is given by

$$\frac{c^3}{3!}\phi_3'''(m)+\frac{c^2}{2!}\phi_2''(m)+c\phi_1'(m)+\phi_0(m)=0.$$

Put $m=-1$, we get

$$6\frac{c^3}{6}+4\frac{c^2}{2}-9c+2=0$$

$$\Rightarrow \quad c^3+2c^2-9c+2=0$$

$$\Rightarrow \quad (c-2)\left(c^2+4c-1\right)=0$$

$$\Rightarrow \quad c=2,-2+\sqrt{5},-2-\sqrt{5}.$$

Substituting the corresponding values of m and c in the equation $=mx+c$, the required asymptotes are $y=-x+2$, $y=-x-2+\sqrt{5}$, $y=-x-2-\sqrt{5}$,

i.e., $$y+x=2,\ y+x+2-\sqrt{5}=0,\ y+x+2+\sqrt{5}=0.$$

7.2.4 Asymptotes by Inspection

If the equation of a curve C of nth degree, is of the form $P_n+P_{n-2}=0$, where P_n is a polynomial of degree n (i.e., P_n contains terms of degree n and may also contain terms of lower degree to n) and P_{n-2} is a polynomial of degree $(n-2)$ at the most, then every linear factor of $P_n=0$ will be an asymptote of the curve C, provided no two linear factors are either parallel or coincident.

EXAMPLE 3 Find the asymptotes of the following curves by the inspection method:

(a) $x^2b^2-y^2a^2=1$.

(b) $xy(x^2-y^2)+3x^2-4y^2+1=0$.

(c) $(x-y)(x+y-3)(2x+5y)+3x+4y+1=0$.

(d) $(x-y-1)(x-y)(x+y)=8x-1$.

Solution:

(a) The equation of the given curve is $\left(\frac{x^2}{a^2}-\frac{y^2}{b^2}\right)-1=0$.

It is of the form $P_n+P_{n-2}=0$.
Here, n = 2 and P_n can be resolved into real linear non-repeated factors.
Thus, asymptotes are given by $P_n=0$,

i.e., $$\frac{x^2}{a^2}-\frac{y^2}{b^2}=0$$

$\Rightarrow$ $$y^2=\frac{b^2x^2}{a^2}$$

Therefore, the required asymptotes are $y=\frac{b}{a}x, -\frac{b}{a}x$.

(b) The equation of the given curve is $xy(x^2-y^2)+(3x^2-4y^2+1)=0$.
It is of the form $P_n+P_{n-2}=0$.
Here, n = 4 and P_n can be resolved into real linear non-repeated factors.
Thus, asymptotes are given by $P_n=0$,

i.e., $xy(x^2-y^2)=0$

$\Rightarrow$ $xy(x-y)(x+y)=0$.

Therefore, the required asymptotes are $x=0, y=0, x-y=0, x+y=0$.

(c) The equation of the given curve is $(x-y)(x+y-3)(2x+5y)+3x+4y+1=0$.
It is of the form $P_n+P_{n-2}=0$.
Here, n = 3 and P_n can be resolved into real linear non-repeated factors.
Thus, asymptotes are given by $P_n=0$,

i.e., $(x-y)(x+y-3)(2x+5y)=0$.

Therefore, the required asymptotes are $x-y=0, x+y-3=0$, $2x+5y=0$.

(d) The equation of the given curve is $(x-y-1)(x-y)(x+y)=8x-1$
It is of the form $P_n+P_{n-2}=0$.
Here, n = 3 and P_n can be resolved into real linear non-repeated factors.
Thus, asymptotes are given by $P_n=0$,

i.e., $(x-y-1)(x-y)(x+y)=0$.

Therefore, the required asymptotes are $x-y-1=0$, $x-y=0$, $x+y=0$.

7.3 INTERSECTION OF A CURVE AND ITS ASYMPTOTES

From equation (7.2), the equation of a curve of degree n is of the form

$$x^n\phi_n\left(\frac{y}{x}\right)+x^{n-1}\phi_{n-1}\left(\frac{y}{x}\right)+\ldots+\phi_0\left(\frac{y}{x}\right)=0.$$

Let $y=mx+c$ be an asymptote to the curve. Then put the value of y in the equation of curve, we get

$$x^n\phi_n\left(m+\frac{c}{x}\right)+x^{n-1}\left(m+\frac{c}{x}\right)+x^{n-2}\left(m+\frac{c}{x}\right)+\ldots=0.$$

Expand each expression by Taylor's theorem, we get

$$x^n\left\{\phi_n(m)+\frac{c}{x}\phi_n'(m)+\frac{c^2}{2!.x^2}\phi_n''(m)+\ldots\right\}+x^{n-1}\left\{\phi_{n-1}(m)+\frac{c}{x}\phi_{n-1}'(m)+\frac{c^2}{2!.x^2}\phi_{n-1}''(m)+\ldots\right\}$$

$$+x^{n-2}\left\{\phi_{n-2}(m)+\frac{c}{x}\phi_{n-2}'(m)+\frac{c^2}{2!.x^2}\phi_{n-2}''(m)+\ldots\right\}+\ldots=0.$$

$$\Rightarrow x^n\phi_n(m)+x^{n-1}\left[c\phi_n'(m)+\phi_{n-1}(m)\right]$$

$$+x^{n-2}\left[\frac{1}{2}c^2\phi_n''(m)+c\phi_{n-1}'(m)+\phi_{n-2}(m)\right]+\ldots=0 \tag{7.11}$$

Since $y = mx + c$ is an asymptote to the given curve, then the value of m and c are given by

$$\phi_n(m) = 0 \text{ and } c\phi_n'(m) + \phi_{n-1}(m) = 0.$$

Then, equation (7.11) reduces to

$$x^{n-2}\left[\frac{1}{2}c^2\phi_n''(m) + c\phi_{n-1}'(m) + \phi_{n-2}(m)\right] + \ldots = 0.$$

which is an equation of degree $(n-2)$ and determines $(n-2)$ values of x. So, we have the following statements:

- *An asymptotes of a curve of degree n cuts the curve in $(n-2)$ points.*
- *A curve of degree n has n number of asymptotes and these asymptotes intersect the curve in $n(n-2)$ points.*
- *A curve of degree n can be expressed as $P_n + P_{n-2} = 0$, where P_{n-2} is of degree $(n-2)$ at the most and $P_n = 0$ has n non-repeated linear factors, then the $n(n-2)$ points of intersection of the curve and its asymptotes lie on the curve $P_{n-2} = 0$.*

EXAMPLE 4 Find the asymptotes of the curve $x^2y - xy^2 + xy + y^2 + x - y = 0$ and show that they cut the curve on three points which lie on the straight line $x + y = 0$.

Solution The given equation of the given curve is

$$x^2y - xy^2 + xy + y^2 + x - y = 0.$$

To find the parallel asymptotes to x and y axes, equate the coefficient of x^2 and y^2, highest degree terms in x and y to zero. We get

$$y = 0 \text{ and } 1 - x = 0.$$

So, x-axis is asymptote to the given curve and the line $x = 1$ is an asymptote to the given curve which is parallel to y-axis.

Now, we find the oblique asymptotes, by putting $x = 1$ and $y = m$ in the given equation, we get

$$\phi_3(m) = m - m^2,\ \phi_2(m) = m + m^2 \text{ and } \phi_1(m) = 1 - m.$$

To find the value of m, let $\phi_3(m) = 0$ which gives

$$m - m^2 = 0$$

$$\Rightarrow \qquad m(1 - m) = 0$$

$$\Rightarrow \qquad m = 0, 1.$$

Thus, the equation $\phi_3(m) = 0$ has distinct roots, the value of c is

$$c = \frac{-\phi_2(m)}{\phi_3'(m)}.$$

So, when $m=0,\ c=0$

and when $m=1,\ c=\dfrac{-\left(m+m^2\right)}{(1-2m)}=2$.

Substituting the corresponding values of m and c in equation $y=mx+c$, the required asymptotes are $y=0, x=1$ and $y=x+2$.

Now, the joint equation of the asymptotes is

$$y(x-1)(x-y+2)=0$$

$$\Rightarrow x^2y-xy^2+xy+y^2-2y=0.$$

The given equation of the curve can be written as

$$\left(x^2y-xy^2+xy+y^2-2y\right)+x+y=0.$$

It is of the form $P_3+P_1=0$ and $P_3=0$ and has three non-repeated linear factors. The points of intersection are $n(n-2)=3(3-2)=3$ and they lie on the straight line $x+y=0$.

EXAMPLE 5 Find the cubic equation of a curve which has the same asymptotes as the curve $x^3-6x^2y+11xy^2-6y^3+x+y+1=0$ and passes through the points $(0,0), \left(0,\dfrac{1}{2}\right)$ and $(2,0)$.

Solution: The equation of the given curve is

$$x^3-6x^2y+11xy^2-6y^3+x+y+1=0.$$

Since the coefficient of x^3 and y^3, highest degree terms in x and y, are constants. Thus, the curve has no asymptotes parallel to x-axis or y-axis.

Now, we find oblique asymptotes, by putting $x=1$ and $y=m$ in the given equation. We get

$$\phi_3(m)=1-6m+11m^2-6m^3,\ \phi_2(m)=0 \text{ and } \phi_1(m)=1+m.$$

To find the value of m, let $\phi_3(m)=0$, which gives

$$1-6m+11m^2-6m^3=0$$

$$\Rightarrow \quad (1-m)(1-2m)(1-3m)=0$$

$$\Rightarrow \quad m=1,\frac{1}{2},\frac{1}{3}.$$

Thus, the equation $\phi_3(m)=0$ has distinct roots, so the value of c is

$$c=\frac{-\phi_2(m)}{\phi_3'(m)}.$$

Since $\phi_2(m)=0$, for all values of m, so c = 0, for all values of m.

Substitute the corresponding values of m and c in equation $y = mx + c$, the required asymptotes are $y = x$, $y = \frac{1}{2}x$ and $y = \frac{1}{3}x$.

Now, the joint equation of the asymptotes is

$$(y - x)\left(y - \frac{1}{2}x\right)\left(y - \frac{1}{3}x\right) = 0$$

$$\Rightarrow \qquad (x - y)(x - 2y)(x - 3y) = 0.$$

The general equation of the curve which has three asymptotes is of the form $P_3 + P_1 = 0$, i.e. $(x - y)(x - 2y)(x - 3y) + ax + by + c = 0$, where a, b and c are constants (as P_1 is of degree 1 and so we let $P_1 = ax + by + c$).

Now, it passes through the points $(0,0)$, $\left(0, \frac{1}{2}\right)$ and $(2,0)$, we get

$$c = 0,$$

$$-\frac{3}{4} + 2b = 0 \Rightarrow b = 3/8,$$

$$27 + 2a = 0 \Rightarrow a = -4.$$

Hence, the required equation of cubic curve is

$$x^3 - 6x^2y + 11xy^2 - 6y^3 - 4x + \frac{3}{8}y = 0.$$

7.4 ASYMPTOTES OF THE POLAR CURVES

Let the equation of the polar curve be $r = f(\theta)$.

To find the asymptotes to the polar curve, we require an equation of a straight line in polar coordinates.

So, first of all, we find an equation of straight line in polar coordinates.

Let MN be a straight line and $P(r, \theta)$ be a point on a straight line MN. Let p be the perpendicular distance from the pole to the line and α be an angle which the perpendicular makes with the positive direction of x-axis,

i.e., $OM = p$ and $\angle MOX = \alpha$.

Since $P(r, \theta)$ is a point on line MN, then we have

$$\angle POX = \theta \Rightarrow \angle POM = \theta - \alpha.$$

Now, in right angled triangle NMO,

$$\frac{OM}{OP} = \cos(\theta - \alpha)$$

$$\Rightarrow \qquad \frac{p}{r} = \cos(\theta - \alpha)$$

$$\Rightarrow \qquad p = r\cos(\theta - \alpha)$$

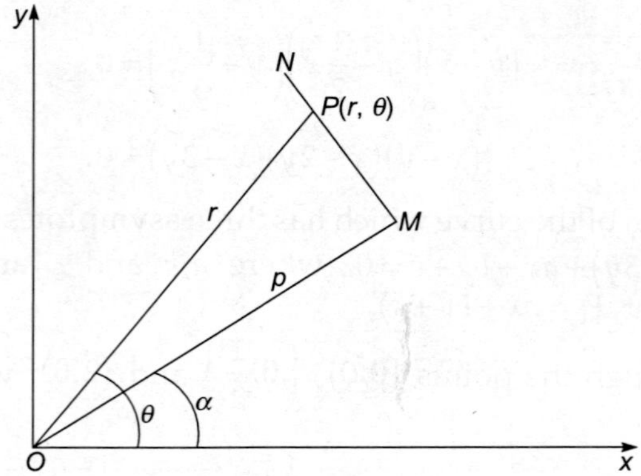

Figure 7.3

Therefore, the equation of straight line in polar coordinates is $p = r\cos(\theta - \alpha)$.

Now, we will find p and α so that p = r cos (θ – α) is an asymptote to the polar curve r = f(θ).

Let $P(r,\theta)$ be a point on the polar curve and PR be a perpendicular from the point to the line MN.

So, we have $PR = TM$ (as shown in Figure 7.4)

$$\Rightarrow \qquad PR = OM - OT = p - OT = p - r\cos(\theta - \alpha)$$

$$\Rightarrow \qquad \frac{PR}{r} = \frac{p}{r} - \cos(\theta - \alpha). \tag{7.12}$$

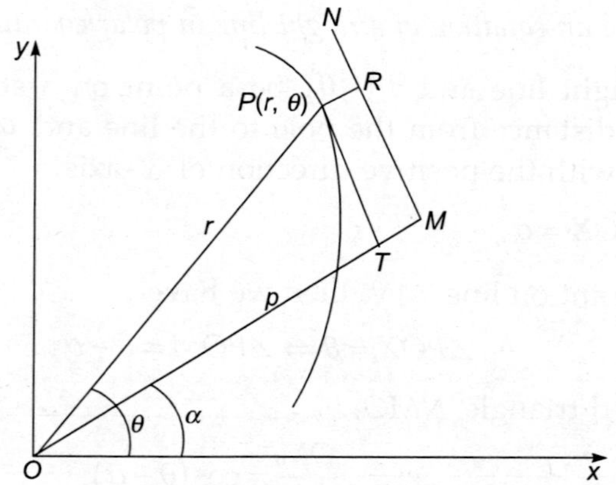

Figure 7.4

The line MN will be an asymptote to the curve $r = f(\theta)$ when PR tends to zero and the point $P(r,\theta)$ tends to infinity along the branch of the curve, i.e., $r \to \infty$.

Let $\theta \to \theta'$ when $r \to \infty$.

Taking limits $r \to \infty$ in (7.12), we get

$$\cos(\theta' - \alpha) = 0$$

$$\Rightarrow \qquad \theta' - \alpha = \frac{\pi}{2}$$

$$\Rightarrow \qquad \alpha = \theta' - \frac{\pi}{2}.$$

Now,
$$p = OM = \lim_{r\to\infty} OT$$

$$= \lim_{r\to\infty} r\cos(\theta - \alpha)$$

$$= \lim_{r\to\infty} \frac{\cos(\theta-\alpha)}{\left(\frac{1}{r}\right)} \left(\frac{0}{0} \text{ form}\right)$$

(as $\theta \to \theta'$, when $r \to \infty$)

$$= \lim_{\substack{u\to 0\\ \theta\to\theta'}} \frac{-\sin(\theta-\alpha)}{\frac{du}{d\theta}}, \text{ where } \left(u = \frac{1}{r}\right)$$

$$= \lim_{\substack{u\to 0\\ \theta\to\theta'}} \frac{-1}{\frac{du}{d\theta}}$$

$$= \lim_{\substack{u\to 0\\ \theta\to\theta'}} \left(-\frac{d\theta}{du}\right).$$

Therefore, the equation of the asymptotes to the polar curve $r = f(\theta)$ is

$$p = r\cos(\theta - \alpha).$$

$$\lim_{\substack{u\to 0\\ \theta\to\theta'}} \left(-\frac{d\theta}{du}\right) = r\cos(\theta-\alpha) = r\cos\left(\theta - \left(\theta' - \frac{\pi}{2}\right)\right) = r\sin(\theta' - \theta).$$

Working Rule

To find the asymptotes of the polar curve $r = f(\theta)$, we summarize all the discussed concepts in the following steps:

1. Substitute $r = \frac{1}{u}$ in the equation of the given curve.
2. Find the value of θ for which $u \to 0$. Let θ' be such value.

3. Find $\left(-\dfrac{d\theta}{du}\right)$ and $\lim\limits_{\substack{u\to 0\\ \theta\to\theta'}}\left(-\dfrac{d\theta}{du}\right)$.

4. Then $\lim\limits_{\substack{u\to 0\\ \theta\to\theta'}}\left(-\dfrac{d\theta}{du}\right) = r\sin(\theta'-\theta)$ will be an asymptote to the polar curve $r = f(\theta)$.

Note: If $\lim\limits_{\theta\to\theta'}\left(-\dfrac{d\theta}{du}\right)$ is infinite, then there will be no asymptote to the given polar curve.

The following result will be helpful to find the asymptotes for polar curve $r = f(\theta)$

1. $\sin\theta = 0 \Rightarrow \theta = n\pi$.
2. $\cos\theta = 0 \Rightarrow \theta = (2n+1)\dfrac{\pi}{2}$.
3. $\sin\theta = \sin\alpha \Rightarrow \theta = n\pi + (-1)^n\alpha$.
4. $\cos\theta = \cos\alpha \Rightarrow \theta = 2n\pi \pm \alpha$.
5. $\tan\theta = \tan\alpha \Rightarrow \theta = n\pi + \alpha$.
6. $\sin(n\pi + \theta) = (-1)^n \sin\theta$.
7. $\cos(n\pi + \theta) = (-1)^n \cos\theta$.
8. $\tan(n\pi + \theta) = \tan\theta$.
9. $\sin(2n+1)\dfrac{\pi}{2} = (-1)^n$.
10. $\cos n\pi = (-1)^n$.
11. $\cos\theta = 1 \Rightarrow \theta = 2n\pi$.

EXAMPLE 6 Find the asymptotes of the curve $r\theta = 1$.

Solution The equation of the curve is $r\theta = 1$.

Substitute $r = \dfrac{1}{u}$ in the given equation, we get

$$\frac{1}{u}\theta = 1 \Rightarrow u = \theta.$$

As $u \to 0$ implies $\theta \to 0$. Then $\theta' = 0$.

Now, $\dfrac{du}{d\theta} = 1 \Rightarrow \dfrac{d\theta}{du} = 1$.

So, $\lim\limits_{\theta\to\theta'}\left(-\dfrac{d\theta}{du}\right) = -1$.

Thus, the asymptote is $-1 = r\sin(0-\theta) = -r\sin\theta$.

Therefore, $r\sin\theta = 1$ is required asymptote of the given curve.

EXAMPLE 7 Find the asymptote of the curve $r\sin\theta = 2\cos 2\theta$.

Solution The equation of the given curve is $r\sin\theta = 2\cos 2\theta$,

i.e., $$r = \frac{2\cos 2\theta}{\sin\theta} \Rightarrow \frac{1}{r} = \frac{\sin\theta}{2\cos 2\theta}.$$

Substitute $r = 1/u$ and then taking limit $u \to 0$, we get

$$\sin\theta = 0$$

$$\Rightarrow \quad \theta' = n\pi.$$

Now, $$\frac{du}{d\theta} = \frac{1}{2}\left(\frac{\cos\theta\cos 2\theta + 2\sin\theta\sin 2\theta}{\cos^2 2\theta}\right).$$

$$\Rightarrow \lim_{\theta\to\theta'}\left(-\frac{d\theta}{du}\right) = -2\left(\frac{\cos^2 2n\pi}{\cos n\pi\cos 2n\pi + 2\sin n\pi\sin 2n\pi}\right)$$

$$= -2\frac{1}{(-1)^n}.$$

Thus, the asymptote is $$-2\frac{1}{(-1)^n} = r\sin(n\pi - \theta)$$

$$= r(-1)^{n+1}\sin\theta.$$

Therefore, $r\sin\theta = 2$ is the required asymptote to the given curve.

EXAMPLE 8 Find the asymptote of the curve $r = a\sec\theta + b\tan\theta$.

Solution The equation of the given curve is

$$r = a\sec\theta + b\tan\theta.$$

$$\Rightarrow \quad r = \frac{a + b\sin\theta}{\cos\theta}$$

$$\Rightarrow \quad \frac{1}{r} = \frac{\cos\theta}{a + b\sin\theta}.$$

Substitute $r = 1/u$ and then taking limit $u \to 0$, we get

$$\cos\theta = 0$$

$$\Rightarrow \quad \theta' = (2n+1)\frac{\pi}{2}.$$

Now, $\dfrac{du}{d\theta}=-\left(\dfrac{a\sin\theta+b}{(a+b\sin\theta)^2}\right)$

$$\Rightarrow \lim_{\theta\to\theta'}\left(-\frac{d\theta}{du}\right)=\frac{\left(a+b\sin\left((2n+1)\frac{\pi}{2}\right)\right)^2}{a\sin\left((2n+1)\frac{\pi}{2}\right)+b}$$

$$=\frac{\left(a+b(-1)^n\right)^2}{\left(a(-1)^n+b\right)}.$$

Thus, the asymptote is $\dfrac{\left(a+b(-1)^n\right)^2}{\left(a(-1)^n+b\right)}=r\sin(\theta'-\theta)$.

$$\Rightarrow \qquad \frac{\left(a+b(-1)^n\right)^2}{\left(a(-1)^n+b\right)}=r\sin\left((2n+1)\frac{\pi}{2}-\theta\right).$$

$$\Rightarrow \qquad \frac{\left(a+b(-1)^n\right)^2}{\left(a(-1)^n+b\right)}=r\sin\left(n\pi+\left(\frac{\pi}{2}-\theta\right)\right).$$

$$\Rightarrow \qquad \frac{\left(a+b(-1)^n\right)^2}{\left(a(-1)^n+b\right)}=r(-1)^n\sin\left(\frac{\pi}{2}-\theta\right)=r(-1)^n\cos\theta.$$

Therefore, $\dfrac{\left(a+b(-1)^n\right)^2}{\left(a(-1)^n+b\right)}=(-1)^n r\cos\theta$ is the required asymptote of the given curve.

EXAMPLE 9 Find the asymptote of the curve $r=\tan\theta$.

Solution: The equation of the given curve is

$$r=\tan\theta$$

$$\Rightarrow \qquad r=\frac{\sin\theta}{\cos\theta}$$

$$\Rightarrow \qquad \frac{1}{r}=\frac{\cos\theta}{\sin\theta}.$$

Substitute $r=1/u$ and then taking limit $u\to 0$, we get

$$\cos\theta=0$$

$$\Rightarrow \qquad \theta'=(2n+1)\frac{\pi}{2}.$$

Now, $$\frac{du}{d\theta} = \operatorname{cosec}^2\theta$$

$$\Rightarrow \quad \lim_{\theta\to\theta'}\left(-\frac{d\theta}{du}\right) = -\sin\left((2n+1)\frac{\pi}{2}\right)^2$$

$$= -(-1)^{2n} = -1.$$

Thus, the asymptote is $-1 = r\sin(\theta' - \theta) = r(-1)^n\cos\theta$ (as in Example 8).

Therefore, $r\cos\theta = (-1)^{1-n}$ is the required asymptote of the given curve.

EXAMPLE 10 Show that the curve $r = \dfrac{a}{1-\cos\theta}$ has no asymptotes.

Solution: The equation of the given curve is

$$r = \frac{a}{1-\cos\theta}$$

$$\Rightarrow \quad \frac{1}{r} = \frac{1}{a}(1-\cos\theta).$$

Substitute $r = 1/u$ and then taking limit $u \to 0$, we get

$$\cos\theta = 1$$

$$\Rightarrow \quad \theta' = 2n\pi.$$

Now, $\dfrac{du}{d\theta} = \dfrac{1}{a}\sin\theta.$

$$\Rightarrow \quad \lim_{\theta\to\theta'}\left(-\frac{d\theta}{du}\right) = -\frac{1}{a}\frac{1}{\sin(2n\pi)} \to -\infty,$$

i.e. $\lim_{\theta\to\theta'}\left(-\dfrac{d\theta}{du}\right)$ does not exist.

Therefore, the give polar curve has no asymptotes.

Exercise

Find all the asymptotes of the following curves:

1. $3x^3 + 2x^2y - 7xy^2 + 2y^3 - 14xy + 7y^2 + 4x + 5y = 0$.
2. $4x^2(y-x) + y(y-2)(x-y) - 4x - 4y + 7 = 0$.
3. $x^3 - x^2y - xy^2 + y^3 + 2x^2 - 4y^2 + 2xy + x + y + 1 = 0$.
4. $x^2y^2(x^2-y^2)^2 - (x^2+y^2)^3 = 0$.

5. $x^3 + 3x^2y - xy^2 - 3y^3 + x^2 - 2xy + 3y^2 + 4x + 7 = 0$.

6. $x^2(x-y)^2 - a^2(x^2 + y^2) = 0$.

7. $x^3 + 4x^2y + 5xy^2 + 2y^3 + 2x^2 + 4xy + 2y^2 - x - 9y + 1 = 0$.

8. $y^4 - 2xy(y^2 - x^2) - x^4 - 3x^3 + 3x^2y + 3xy^2 - 3y^3 - 2x^2 + 2y^2 - 1 = 0$.

9. $2x^3 + 3xy(x-y) - 2y^3 + 3x^2 - 3y^2 + y = 3$.

10. $(x+y)^2(2x^2 + xy - 3y^2) + x^3 + y^3 + xy(x+y) + x + 3y + 1 = 0$.

Find all the asymptotes of the following curves by the inspection method:

11. $xy(x^2 - y^2) + 2(x^2 + 2y^2 + 1) = 0$.

12. $2x^3 + 3x^2y - 2xy^2 - 3y^3 + 6x + y + 3 = 0$.

13. $(x-y)(x+y)(x+2y-1) = 3x + 4y + 5$.

14. Find the asymptotes of the curve

$$(x^2 - y^2)(y^2 - 4x^2) + 6x^3 - 5x^2y - 3xy^2 + 2y^3 - x^2 + 3xy - 1 = 0$$

and show that the asymptotes cut the curve again at eight points which lie on a circle of radius unity.

15. Show that the asymptotes of the curve $x^3 - 2y^3 + xy(2x - y) + y(x - y) + 1 = 0$ cut the curve at three points which lie on the straight line $x - y + 1 = 0$.

16. Find the equation of the straight line on which lie the three points of intersection of the curve $x^3 + 2x^2y - xy^2 - 2y^3 + 4y^2 + 2xy + y - 1 = 0$. Also find its asymptotes.

17. Find asymptotes of the curve $x^4 - 5x^2y^2 + 4y^4 + x^2 - y^2 + x + y + 1 = 0$ and show that the eight points of intersection of the curve and its asymptotes lie on a rectangular hyperbola.

18. Find asymptotes of the following polar curves:

(i) $r = \dfrac{3a \sin\theta \cos\theta}{\sin^3\theta + \cos^3\theta}$

(ii) $r \sin 2\theta = a$

(iii) $r = a \operatorname{cosec}\theta + b$

(iv) $r = \dfrac{a\theta}{\theta - 1}$

(v) $r\theta = a$

(vi) $r \ln\theta = a$

(vii) $r = a(\sec\theta + \tan\theta)$

(viii) $r = a(\sec\theta + \cos\theta)$

(ix) $r = a + b \cot n\theta$

(x) $r \sin n\theta = a$.

8 CHAPTER

Singular Points and Curve Tracing

In this chapter, we will study the techniques of curve tracing by using the concept of differential calculus. In curve tracing, the concept of singular points plays an important role. So, we begin this chapter with the study of singular points.

8.1 SINGULAR POINTS

Definition: A *singular point (or multiple point)* of the curve $f(x,y)=0$ is the point (x,y) on the curve through which pass m branches of the curve.

Equivalently, *a singular point (or multiple point)* of the curve $f(x,y)=0$ is the point (x,y) on the curve where more than one tangent can be drawn.

A multiple point through which pass two branches of the curve is known as *double point.*

Note:

1. A point (x,y) on the curve $f(x,y)=0$ where exactly one tangent can be drawn is known as an ordinary point.
2. A multiple point (x,y) through which pass m branches of curve $f(x,y)=0$ is known as multiple point of order m.
3. If a point (x,y) on the curve $f(x,y)=0$ has multiple point of order m, then the curve has m tangents at the multiple point (x,y).

8.1.1 Types of Double Points

Since, the curve $f(x,y)=0$ has two tangents at the double point (x,y). So, we define three types of double points:

1. **Node.** A double point $P(x,y)$ on the curve is called a node, if two real branches of the curve pass through P and the tangents at P to the curve are real and distinct (or real and not coincident) (Figure 8.1(a)).
2. **Cusp.** A double point $P(x,y)$ on the curve is called a cusp, if two real branches of the curve pass through P and the tangents at P to the curve are real, but coincident (Figure 8.1(b)).

3. **Conjugate Point.** A double point $P(x,y)$ on the curve is called a conjugate point (or an isolated point), if there are no real points on the curve in the neighbourhood of the point P, i.e., the tangents at a conjugate point are imaginary (Figure 8.1(c))).

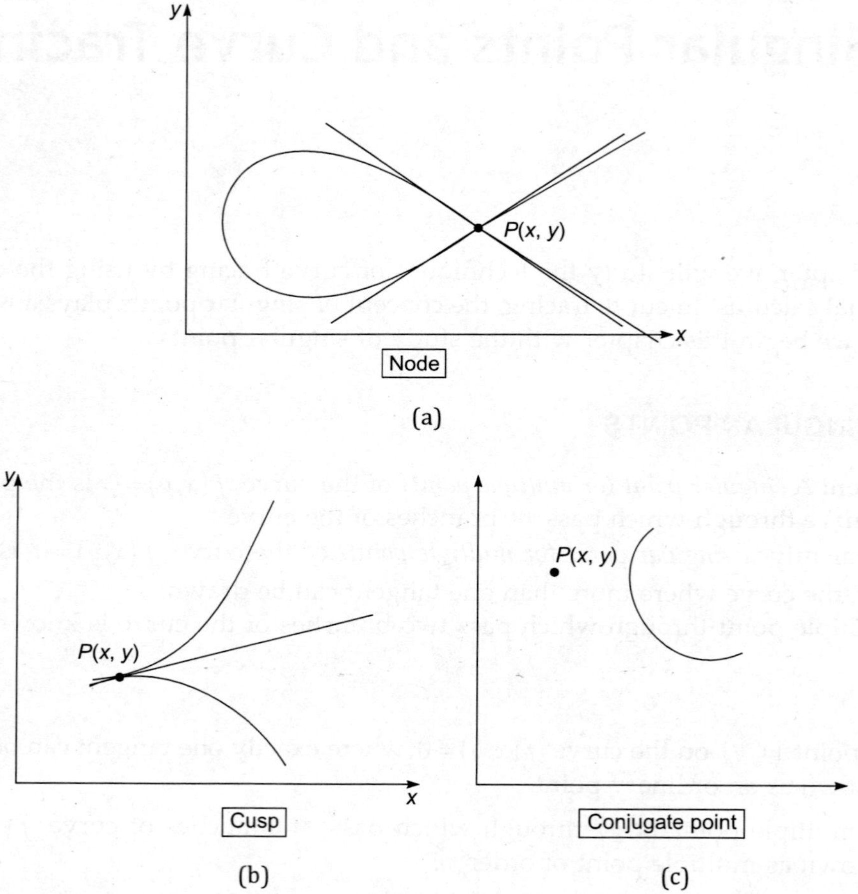

Figure 8.1

Remark: At the conjugate point $P(x,y)$ on the curve, there are some possibilities that the tangents come out to be real. In such cases, no conclusion is found about the nature of the curve in the neighbourhood of the conjugate point.

8.1.2 Tangents at the Origin

To study the position and nature of multiple points, it is necessary to find the tangents at the origin. We can find the tangents at the origin to the given curve by the following proposition:

"If a rational algebraic curve of degree n passes through the origin, then the equation of tangents at the origin is obtained by equating the lowest degree term from the given equation of the curve to zero."

EXAMPLE 1 Find the equations of tangents at the origin to the following curves:

(a) $a^2y^2 + x^4 = a^2x^2$

(b) $x^4 + y^4 + 2x^3 - x^2 - 2xy - y^2 = 0$

(c) $x^4 + y^4 = 2a^2y^3 - 2b^2x^2y$

(d) $x^3 + x^2y + b^2x^3 = a\left(x^2 + y^2\right)$

Solution:

(a) The given equation of curve is $a^2y^2 + x^4 = a^2x^2$

$$\Rightarrow \quad a^2y^2 + x^4 - a^2x^2 = 0.$$

Equate the lowest degree term to zero, we get

$$a^2y^2 - a^2x^2 = 0$$

$$\Rightarrow \quad y^2 = x^2$$

$$\Rightarrow \quad y = \pm x.$$

Thus, the equations of tangent to the given curve at the origin are $y = x$ and $y = -x$.

(b) The given equation of the curve is $x^4 + y^4 + 2x^3 - x^2 - 2xy - y^2 = 0$

Equate the lowest degree term to zero, we get

$$-x^2 - 2xy - y^2 = 0.$$

$$\Rightarrow \quad (x + y)^2 = 0.$$

Thus, the equations of tangent to the given curve at the origin are $x + y = 0$ and $x + y = 0$, i.e., the given curve has two tangents at the origin, but they are coincident.

(c) The given equation of the curve is $x^4 + y^4 = 2a^2y^3 - 2b^2x^2y$.

$$\Rightarrow \quad x^4 + y^4 - 2a^2y^3 + 2b^2x^2y = 0.$$

Equate the lowest degree term to zero, we get

$$-2a^2y^3 + 2b^2x^2y = 0.$$

$$\Rightarrow \quad -y\left(a^2y^2 - b^2x^2\right) = 0$$

$$\Rightarrow \quad y = 0,\ y = \pm\frac{b}{a}x.$$

Thus, the equations of tangent to the given curve at the origin are $y=0, y=\frac{b}{a}x$ and $y=-\frac{b}{a}x$.

(d) The given equation of the curve is $x^3+x^2y+b^2x^3=a\left(x^2+y^2\right)$.

$$\Rightarrow \qquad x^3+x^2y+b^2x^3-ax^2-ay^2=0.$$

Equate the lowest degree term to zero, we get

$$-ax^2-ay^2=0$$

$$\Rightarrow \qquad y^2=-x^2$$

$$\Rightarrow \qquad y=\pm ix.$$

Thus, the equations of tangent to the given curve at the origin are $y=ix$ and $y=-ix$.

EXAMPLE 2 Find the equation of tangent at $(-1,-2)$ to the curve $x^3+2x^2+2xy-y^2+5x-2y=0$.

Solution: Let the origin be shifted to the point $(-1,-2)$, then put $x=X+1$ and $y=Y+2$, where (X,Y) is the point in the new XY coordinate plane.

Thus, the equation of given curve becomes

$$(X+1)^3+2(X+1)^2+2(X+1)(Y+2)-(Y+2)^2+5(X+1)-2(Y+2)=0.$$

$$\Rightarrow \qquad X^3+5X^2-Y^2+2XY+16X-4Y+4=0.$$

Equate the lowest degree term to zero, we get

$$16X-4Y=0.$$

$$\Rightarrow \qquad Y=4X$$

$$\Rightarrow \qquad y-2=4(x-1)$$

$$\Rightarrow \qquad y=2(2x-1).$$

Thus, the tangent to the given curve at the point $(-1,-2)$ is $y=2(2x-1)$.

8.1.3 Position of Multiple Points

Let the equation of the curve be $f(x,y)=0$.

Differentiating the above equations of the curve w.r.t. x, we get

$$\frac{\partial f}{\partial x}+\frac{\partial f}{\partial y}\cdot\frac{dy}{dx}=0. \qquad (8.1)$$

Since the curve has at least two branches and so, at least two tangents at the multiple point. Thus, $\frac{dy}{dx}$ has at least two values at the multiple point. But the equation (8.1) is of first degree in $\frac{dy}{dx}$ and it gives exactly one value of $\frac{dy}{dx}$.

Therefore, equation (8.1) can be satisfied by more than one value of $\frac{dy}{dx}$ if and only if

$$\frac{\partial f}{\partial x}=0 \quad \text{and} \quad \frac{\partial f}{\partial y}=0.$$

Thus, the necessary and sufficient conditions for a point (x,y) on the curve $f(x,y)=0$ to be a multiple point are

$$\frac{\partial f}{\partial x}=0 \quad \text{and} \quad \frac{\partial f}{\partial y}=0.$$

Working Rule: To find the multiple points (x,y) on the curve $f(x,y)=0$, the following steps are involved:

1. Find the value of $\frac{\partial f}{\partial x}$ and $\frac{\partial f}{\partial y}$.
2. Solve simultaneously, $\frac{\partial f}{\partial x}=0$, $\frac{\partial f}{\partial y}=0$ and $f(x,y)=0$.
3. The obtained solution will be the required multiple points.

8.1.4 Nature of Double Points

Let (x,y) be the multiple point on the curve $f(x,y)=0$.

Now, we will find the nature of the multiple point.

So, differentiate (8.1) w.r.t. x, we get

$$\frac{\partial}{\partial x}\left(\frac{\partial f}{\partial x}+\frac{\partial f}{\partial y}.\frac{dy}{dx}\right)+\frac{\partial}{\partial y}\left(\frac{\partial f}{\partial x}+\frac{\partial f}{\partial y}.\frac{dy}{dx}\right).\frac{dy}{dx}=0.$$

$$\Rightarrow \frac{\partial^2 f}{\partial x^2}+\left(\frac{\partial^2 f}{\partial x\partial y}\frac{dy}{dx}+\frac{\partial f}{\partial y}.\frac{\partial}{\partial x}\left(\frac{dy}{dx}\right)\right)+\frac{\partial^2 f}{\partial y\partial x}.\frac{dy}{dx}+\frac{\partial^2 f}{\partial y^2}\left(\frac{dy}{dx}\right)^2+\left(\frac{\partial f}{\partial y}.\frac{\partial}{\partial y}\left(\frac{dy}{dx}\right)\right).\frac{dy}{dx}=0. \tag{8.2}$$

Since at multiple points, $\frac{\partial f}{\partial x}=0$, $\frac{\partial f}{\partial y}=0$ and $f(x,y)=0$, then by substituting these values in (8.2), we have

$$\Rightarrow \quad \frac{\partial^2 f}{\partial x^2}+2\frac{\partial^2 f}{\partial x\partial y}\frac{dy}{dx}+\frac{\partial^2 f}{\partial y^2}\left(\frac{dy}{dx}\right)^2=0. \tag{8.3}$$

Thus, if $\frac{\partial^2 f}{\partial y^2},\frac{\partial^2 f}{\partial x^2}$ and $\frac{\partial^2 f}{\partial x\partial y}$ all are not equal to zero at the multiple point (x,y), then equation (8.3) is a quadratic equation in $\frac{dy}{dx}$ and so, it gives two values of $\frac{dy}{dx}$. Therefore, there are two tangents at that multiple point (x,y) and so, the multiple point is of order two, i.e. it is a double point.

Hence, the nature of double point (x,y) on the curve $f(x,y)=0$ is given as:

1. If $\left(\dfrac{\partial^2 f}{\partial x \partial y}\right)^2 - \dfrac{\partial^2 f}{\partial x^2} \cdot \dfrac{\partial^2 f}{\partial y^2} > 0$ at the double point (x,y), then the value of $\dfrac{dy}{dx}$ are real and distinct, i.e., the tangents at that double point are real and distinct. Therefore, the double point is node.

2. If $\left(\dfrac{\partial^2 f}{\partial x \partial y}\right)^2 - \dfrac{\partial^2 f}{\partial x^2} \cdot \dfrac{\partial^2 f}{\partial y^2} = 0$ at the double point (x,y), then the values of $\dfrac{dy}{dx}$ are real and equal, i.e., the tangents at that double point are real but coincident. Therefore, the double point is a cusp.

3. If $\left(\dfrac{\partial^2 f}{\partial x \partial y}\right)^2 - \dfrac{\partial^2 f}{\partial x^2} \cdot \dfrac{\partial^2 f}{\partial y^2} < 0$ at the double point (x,y), then the values of $\dfrac{dy}{dx}$ are imaginary, i.e., the tangents at that double point are imaginary. Therefore, the double point is a conjugate point.

Remark: If the value of $\dfrac{\partial^2 f}{\partial y^2}, \dfrac{\partial^2 f}{\partial x^2}$ and $\dfrac{\partial^2 f}{\partial x \partial y}$ are zero at the multiple point (x,y), then equation (8.3) becomes an identity. Thus, the order of multiple point is greater than two. So, we need to again differentiate equation (8.3) w.r.t. x, to identify if the nature of the multiple point is of order greater than two.

Notation. The first and second order partial derivatives $\dfrac{\partial f}{\partial x}, \dfrac{\partial f}{\partial y}, \dfrac{\partial^2 f}{\partial x^2}, \dfrac{\partial^2 f}{\partial x \partial y}$ and $\dfrac{\partial^2 f}{\partial y^2}$ can also be denoted as f_x, f_y, f_{xx}, f_{xy} and f_{yy}, respectively.

EXAMPLE 3 Find the position and nature of the multiple points for the following curves:

(a) $x^3 + a(y^2 - x^2) = 0$

(b) $x^3 - y^2 - 7x^2 + 4y + 15x - 13 = 0$

(c) $x^3 + 2x^2 + 2xy - y^2 + 5x - 2y = 0$

(d) $x^4 - 4ax^3 + 2ay^3 + 4a^2x^2 - 3a^2y^2 - a^4 = 0$

Solutions:

(a) The given equation of the curve is $x^3 + a(y^2 - x^2) = 0$.

Let $f(x,y) = x^3 + a(y^2 - x^2)$

$\Rightarrow f_x = 3x^2 - 2ax$ and $f_y = 2ay$

Now, $f_x = 0 \Rightarrow 3x^2 - 2ax = 0 \Rightarrow x(3x - 2a) = 0 \Rightarrow x = 0, \frac{2a}{3}$ and

$f_y = 0 \Rightarrow 2ay = 0 \Rightarrow y = 0$.

But, the point $\left(0, \frac{2a}{3}\right)$ does not satisfy the given equation of the curve, so the point $(0,0)$ is the only multiple point on the curve $f(x,y) = 0$.

Now, since the origin is a multiple point, the equation of tangents at origin is obtained by equating the lowest degree term from the given equation to zero,

i.e., $$a\left(y^2 - x^2\right) = 0.$$

$\Rightarrow$ $$y = \pm x.$$

Thus, we get two tangents at the origin, hence, the origin is a double point. Also, the tangents at $(0,0)$ are real and distinct, therefore the origin is node.

(b) The given equation of curve is $x^3 - y^2 - 7x^2 + 4y + 15x - 13 = 0$.

Let $f(x,y) = x^3 - y^2 - 7x^2 + 4y + 15x - 13$.

$\Rightarrow f_x = 3x^2 - 14x + 15$ and $f_y = -2y + 4$.

Now, $f_x = 0 \Rightarrow 3x^2 - 14x + 15 = 0 \Rightarrow (x-3)\left(x - \frac{5}{3}\right) = 0 \Rightarrow x = 3, \frac{5}{3}$ and

$f_y = 0 \Rightarrow -2y + 4 = 0 \Rightarrow y = 2$.

But the point $\left(\frac{5}{3}, 2\right)$ does not satisfy the given equation of the curve, so the point $(3,2)$ is the only multiple point.

Now, $f_{xx} = 6x - 14$, $f_{xy} = 0$ and $f_{yy} = -2$.

At $(3,2)$, $f_{xx} = 4$, $f_{xy} = 0$ and $f_{yy} = -2$ are not equal to zero. Thus, the multiple point $(3,2)$ is a double point.

Now, at $(3,2)$, $f_{xy}^2 - f_{xx}f_{yy} = (0)^2 - (4)(-2) = 8 > 0$.

Therefore, the double point $(3,2)$ is a node.

(c) The given equation of the curve is $x^3 + 2x^2 + 2xy - y^2 + 5x - 2y = 0$.

Let $f(x,y) = x^3 + 2x^2 + 2xy - y^2 + 5x - 2y$.

$\Rightarrow f_x = 3x^2 + 4x + 2y + 5$ and $f_y = 2x - 2y - 2$.

Now, $f_x = 0 \Rightarrow 3x^2 + 4x + 2y + 5 = 0$ and $f_y = 0 \Rightarrow 2x - 2y - 2 = 0 \Rightarrow y = x - 1$.

So, put the value of $y = x - 1$ in $f_x = 0$, we get

$$3x^2 + 4x + 2(x-1) + 5 = 0.$$

$$\Rightarrow \qquad 3x^2 + 6x + 3 = 0$$

$$\Rightarrow \qquad (x+1)^2 = 0$$

$$\Rightarrow \qquad x = -1.$$

So, $\qquad y = x - 1 \Rightarrow y = -2$.

Thus, $(-1,-2)$ satisfies the given equation of the curve and it is the only multiple point on the given curve.

Now, $f_{xx} = 6x + 4$, $f_{xy} = 2$ and $f_{yy} = -2$.

At $(-1,-2)$, $f_{xx} = -2$, $f_{xy} = 2$ and $f_{yy} = -2$ are not equal to zero. Thus, the multiple point $(-1,-2)$ is a double point.

Now, at $(-1,-2)$, $f_{xy}^2 - f_{xx} \cdot f_{yy} = (2)^2 - (-2)(-2) = 0$.
Therefore, the double point $(-1,-2)$ is a cusp.

(d) The given equation of the curve is $x^4 - 4ax^3 + 2ay^3 + 4a^2x^2 - 3a^2y^2 - a^4 = 0$.

Let $f(x,y) = x^4 - 4ax^3 + 2ay^3 + 4a^2x^2 - 3a^2y^2 - a^4$.

$\Rightarrow f_x = 4x^3 - 12ax^2 + 8a^2x$ and $f_y = 6ay^2 - 6a^2y$.

Now, $f_x = 0 \Rightarrow 4x^3 - 12ax^2 + 8a^2x = 0 \Rightarrow 4x\left(x^2 - 3ax + 2a^2\right) = 0$

$$\Rightarrow \qquad x(x-a)(x-2a) = 0$$

$$\Rightarrow \qquad x = 0,\ a,\ 2a.$$

and $f_y = 0 \Rightarrow 6ay^2 - 6a^2y = 0 \Rightarrow 6ay(y-a) = 0 \Rightarrow y = 0, a$.

So, the points are $(0,0),(0,a),(a,0),(a,a),(2a,0)$ and $(2a,a)$. But the points $(0,0)$, $(0,a)$, (a,a) and $(2a,0)$ do not satisfy the given equation of the curve. So, the point $(a,0)$ is the only multiple point on the given curve.

Now, $f_{xx} = 12x^2 - 24ax + 8a^2$, $f_{xy} = 0$ and $f_{yy} = 12ay - 6a^2$.
At $(a,0)$, $f_{xx} = -4a^2$, $f_{xy} = 0$ and $f_{yy} = -6a^2$ are not equal to zero. Thus, the multiple point $(a,0)$ is a double point.

Now, at $(a,0)$, $f_{xy}^2 - f_{xx} \cdot f_{yy} = 0 - \left(-4a^2\right)\left(-6a^2\right) = -24a^4 < 0$, for all values of a.
Therefore, the double point $(a,0)$ is a conjugate point.

EXAMPLE 4 Find the position and nature of the multiple points (x,y) on the curve $x^2y^2-(a+y)^2(b^2-y^2)=0$.

Solution: The equation of the given curve is $x^2y^2-(a+y)^2(b^2-y^2)=0$.

Let $f(x,y)=x^2y^2-(a+y)^2(b^2-y^2)$

$\Rightarrow f_x=2xy^2$ and $f_y=2x^2y-2(a+y)(b^2-y^2)+2y(a+y)^2$.

Now, $f_x=0\Rightarrow 2xy^2=0 \Rightarrow x=0$ or $y=0$

and $f_y=0\Rightarrow 2x^2y-2(a+y)(b^2-y^2)+2y(a+y)^2=0$.

If $y=0$, then $f_y=0\Rightarrow -2ab^2=0$, which is not possible.

So, $x=0$. Substitute it in $f_y=0$, we get

$$-2(a+y)(b^2-y^2)+2y(a+y)^2=0.$$

$$\Rightarrow \quad 2(a+y)\left(y(a+y)-(b^2-y^2)\right)=0$$

$$\Rightarrow \quad (a+y)(2y^2+ay-b^2)=0$$

$$\Rightarrow \quad y=-a,\frac{\left(-a\pm\sqrt{a^2+8b^2}\right)}{4}.$$

But, the point $\left(0,\frac{\left(-a\pm\sqrt{a^2+8b^2}\right)}{4}\right)$ does not satisfy the given equation of the curve.

So, the point $(0,-a)$ is the only multiple point.

Now, $f_{xx}=2y^2$, $f_{xy}=4xy$ and $f_{yy}=2x^2+2(a+y)^2+4y(a+y)-2(b^2-y^2)+4y(a+y)$.

At $(0,-a)$, $f_{xx}=2a^2$, $f_{xy}=0$ and $f_{yy}=-2(b^2-a^2)$ all are not equal to zero. Thus, the multiple point $(a,0)$ is a double point.

Now, at $(0,-a)$, $f_{xy}^2-f_{xx}\cdot f_{yy}=4a^2(b^2-a^2)$.

So, if $b>a$, then $f_{xy}^2-f_{xx}\cdot f_{yy}>0$. Therefore, the double point $(0,-a)$ is a node.

If $b=a$, then $f_{xy}^2-f_{xx}\cdot f_{yy}=0$. Therefore, the double point $(0,-a)$ is a cusp.

If $b<a$, then $f_{xy}^2-f_{xx}\cdot f_{yy}<0$. Therefore, the double point $(0,-a)$ is a conjugate point.

EXAMPLE 5 Show that at the point $(1,-1)$, there is a cusp on the curve:

$$x^3+y^3+xy^2-4x^2+y^2+4x+y-1=0.$$

Solution: The equation of the given curve is

$$x^3+y^3+xy^2-4x^2+y^2+4x+y-1=0.$$

Let $f(x,y)=x^3+y^3+xy^2-4x^2+y^2+4x+y-1$.

$\Rightarrow f_x=3x^2+y^2-8x+4$ and $f_y=3y^2+2xy+2y+1$

At $(1,-1)$, $f_x=0$ and $f_y=0$.

Thus, $(1,-1)$ is the multiple point on the given curve.

Now, $f_{xx}=6x-8$, $f_{xy}=2y$ and $f_{yy}=6y+2x+2$.

At $(1,-1)$, $f_{xx}=-2$, $f_{xy}=-2$ and $f_{yy}=-2$ are not equal to zero. Thus, the multiple point $(1,-1)$ is a double point.

Now, at $(1,-1)$, $f_{xy}^2-f_{xx}\cdot f_{yy}=(-2)^2-(-2)(-2)=0$.

Therefore, the double point $(1,-1)$ is a cusp.

8.1.5 Classification of Cusps

We know that, the curve has two branches at the double point and if the double point is cusp, then both the branches of the curve have a common tangent. But, at the cusp, these two branches of the curve have different positions in coordinate plane relative to the common tangent and the common normal. Such as, both the branches of the curve can either lie on the same side or opposite sides of the tangent at the cusp. Similarly, both the branches of the curve can either lie on the same side or opposite sides of the normal at the cusp.

Regarding this, we give the classification of cusps relative to the common tangent and common normal as follows:

1. **Single Cusp**: If both the branches of the curve at the cusp lie entirely on the same side of the normal, then the cusp is known as *single cusp.*
2. **Double Cusp**: If both the branches of the curve at the cusp lie opposite sides of the normal or at least one of the branch of the curve can be extended to the both sides of the normal, then the cusp is known as *double cusp.*
3. **First Species**: If both the branches of the curve at the cusp lie on the opposite sides of the tangent, then the cusp is known as the *first species (or ceratoid).*
4. **Second Species**: If both the branches of the curve at the cusp lie entirely on the same side of the tangent, then the cusp is known as the *second species (or ramphoid).*

Now, we give further classification of cusps:

1. **Single Cusp of First Species**: If both the branches of the curve at the cusp lie entirely on the same side of the normal, but are on opposite sides of the tangent. Then, the cusp is known as the *single cusp of the first species* (Figure 8.2(a)).

2. **Single Cusp of Second Species**: If both the branches of the curve at the cusp lie entirely on the same side of the normal and also on the same side of the tangent. Then, the cusp is known as the *single cusp of the second species* (Figure 8.2(b)).

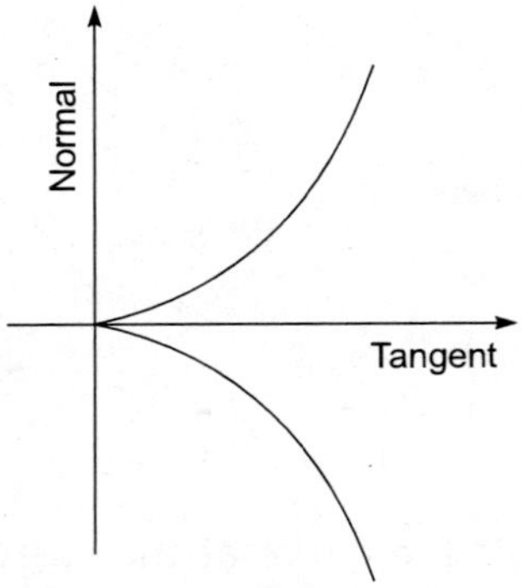

Figure 8.2(a)

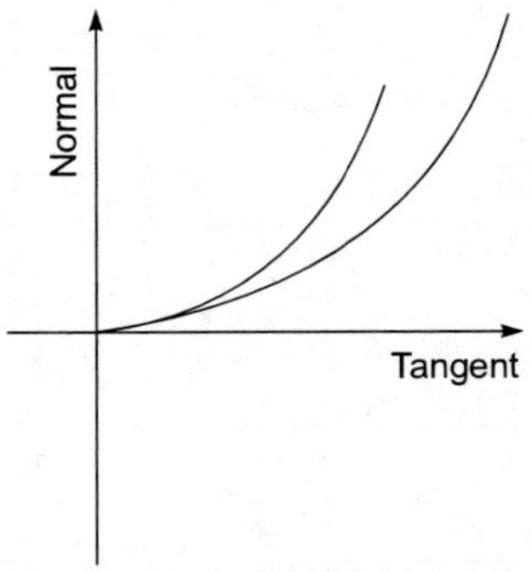

Figure 8.2(b)

3. **Double Cusp of First Species**: If both the branches of the curve at the cusp lie on the opposite sides of the normal and also on the opposite sides of the tangent. Then, the cusp is known as the *double cusp of the first species* (Figure 8.2(c)).

4. **Double Cusp of Second Species**: If both the branches of the curve at the cusp lie on the opposite sides of the normal, but are on the same sides of the tangent. Then, the cusp is known as the *double cusp of the second species* (Figure 8.2(d)).

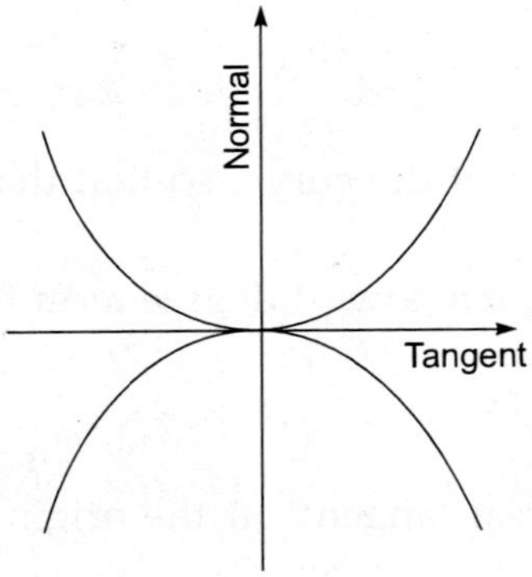

Figure 8.2(c)

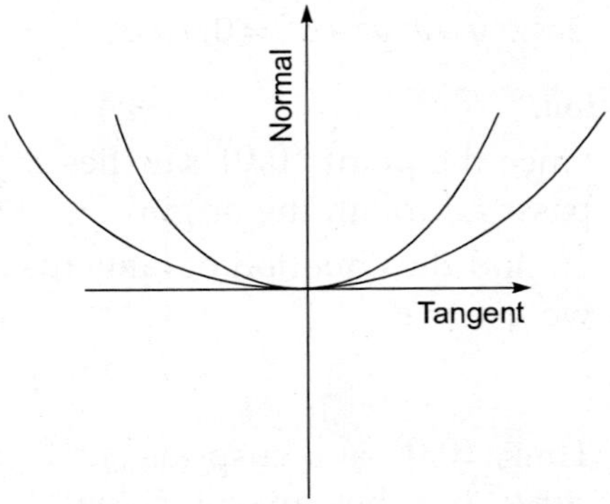

Figure 8.2(d)

5. **Double Cusp with Change of Species**: If the branches of the curve at the cusp lie on the opposite sides of the normal, but the curve has first species on the right of the normal and the second species on the left of the normal. Then, the cusp is known as the *double cusp with change of species* or *the cusp is said to have osco-inflexion point* (Figure 8.2(e)).

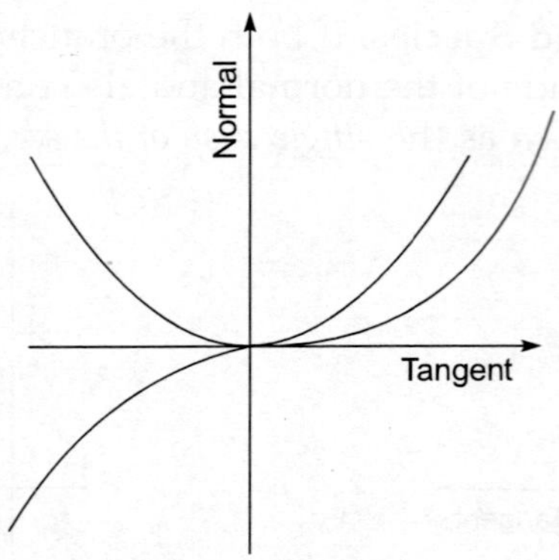

Figure 8.2(e)

Remark: At the osco-inflexion point, the branches of the curve at the cusp lie on the opposite sides of the normal and also, the curve can have first species on the left of the normal and second species on the right of the normal.

EXAMPLE 6 Find the nature and position of the cusps for the following curves:

(a) $x^3 - x^2y + y^2 = 0$.

(b) $y^2 + 4x^4 - 4x^2y - x^5 = 0$.

(c) $x^6 - ayx^4 - a^3x^2y + a^4y^2 = 0$.

(d) $2a^2x^2y - a^3y^2 - x^5 = 0, a > 0$.

Solution:

(a) Since the point $(0,0)$ satisfies the given equation of the curve, so that the curve passes through the origin.
To find the equation of tangents at the origin, equate lowest degree term to zero, we get

$$y^2 = 0.$$

Thus, $(0,0)$ is a cusp (as $y = 0$ and $y = 0$ are real tangents at the origin of the curve, but they are coincident).
Now, solve the given equation of the curve for y as a quadratic in y, we get

$$y = \frac{x^2 \pm \sqrt{x^4 - 4x^3}}{2} = x^2\left(\frac{1 \pm \sqrt{1 - \frac{4}{x}}}{2}\right).$$

Thus, the value of y is real, when $\frac{4}{x} < 1$, i.e., $x > 4$.

Therefore, for each value of $x > 4$, the two values of y are of positive sign. Hence, the origin is a single cusp of the second species.

(b) Since the point $(0,0)$ satisfies the given equation of the curve, so that the curve passes through the origin.

To find the equation of tangents at the origin, equate lowest degree term to zero, we get

$$y^2 = 0.$$

Thus, $(0,0)$ is a cusp (as $y=0$ and $y=0$ are real tangents at the origin of the curve, but they are coincident).

Now, we the given equation of the curve can be written as

$$\left(y-2x^2\right)^2 = x^5.$$

This implies that x can never be negative as left hand side is always positive. Thus, the given curve lies completely on the right side of the normal at the cusp.

Therefore, $(0,0)$ is a single cusp.

Now, solve the given equation for y, we get

$$y = 2x^2 \pm x^{5/2}.$$

Since $2x^2 < x^{\frac{5}{2}}$, for small values of x, the two values of y are of opposite sign. Hence, the origin is a single cusp of the first species.

(c) Let $f(x,y) = x^6 - ayx^4 - a^3x^2y + a^4y^2$.

$\Rightarrow f_x = 6x^5 - 4ayx^3 - 2a^3xy$ and $f_y = -ax^4 - a^3x^2 + 2a^4y$.

On solving, the equations simultaneously $f_x = 0$, $f_y = 0$ and $f(x,y) = 0$, we obtained that $(0,0)$ is the only multiple point.

To find the tangent at the origin, equate the lowest degree term to zero, we get

$$y^2 = 0.$$

Thus, $(0,0)$ is a cusp (as $y=0$ and $y=0$ are real tangents at the origin of the curve, but they are coincident).

Now, solve the given equation of the curve for y as a quadratic in y, we get

$$y = \frac{ax^4 + a^3x^2 \pm \sqrt{\left(ax^4 + a^3x^2\right)^2 - 4a^4x^6}}{2a^4}.$$

$$\Rightarrow \qquad y = \frac{ax^4 + a^3x^2 \pm \sqrt{\left(ax^4 - a^3x^2\right)^2}}{2a^4}$$

$$\Rightarrow \qquad y=\frac{(ax^4+a^3x^2)\pm(ax^4-a^3x^2)}{2a^4}.$$

Now, we observe that the value of y is positive for all values of x (negative or positive). Thus, both the branches of the curve at the cusp lie on the opposite sides of the normal, but entirely above the x-axis (which is a cuspidal tangent).

Hence, the origin is double cusp of the second species.

(d) Let $f(x,y)=a^2y^2-2a^2x^2y+x^5$.

Since the point $(0,0)$ satisfies the given equation of the curve, so that the curve passes through the origin.

To find the equation of tangents at the origin, equate lowest degree term to zero, we get

$$y^2=0.$$

Thus, $(0,0)$ is a cusp (as $y=0$ and $y=0$ are real tangent at origin to the curve, but they are coincident).

Now, solve the given equation of the curve for y as a quadratic in y, we get

$$y=\frac{2a^2x^2\pm\sqrt{4a^4x^4-4a^3x^5}}{2a^3}.$$

$$\Rightarrow \qquad y=\frac{x^2}{a}\left(1\pm\sqrt{\left(1-\frac{x}{a}\right)}\right).$$

Now, we observe that the value of y is real for all small values of x (negative or positive). Thus, both the branches of the curve at the cusp lie on both sides of the origin.

Therefore, the origin is a double cusp.

Now, to obtain the nature of cusp, we consider the following two cases:

(I) If x is small but positive, then the two values of y are positive and in this region the curve lies entirely above the x-axis $\left(\text{as } 1>\sqrt{\left(1-\frac{x}{a}\right)}\right)$.

(II) If x is small but negative, then the two values of y are of opposite signs and in this region the curve lies above as well as below the x-axis in the left of normal $\left(\text{as } 1<\sqrt{\left(1-\frac{x}{a}\right)}\right)$.

Hence, we conclude that the origin is a point of osco-inflexion.

8.2 CURVE TRACING

In this section, we will study that, how we can find the appropriate shape of a curve without plotting large number of points.

For Cartesian Curve $y = f(x)$.

The following steps are involved in tracing a Cartesian curve:

1. **Symmetry:**

   (a) ***Symmetric about x-axis:*** If the equation of curve has only even powers of y or the equation of a curve remains unchanged when y is replaced by $-y$, then the curve is symmetrical about x-axis.

   (b) ***Symmetric about y-axis:*** If the equation of curve has only even powers of x or the equation of a curve remains unchanged when x is replaced by $-x$, then the curve is symmetrical about y-axis.

   (c) ***Symmetric in opposite quadrants:*** If the equation of a curve remains unchanged when x is replaced by $-x$ and y is replaced by $-y$, then the curve is symmetrical in opposite quadrants.

   (d) ***Symmetric about the line $y = x$:*** If the equation of a curve remains unchanged, on interchanging of x and y, then the curve is symmetric about the line $y = x$.

   (e) ***Symmetric about the line $y = -x$:*** If the equation of a curve remains unchanged when x is interchanged by $-y$ and when y is interchanged by $-x$, then the curve is symmetric about the line $y = -x$.

2. **Origin:** If the equation of a curve does not have a constant term, then the curve passes through the origin.

   If the curve passes through the origin, then find the equation of tangents at the origin. Now, there are two possibilities as follows:

   (a) If there is only one tangent at the origin, then find whether the curve lies below or above the tangent in the neighbourhood of origin.

   (b) If there are two tangents, then the origin is a double point. Find the nature of a double point.

3. **Points of Intersection with the Coordinate Axes:** Find the points, where the curve meets the coordinate axes.

   (a) To find the points, where the curve meets the x-axis, by putting $y = 0$ in the given equation of a curve.

   (b) To find the points, where the curve meets the y-axis, by putting $y = 0$ in the given equation of a curve.

   Also find the tangents at these points, if necessary and position of the curve relative to these points.

(c) If a line $y = x$ or $y = -x$ is a line of symmetry, then find the points of intersection of the line and the given curve and the tangents at these points.

4. **Asymptotes:** Find all the asymptotes (parallel or oblique) of the curve.

5. **Special Points:** Find the points on a curve, where $\frac{dy}{dx} = 0$ or ∞, i.e., points, where the tangent to the curve is parallel to the coordinate axes. At these points, the behaviour of the curve changes from increasing to decreasing or vice versa.

6. **Points of Inflexion:** Find the point of inflexion, by putting $\frac{d^2y}{dx^2} = 0$, (but at the point of inflexion, $\frac{d^3y}{dx^3} \neq 0$).

7. **Region:** Find out the region of *xy*-plane, where the curve does not lie. Such a region is obtained, by solving the given equation of a curve for one variable in terms of other. Then, the curve will not lie for those sets of values of one variable which make the other variable imaginary or undefined.

EXAMPLE 7 Trace the following Cartesian curves:

(i) $x\left(x^2 + y^2\right) = a\left(x^2 - y^2\right)$

(ii) $y^2(x - a) = x^2(x + a)$

(iii) $x^3 + y^3 = 3ax^2$

(iv) $y^2\left(a^2 + x^2\right) = x^2\left(a^2 - x^2\right)$

(v) $y^2x^2 = x^2 + a^2$.

Solution: (i) The given equation of the curve is

$$x\left(x^2 + y^2\right) = a\left(x^2 - y^2\right).$$

$\Rightarrow x^3 + xy^2 - ax^2 + ay^2 = 0$

$\Rightarrow y^2(a + x) = x^2(a - x)$.

1. **Symmetry:** Since the curve has only even powers in *y*, the given curve is symmetric about *x*-axis.

2. **Origin:** Since the curve does not have any constant term, so the curve passes through the origin.

   Now, we find the equation of the tangents to the curve at the origin, by equating lowest degree term from the given equation to zero,

   i.e., $$ay^2 - ax^2 = 0 \Rightarrow y = \pm x.$$

Thus, the curve has two tangents $y = \pm x$ at the origin. So, the origin is a double point and since the tangents are real and distinct, the origin is a node.

3. **Points of Intersection with the Coordinate Axes:** First, put $y = 0$ in the given equation of a curve, we get

$$x^3 - ax^2 = 0 \Rightarrow x^2(x-a) = 0 \Rightarrow x = 0, a.$$

Again, put $x = 0$ in the given equation of a curve, we get

$$ay^2 = 0 \Rightarrow y = 0.$$

Thus, the curve intersects x-axis at the points $(0,0)$ and $(a,0)$ and y-axis at $(0,0)$, i.e., the curve passes through $(0,0)$ and $(a,0)$.

Now, we find the equation of tangents at these points.

At $(0,0)$, the equation of tangents and its nature is given in (2).

At $(a,0)$, let $X = x - a$ and $Y = y$.

$\Rightarrow x = X + a$ and $Y = y$.

Substituting in the given equation, we get

$$Y^2(X + 2a) = (X + a)^2(-X).$$

Equating the lowest degree term to zero, we get

$$a^2X = 0 \Rightarrow X = 0 \Rightarrow x = a.$$

Thus, the equation of tangent at the point $(a,0)$ is $x = a$. So, $(a,0)$ is an ordinary point.

4. **Asymptotes**

To find parallel asymptotes

Since the coefficient of x^3, highest degree term in x, is constant. So, there is no asymptote parallel to x-axis.

Now, the coefficient of y^2, highest degree term in y to zero, we get

$$x + a = 0 \Rightarrow x = -a.$$

Thus, $x = -a$ is an asymptote of the given curve parallel to y-axis.

To find oblique asymptotes

Put $x = 1$ and $y = m$ in the given equation of the curve, we get

$\phi_3(m) = 1 + m^2$, $\phi_2(m) = am^2 - a$ and $\phi_1(m) = 0$.

To find the value of m, let $\phi_3(m)=0$ which gives

$$m^2+1=0 \Rightarrow m^2=-1.$$

Thus, m is imaginary and so there are no oblique asymptotes of the curve. Therefore, the only asymptote of the given curve is $x=-a$.

5. **Special Points**

The given equation of the curve can be written as

$$y^2=\frac{x^2(a-x)}{a+x}.$$

Now, differentiate w.r.t. x to find the value $\frac{dy}{dx}$ and then, equate it to zero, we get

$$x^2+ax-a^2=0$$

$$\Rightarrow \qquad x=\frac{-a\pm\sqrt{a^2+4a^2}}{2}=\frac{-a\pm a\sqrt{5}}{2}$$

$$\Rightarrow \qquad x=0.6a,-1.6a.$$

6. **Region**

Since $y^2=\frac{x^2(a-x)}{a+x}$, then the region of existence as follows:

Region	**Sign of R.H.S**	**Existence**
$x<-a$	Negative	No
$-a<x<0$	Positive	Yes
$0<x<a$	Positive	Yes
$x>a$	Negative	No

So, we conclude that the region of existence is $-a<x<a$ and so, the point $x=-1.6a$ is not considered as it lies outside the region of existence.

Now, $\frac{dy}{dx}>0$, for all $x\in(-a,0.6a)$ and $\frac{dy}{dx}<0$, for all $x\in(0.6a,a]$. Thus, y is strictly increasing for $x\in(-a,0.6a)$ and strictly decreasing for $x\in(0.6a,a]$. But $y=0$, when $x=a$.

Therefore, the point $x=0.6a$ behaves as a turning point.

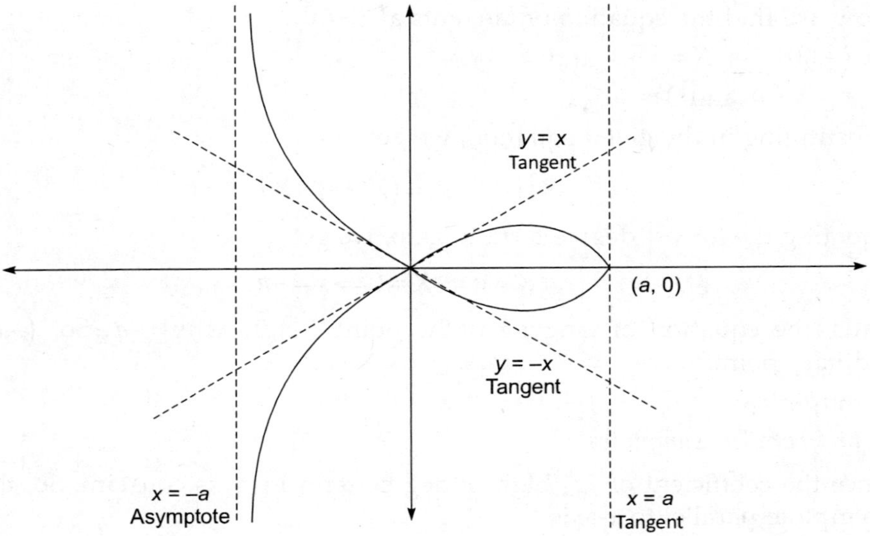

Figure 8.3

(ii) The given equation of the curve is

$$y^2(x-a) = x^2(x+a).$$

1. **Symmetry:** Since the curve has only even powers in y, thus the given curve is symmetric about x-axis.

2. **Origin:** Since the curve does not have any constant term, the curve passes through the origin.
   Now, we find the equation of the tangents to the curve at the origin, by equating lowest degree term from the given equation to zero,

   i.e., $$-ay^2 - ax^2 = 0.$$

   Then, the values of y are not real. So, the origin is a conjugate point.

3. **Points of Intersection with the Coordinate Axes:** First, putting $y = 0$ in the given equation of a curve, we get

   $$x^3 + ax^2 = 0 \Rightarrow x^2(x+a) = 0 \Rightarrow x = 0, -a.$$

   Again, putting $x = 0$ in the given equation of a curve, we get

   $$-ay^2 = 0 \Rightarrow y = 0.$$

   Thus, the curve intersects x-axis at the points $(0,0)$ and $(-a,0)$ and y-axis at $(0,0)$, i.e., the curve passes through $(0,0)$ and $(-a,0)$. But the origin is a conjugate point.

Now, we find the equation of tangents at $(-a,0)$.
At $(-a,0)$, let $X = x + a$ and $Y = y$.
$\Rightarrow x = X - a$ and $Y = y$.
Substituting in the given equation, we get

$$Y^2(X-2a) = (X-a)^2(X).$$

Equating the lowest degree term to zero, we get

$$a^2X = 0 \Rightarrow X = 0 \Rightarrow x = -a.$$

Thus, the equation of tangent at the point $(-a,0)$ is $x = -a$. So, $(-a,0)$ is an ordinary point.

4. **Asymptotes**

To find parallel asymptotes

Since the coefficient of x^3, highest degree term in x, is constant. So, there is no asymptote parallel to x-axis.

Now, the coefficient of y^2, highest degree term in y to zero, we get

$$x - a = 0 \Rightarrow x = a.$$

Thus, $x = -a$ is an asymptote of the given curve parallel to y-axis.

To find oblique asymptotes

Put $x = 1$ and $y = m$ in given equation of the curve, we get
$\phi_3(m) = m^2 - 1$, $\phi_2(m) = -am^2 - a$ and $\phi_1(m) = 0$.
To find the value of m, let $\phi_3(m) = 0$ which gives

$$m^2 - 1 = 0 \Rightarrow m^2 = \pm 1.$$

Thus, the values of m are real and distinct. Thus, the value of c is given by

$$c = -\frac{\phi_2(m)}{\phi_3'(m)} = -\frac{-(-m^2a - a)}{2m} = \frac{a(m^2+1)}{2m}.$$

for $m = 1$, $c = a$ and
for $m = -1$, $c = -a$.
Therefore, the asymptotes of the given curve are:

$$x - a = 0,\ y = x + a \text{ and } y = -(x+a).$$

5. **Special Points**

The given equation of the curve can be written as

$$y^2 = \frac{x^2(a+x)}{x-a}.$$

Now, differentiate w.r.t. x to find the value $\frac{dy}{dx}$ and then equating it to zero, we get

$$x^2 - ax - a^2 = 0.$$

$$\Rightarrow \qquad x = \frac{a \pm \sqrt{a^2 + 4a^2}}{2} = \frac{a \pm a\sqrt{5}}{2}$$

$\Rightarrow x = 1.6189a, -0.618a.$

6. **Region**

Since $y^2 = \frac{x^2(x+a)}{x-a}$, then the region of existence is as follows:

Region	Sign of R.H.S.	Existence
$x < -a$	Positive	Yes
$-a < x < 0$	Negative	No
$0 < x < a$	Negative	No
$x > a$	Positive	Yes

So, we conclude that the region of existence is $x < -a$ and $x > a$. So, the point $x = -0.6a$ is not considered as it lies outside the region of existence.

Now, $\frac{dy}{dx} > 0$, for all $x \in (-\infty, -a] \cup (1.6189a, \infty)$ and $\frac{dy}{dx} < 0$, for all $x \in (a, 1.6189a)$, therefore, y is strictly increasing for $x \in (-\infty, -a] \cup (1.6189a, \infty)$ and strictly decreasing for $x \in (a, 1.6189a)$.

Therefore, the point $x = 1.6189a$ behaves as a turning point.

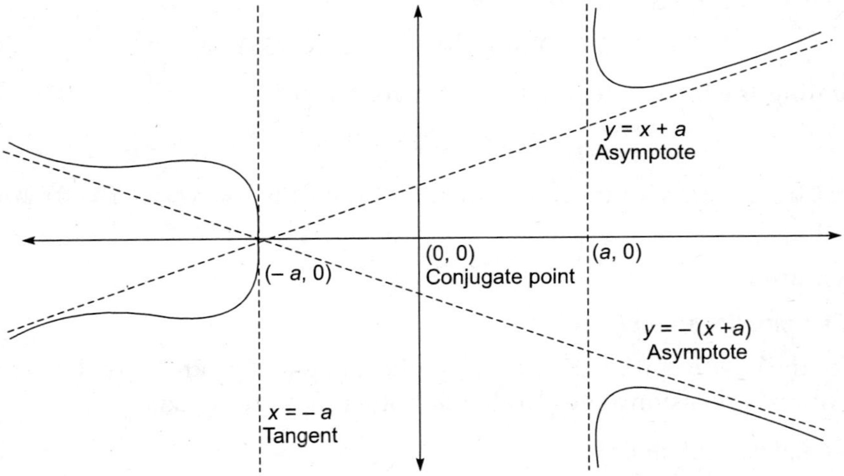

Figure 8.4

(iii) The given equation of the curve is

$$x^3 + y^3 = 3ax^2.$$

1. **Symmetry:** Since the curve does not have even powers in y and in x, thus the given curve is not symmetric about both the y-axis and x-axis.
   Also, the given curve is not symmetric about the line $y = x$.
2. **Origin:** Since the curve does not have any constant term, so the curve passes through the origin.
   Now, we find the equation of the tangents to the curve at the origin, by equating the lowest degree term from the given equation to zero,
   i.e., $$x^2 = 0.$$
   Then, the values of x are real and equal. So, the origin is a cusp.
3. **Points of Intersection with the Coordinate Axes:** First, putting $y = 0$ in the given equation of a curve, we get
   $$x^3 = 3ax^2 \Rightarrow x^2(x - 3a) = 0 \Rightarrow x = 0, 3a.$$
   Again, putting $x = 0$ in the given equation of a curve, we get
   $$y^3 = 0 \Rightarrow y = 0.$$
   Thus, the curve intersects x-axis at the points $(0,0)$ and $(3a,0)$ and y-axis at $(0,0)$, i.e., the curve passes through $(0,0)$ and $(3a,0)$ and the origin is a cusp.
   Now, we find the equation of tangents at $(3a,0)$.
   At $(3a,0)$, let $X = x - 3a$ and $Y = y$.
   $\Rightarrow x = X + 3a$ and $Y = y$.
   Substitute in the given equation, we get
   $$(X + 3a)^3 + Y^3 = 3a(X + 3a)^2.$$
   Equating the lowest degree term to zero, we get
   $$27aX - 18aX = 0 \Rightarrow X = 0 \Rightarrow x = 3a.$$
   Thus, the equation of tangent at the point $(3a,0)$ is $x = 3a$. So, $(3a,0)$ is an ordinary point.
4. **Asymptotes**

   To find parallel asymptotes

   Since the coefficient of x^3 and y^3, highest degree term in x and y are constants. So, there is no asymptote parallel to both x-axis and y-axis.

   To find oblique asymptotes

   Putting $x = 1$ and $y = m$ in the given equation of the curve, we get

$$\phi_3(m) = m^3 + 1,\ \phi_2(m) = -3a \text{ and } \phi_1(m) = 0.$$

To find the value of m, let $\phi_3(m) = 0$ which gives

$$m^3 + 1 = 0 \Rightarrow (m+1)(m^2 - m + 1) = 0.$$

$$m = -1 \text{ (other are imaginary).}$$

Thus, the values of m are real and distinct. Thus, the value of c is given by

$$c = -\frac{\phi_2(m)}{\phi_3'(m)} = \frac{3a}{3m^2}.$$

For $m = -1$, $c = a$.

Therefore, the only asymptote of the given curve is $y = -x + a$ and it intersects the curve at the point $\left(\frac{a}{3}, \frac{2a}{3}\right)$.

5. **Special Points**

The given equation of the curve can be written as

$$y^3 = 3ax^2 - x^3.$$

Now, differentiate w.r.t. x to find the value $\frac{dy}{dx}$ and then, equate it to zero, we get

$x(2a - x) = 0$

$\Rightarrow x = 2a, 0.$

6. **Region**

Since $y^3 = 3ax^2 - x^3$, i.e. $y = \{3ax^2 - x^3\}^{\frac{1}{3}}$, then the region of existence as follows:

Region	**Sign of R.H.S.**	**Existence**
$x < -3a$	Positive	Yes
$-3a < x < 0$	Positive	Yes
$0 < x < 3a$	Positive	Yes
$x > 3a$	Negative	Yes

So, we conclude that the region of existence is $(-\infty, \infty)$.

Now, $\frac{dy}{dx} > 0$, for all $x \in (-\infty, 0] \cup (0, 2a)$ and $\frac{dy}{dx} < 0$, for all $x \in (2a, \infty)$, therefore y is strictly increasing for $x \in (-\infty, 0] \cup (0, 2a)$ and strictly decreasing for $x \in (2a, \infty)$, But $y = 0$, when $x = 3a$.

Therefore, the points $x = 0$ and $2a$ behave as a turning points.

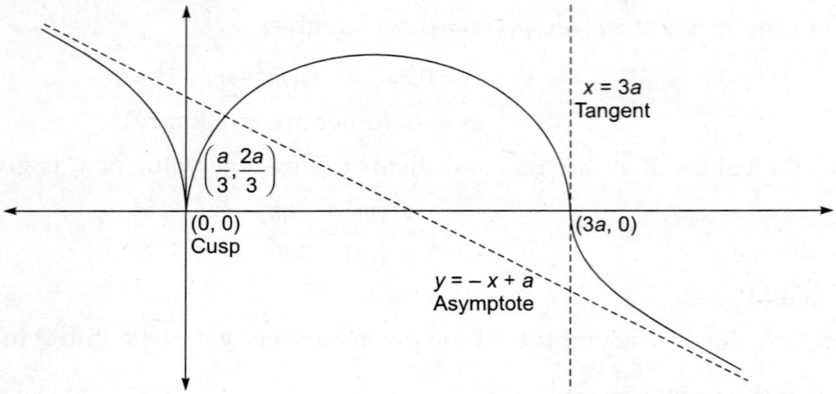

Figure 8.5

(iv) The given equation of the curve is

$$y^2\left(a^2+x^2\right)=x^2\left(a^2-x^2\right)$$

1. **Symmetry:** Since the curve has even powers in y and in x, the given curve is symmetric about both the y-axis and x-axis.
   Also, the given curve is also symmetric in opposite quadrants.
2. **Origin:** Since the curve does not have any constant term, so the curve passes through the origin.
   Now, we find the equation of the tangents to the curve at the origin, by equating lowest degree term from the given equation to zero,
   i.e., $$a^2y^2-a^2x^2=0 \Rightarrow y=\pm x.$$
   Then, the two tangents at origin are real and distinct. So, the origin is a node.
3. **Points of Intersection with the Coordinate Axes:** Firstly, putting $y = 0$ in the given equation of a curve, we get
   $$x^2(a^2 - x^2) = 0 \Rightarrow x = 0, \pm a.$$
   Again, put $x = 0$ in the given equation of a curve, we get
   $$y^2 = 0 \Rightarrow y = 0.$$
   Thus, the curve intersects x-axis at the points $(0,0)$, $(a,0)$, $(-a,0)$ and y-axis at $(0,0)$, i.e., the curve passes through $(0,0)$, $(a,0)$ and $(-a,0)$ and the origin is a node.
   Now, we find the equation of tangents at $(a,0)$ and $(-a,0)$.
   At $(a,0)$, let $X = x - a$ and $Y = y$.

$\Rightarrow x = X + a$ and $Y = y$.

Substituting in given equation, we get

$$Y^2\left(a^2+(X+a)^2\right)=(X+a)^2\left(a^2-(X+a)^2\right).$$

Equate the lowest degree term to zero, we get

$$X = 0 \Rightarrow x = a.$$

Again, at $(-a,0)$, let $X = x + a$ and $Y = y$.

$\Rightarrow x = X - a$ and $Y = y$.

Substitute in given equation, we get

$$Y^2\left(a^2+(X-a)^2\right)=(X-a)^2\left(a^2-(X-a)^2\right).$$

Equate the lowest degree term to zero, we get

$$X = 0 \Rightarrow x = -a.$$

Thus, the equation of tangent at the point $(a,0)$ and $(-a,0)$ are $x = a$ and $x = -a$, respectively. So, $(a,0)$ and $(-a,0)$ are ordinary points.

4. **Asymptotes**

   To find parallel asymptotes

   Since the coefficient of x^4 and y^2, highest degree term in x and y are constants. So, there is no asymptote parallel to both x-axis and y-axis.

   To find oblique asymptotes

   Put $x = 1$ and $y = m$ in the given equation of the curve, we get

   $\phi_4(m) = m^2 + 1$, $\phi_3(m) = 0$, $\phi_2(m) = m^2a^2 - a^2$ and $\phi_1(m) = 0$.

   To find the value of m, let $\phi_4(m) = 0$ which gives

$$m^2 + 1 = 0 \Rightarrow m^2 = -1.$$

   Thus, m is imaginary and so, there are no oblique asymptotes of the curve. Therefore, the given curve has no asymptotes.

5. **Special Points**

   The given equation of the curve can be written as

$$y^2 = \frac{x^2(a^2 - x^2)}{(a^2 + x^2)}.$$

   Now, differentiate w.r.t. x to find the value $\dfrac{dy}{dx}$ and then, equate it to zero, we get

$$a^4 - 2a^2x^2 - x^4 = 0.$$

$$\Rightarrow x = \sqrt{\left(-1+\sqrt{2}\right)}a \approx 0.64a, \; -\sqrt{\left(-1+\sqrt{2}\right)}a \approx -0.64a.$$

6. **Region**

Since $y^2 = \dfrac{x^2(a^2 - x^2)}{(a^2 + x^2)}$, then the region of existence as follows:

Region	Sign of R.H.S	Existence
$x < -a$	Negative	No
$-a < x < 0$	Positive	Yes
$0 < x < a$	Positive	Yes
$x > a$	Negative	No

So, we conclude that the region of existence is $(-a,a)$.

Now, $\dfrac{dy}{dx} > 0$, for all $x \in (-0.64a, 0.64a)$ and $\dfrac{dy}{dx} < 0$, for all $x \in (-a, -0.64a) \cup (0.64a, a)$.

Therefore y is strictly increasing for $x \in (-0.64a, 0.64a)$ and strictly decreasing for $x \in (-a, -0.64a) \cup (0.64a, a)$. But $y = 0$, when $x = \pm a$.
Hence, the points $x = 0.64a$ and $-0.64a$ behave as a turning points.

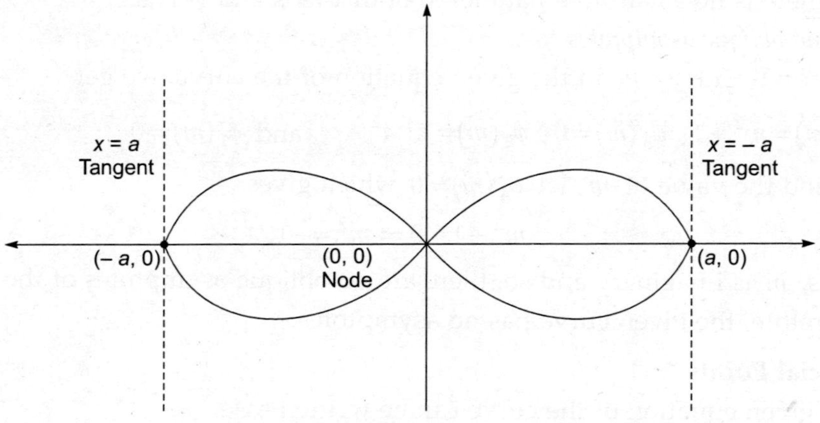

Figure 8.6

(v) The given equation of the curve is

$$y^2x^2 = x^2 + a^2.$$

1. **Symmetry:** Since the curve has even powers in y and in x, thus the given curve is symmetric about both the y-axis and x-axis.

Also, the given curve is also symmetric in opposite quadrants.

2. **Origin:** Since the curve has constant term, so the curve does not pass through the origin.

3. **Points of Intersection with the Coordinate Axes:** First, put $y=0$ in the given equation of a curve, we get

$$x^2+a^2=0.$$

which implies x is not real.

Again, put $x=0$ in the given equation of a curve, we get

$a^2=0$, which is not possible.

Thus, the curve does not intersect with x-axis and y-axis.

4. **Asymptotes**

To find parallel asymptotes

Now, the coefficient of y^2, highest degree term in y to zero, we get

$$x^2=0 \Rightarrow x=0.$$

Thus, y-axis is an asymptote of the given curve.

Now, the coefficient of x^2, highest degree term in x to zero, we get

$$y^2-1=0 \Rightarrow y=\pm 1.$$

Thus, $y=\pm 1$ are two asymptotes of the given curve parallel to x-axis.

To find oblique asymptotes

Put $x=1$ and $y=m$ in given equation of the curve, we get

$\phi_4(m)=m^2$, $\phi_3(m)=0$, $\phi_2(m)=1$ and $\phi_1(m)=0$.

To find the value of m, let $\phi_4(m)=0$ which gives

$$m^2=0 \Rightarrow m=0.$$

Thus, the values of m are real and distinct. Thus, the value of c is given by

$$c=-\frac{\phi_3(m)}{\phi_4'(m)}=0.$$

Therefore, the asymptotes of the given curve are $y=\pm 1$ and y-axis.

5. **Region**

Since $x^2=\dfrac{a^2}{y^2-1}$, then the region of existence is as follows:

Region	Sign of R.H.S	Existence
$y<-1$	Positive	Yes
$-1<y<0$	Negative	No
$0<y<1$	Negative	No
$y>1$	Positive	Yes

So, we conclude that the region of existence is $y<-1$ and $y>1$.

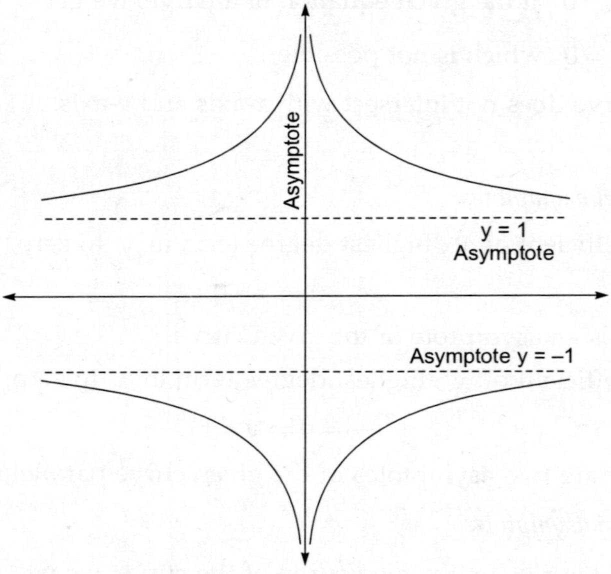

Figure 8.7

For Polar Curve $r = f(\theta)$.

The following important steps are involved in tracing a polar curve:

1. **Symmetry**

   (a) ***Symmetric about the initial line $\theta = 0$:*** If the equation of a curve remains unchanged when θ is replaced by $-\theta$, then the curve is symmetrical about the initial line $\theta = 0$.

   (b) ***Symmetric about the vertical line $\theta = \frac{\pi}{2}$:*** If the equation of a curve remains unchanged when θ is replaced by $\pi-\theta$ or when θ is replaced by $-\theta$ and r is replaced by $-r$, then the curve is symmetrical about the vertical line $\theta = \frac{\pi}{2}$.

(c) ***Symmetric about the pole:*** If the equation of a curve remains unchanged when r is replaced by $-r$, then the curve is symmetrical about the pole.

(d) ***Symmetric about the line*** $\theta = \frac{\pi}{4}$: If the equation of a curve remains unchanged when θ is interchanged by $\frac{\pi}{2} - \theta$, then the curve is symmetric about the line $\theta = \frac{\pi}{4}$.

(e) ***Symmetric about the line*** $\theta = \frac{3\pi}{4}$: If the equation of a curve remains unchanged when θ is interchanged by $\frac{3\pi}{2} - \theta$, then the curve is symmetric about the line $\theta = \frac{3\pi}{4}$.

2. **Origin:** If we find some real value of θ, by putting $r = 0$ in the given equation of curve, then the curve passes through the origin.
   If the curve passes through the origin, then find the equation of tangents at the origin.
3. **Points of Intersection with the Coordinate Axes:** Find the points, where the curve intersects with the initial line $\theta = 0$ and the vertical line $\theta = \frac{\pi}{2}$.
4. **Angle between the radial vector and the tangent:** Find the angle ϕ between the radial vector and the tangent to the curve which is given by

$$\tan\phi = r\frac{d\theta}{dr}$$

   and also the points where $\phi = 0$ or $\frac{\pi}{2}$.
5. **Asymptotes:** Find all the asymptotes of the polar curve.
6. **Special Points:** Solve the equation of the given curve for r and trace the variation of r as θ varies. Also, if $\frac{dr}{d\theta} > 0$, then r is increasing as θ is increasing and of $\frac{dr}{d\theta} < 0$, then r is decreasing as θ is increasing.
7. **Region:** Find out the region of, where the curve does not lie. Such a region is obtained, by solving the given equation of a curve for r in terms of θ. Then, the curve will not lie for those sets of values of θ which make the r imaginary.

EXAMPLE 8 *Trace the following polar curves:*

(i) $r = a(1 + \cos\theta)$ *(Cardioid).*

(ii) $r^2 = a^2 \cos 2\theta$.

(iii) $r = a\cos 2\theta$, $a > 0$.

Solution: (i) The given equation of a curve is

$$r = a(1+\cos\theta)$$

Let $f(\theta) = a(1+\cos\theta)$.

1. **Symmetry:** *Since* $f(\theta) = f(-\theta)$, then the given curve is symmetric about the initial line $\theta = 0$.

   But $f(\theta) \neq f(\pi-\theta)$, so the curve is not symmetric about the vertical line $\theta = \frac{\pi}{2}$.

2. **Origin:** Put $r = 0$ in the given equation we, get

   $$1+\cos\theta = 0 \Rightarrow \cos\theta = -1 \Rightarrow \theta = \pi.$$

   Thus, the curve passes through the pole and the tangent at the pole is $\theta = \pi$.

3. **Points of Intersection with the Coordinate Axes:** The curve intersects the initial line $\theta = 0$ at $r = 2a$ and the vertical line $\theta = \pm\frac{\pi}{2}$ at $r = a$.

   Thus, the points of intersection are $(2a, 0)$, $\left(a, \frac{\pi}{2}\right)$ and $\left(a, -\frac{\pi}{2}\right)$.

4. **Angle between the radial vector and the tangent:** The angle ϕ between the radial vector and the tangent is given by

   $$\tan\phi = r\frac{d\theta}{dr} = \frac{r}{\frac{dr}{d\theta}}$$

   $$= a(1+\cos\theta).\left(\frac{-1}{a\sin\theta}\right)$$

   $$= -\frac{2\cos^2\frac{\theta}{2}}{2\cos\frac{\theta}{2}\sin\frac{\theta}{2}}$$

   $$= -\cot\frac{\theta}{2}$$

   $$= \tan\left(\frac{\pi}{2}+\frac{\theta}{2}\right)$$

   $$\Rightarrow \qquad \phi = \frac{\pi}{2}+\frac{\theta}{2}.$$

   So, when $\theta = 0$, then $\phi = \frac{\pi}{2}$ and $r = 2a$.

   Thus, the tangent at the point $(2a, 0)$ is perpendicular to initial line $\theta = 0$.

5. **Asymptotes:** There is no value of θ for which $r \to \infty$. Thus, the given curve has no asymptotes.

6. **Special Points:** Since the curve is symmetric about the initial line $\theta = 0$, so we consider only the values of θ for $0 \le \theta \le \pi$.

θ	0	$\frac{\pi}{4}$	$\frac{\pi}{2}$	$\frac{3\pi}{4}$	π
r	$2a$	$a\left(1+\frac{1}{\sqrt{2}}\right)=1.7a$	a	$0.3a$	0

7. **Region:** When θ increases from 0 to $\pi/2$, then r decreases from $2a$ to a and θ increases from $\frac{\pi}{2}$ to π, then r decreases from a to 0.

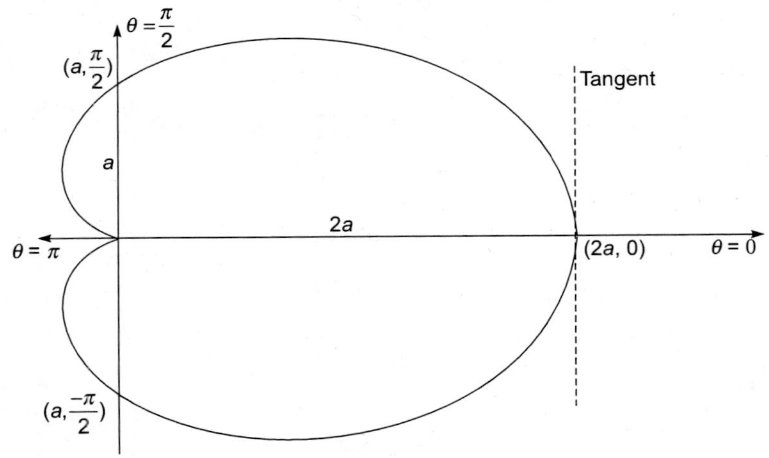

Figure 8.8

(ii) *The given equation of a curve is*

$$r^2 = a^2 \cos 2\theta.$$

1. **Symmetry:** Since the given equation of the curve remains unchanged when θ is replaced by $-\theta$, so the curve is symmetric about the initial line $\theta = 0$.
   Again, since the given equation of the curve remains unchanged when θ is replaced by $\pi - \theta$, so the curve is symmetric about the vertical line $\theta = \frac{\pi}{2}$.

Also, the given equation remains unchanged when r is replaced by $-r$. So, the curve is symmetric about the pole.

2. **Origin:** Put $r=0$ in the given equation we, get

$$\cos 2\theta = 0 \Rightarrow 2\theta = \frac{n\pi}{2} \Rightarrow \theta = \frac{n\pi}{4}.$$

Thus, the curve passes through the pole and the tangent at the pole is $\theta = \frac{n\pi}{4}$, where $n=1$ and 3 (as for other values of n give same value).

3. **Points of Intersection with the Coordinate Axes:** The curve intersects the initial line $\theta=0$ at $r=\pm a$.

Thus, the points of intersection are $(\pm a, 0)$ (as on the vertical line $\theta = \pm\frac{\pi}{2}$, r is not real).

4. **Angle between the radial vector and the tangent:** The angle ϕ between the radial vector and the tangent is given by

$$\tan\phi = r\frac{d\theta}{dr} = \frac{r}{\frac{dr}{d\theta}}$$

$$= r \cdot \left(\frac{-r}{a^2 \sin 2\theta}\right)$$

$$= -\frac{r^2}{a^2 \sin 2\theta}$$

$$= -\cot 2\theta$$

$$= \tan\left(\frac{\pi}{2} + 2\theta\right)$$

$$\Rightarrow \qquad \phi = \frac{\pi}{2} + 2\theta.$$

So, when $\theta = 0$, then $\phi = \frac{\pi}{2}$ and $r = \pm a$.

Thus, the tangent at the point $(\pm a, 0)$ is perpendicular to initial line $\theta = 0$.

5. **Asymptotes:** There is no value of θ for which $r \to \infty$. Thus, the given curve has no asymptotes.

6. **Special Points:** Since the curve is symmetric about the initial line $\theta=0$ and the vertical line $\theta = \frac{\pi}{2}$ and about the pole, so we consider only the values of θ for $0 \le \theta \le \frac{\pi}{2}$.

θ	0	$\frac{\pi}{6}$	$\frac{\pi}{4}$	$\frac{\pi}{2}$
r	$\pm a$	$\pm 0.7a$	0	Not real

7. **Region:** When θ increases from 0 to $\frac{\pi}{6}$, then r decreases from a to $0.7a$ and θ increases from $\frac{\pi}{6}$ to $\frac{\pi}{4}$, then r decreases from $0.7a$ to 0.

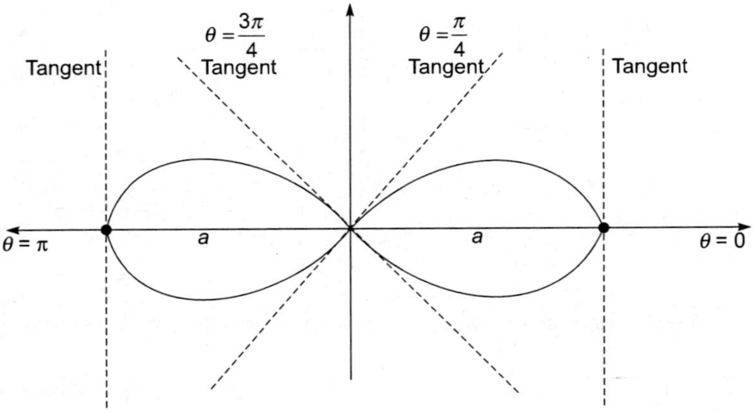

Figure 8.9

(iii) The given equation of the curve is

$$r = a\cos 2\theta.$$

Let $f(\theta) = a\cos 2\theta$.

1. **Symmetry:** Since $f(\theta) = f(-\theta)$, the given curve is symmetric about the initial line $\theta = 0$.

   Again, since $f(\theta) = f(\pi - \theta)$, the curve is symmetric about the vertical line $\theta = \frac{\pi}{2}$.

2. **Origin:** Put $r = 0$ in the given equation we, get

$$\cos 2\theta = 0 \Rightarrow 2\theta = \frac{n\pi}{2} \Rightarrow \theta = \frac{n\pi}{4}.$$

   Thus, the curve passes through the pole and the tangent at the pole is $\theta = \frac{n\pi}{4}$, where $n = 1$ and 3 (as for other values of n give same value).

3. **Points of Intersection with the Coordinate Axes:** The curve intersects the initial line $\theta = 0$ at $r = a$ and the vertical line $\theta = \frac{\pi}{2}$ at $r = -a$.

Thus, the points of intersection are $(a,0)$ and $(-a,0)$.

4. **Angle between the radial vector and the tangent:** The angle ϕ between the radial vector and the tangent is given by

$$\tan\phi = r\frac{d\theta}{dr} = \frac{r}{\frac{dr}{d\theta}}$$

$$= r\cdot\left(\frac{-1}{2a\sin 2\theta}\right)$$

$$= -\frac{\cos 2\theta}{2\sin 2\theta}$$

$$= -\frac{1}{2}\cot 2\theta$$

$$= \frac{1}{2}\tan\left(\frac{\pi}{2}+2\theta\right)$$

So, when $\theta = 0$, then $\phi = \frac{\pi}{2}$ and $r = a$ and when $\theta = \pi$, then $\phi = \frac{\pi}{2}$ and $r = a$.

Thus, the tangent at the point $(a,0)$ and $(-a,0)$ is perpendicular to initial line $\theta = 0$ and $\theta = \pi$, respectively.

5. **Asymptotes:** There is no value of θ for which $r \to \infty$. Thus, the given curve has no asymptotes.
6. **Special Points:** We consider only the values of θ for $0 \le \theta \le 2\pi$.

θ	0	$\frac{\pi}{4}$	$\frac{\pi}{2}$	$\frac{3\pi}{4}$	π	$\frac{5\pi}{4}$	$\frac{3\pi}{2}$	$\frac{7\pi}{4}$	2π
r	a	0	$-a$	0	a	0	$-a$	0	a

7. **Region:**

When θ increases from 0 to $\frac{\pi}{4}$, r decreases from a to 0.

When θ increases from $\frac{\pi}{4}$ to $\frac{\pi}{2}$, r decreases from 0 to $-a$.

When θ increases from $\frac{\pi}{2}$ to $\frac{3\pi}{4}$, r increases from $-a$ to 0.

When θ increases from $\frac{3\pi}{4}$ to π, r increases from 0 to a.

Since the curve is symmetrical about the initial line $\theta = 0$, so further variation of r as θ increases, are same.

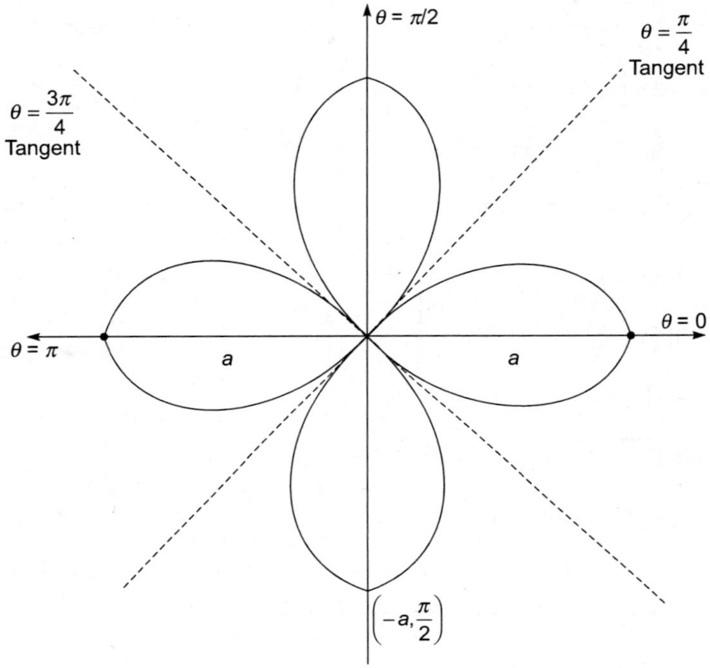

Figure 8.10

Exercise

Find the equations of tangents at the origin to the following curves (1-5):

1. $x^3 + y^3 = 3axy$.
2. $x^5 + 2a^2x^2y - a^3y^3 = 0$.
3. $y^5 + ax^2y^2 - ay^3x - 2a^2x^2y + a^2x^3 = 0$.
4. $x^4 + y^4 = a^2\left(x^2 - y^2\right)$
5. $y^2 = x^3 + x + 2x^2$
6. Find the equations of the tangents to the curve

$$y^4 + 4cy^3 + 2cx^3 + 4c^2y^2 + 3c^2x^2 - c^4 = 0$$

at the point $(-c, -2c)$ and show that it is node.

7. Find the equations of the tangents to the curve

$$x^3 + xy^2 + 2ax^2 + 2ay^2 + 4axy + a^2(5x + 8y + 8a) = 0$$

at the point $(-a, -2a)$ and show that it is a node.

Find the position and nature of the multiple points for the following curves:

8. $x^3 + y^3 = 3axy$.

9. $xy^2 - ax^2 + 2a^2x - a^3 = 0$.

10. $y^2\left(a^2 + x^2\right) = x^2\left(a^2 - x^2\right)$.

11. $x^4 - 2ay^3 - 3a^2y^2 - 2a^2x^2 + a^4 = 0$.

12. $(x+y)^3 - \sqrt{2}(y-x+2)^2 = 0$.

13. $y^2 = (x-1)(x-2)^2$.

14. $y^2 = (x-a)^2(x-b)$.

15. $y^2 = x^2(x-1)$.

16. $y^2 = x(x-a)^2$.

17. $(2y+x+1)^2 - 4(1-x)^5 = 0$.

Find the nature and position of the cusp for the following curves:

18. $y^2 = x^3$

19. $y^2 = x^4$

20. $\left(y - 4x^2\right)^2 - x^7 = 0$

21. $y^2 - 2yx^2 - yx^3 = x^3$

22. $\left(y - 2x^2\right)^2 = x^3$

23. $y^2 + 3x^4 - 4x^2y - x^5 = 0$

24. $y^2(3a - x) = (x-a)^2$

25. $x^5 + 16x^2y - 64y^2 = 0$.

26. $x^4 - 2x^2y - xy^2 - 2x^2 - 2xy + y^2 - x + 2y + 1 = 0$

27. $x^5 - ax^3y - a^2x^2y + a^2y^2 = 0$.

Trace the following cartesian curves:

28. $9ay^2 = x(x-3a)^2$

29. $y(x^2+4a^2) = 8a^3$

30. $y^2x = a^2(a-x)$

31. $y^2(2a-x) = x^3$

32. $y^2(a^2-x^2) = x^4$

33. $y^2x^2 = x^2 - a^2$

34. $y^2(a^2-x^2) = a^3x$

35. $y(1-x^2) = x^2$

36. $x^3 + y^3 = 3axy$

37. $y^2 = x(16-x^2)$

38. $y(a^2+x^2) = a^2x$

39. $y^2(x^2-1) = 2x-1$

Trace the following polar curves:

40. $r = a(1-\cos\theta)$

41. $r = a\sin 3\theta$

42. $r = \sin 2\theta$

43. $r = \dfrac{a\theta^2}{1+\theta^2}$.

9 CHAPTER

Integral Calculus

In this chapter, first we will discuss reduction formulae, derivation and illustration of reduction formulae. Further, we will discuss the applications of integral calculus which includes finding area of irregular plane curve, arc length, arc length of parametric curves, area of surface revolution, volume of slicing, disks and washers methods, volume of cylindrical shells.

9.1 REDUCTION FORMULAE

Reduction formulae reduce a given integral to a known integral of lower order from repeated application of integration by parts. In this section, we discuss reduction formulae of only trigonometric functions like $\sin^n x, \cos^n x, \tan^n x, \cot^n x, \sec^n x, \csc^n x, \sin^n x\cos^n x$.

Reduction Formula for $\int \sin^n x dx; n > 0$.

Let $I = \int \sin^n x\,dx = \int \sin^{n-1} x \sin x dx$.

Now, using integration by parts, we get

$$I = \sin^{n-1} x(-\cos x) - \int(-\cos x)(n-1)\sin^{n-2} x\cos x\,dx$$

$$= \sin^{n-1} x(-\cos x) + (n-1)\int \sin^{n-2} x\cos^2 x\,dx$$

$$= -\cos x\sin^{n-1} x + (n-1)\int \sin^{n-2} x(1-\sin^2 x)dx$$

$$= -\cos x\sin^{n-1} x + (n-1)\int \sin^{n-2} xdx - (n-1)\int \sin^n xdx.$$

$$\Rightarrow \quad (1+n-1)I = -\cos x\sin^{n-1} x + (n-1)\int \sin^{n-2} xdx.$$

$$\Rightarrow \quad nI = -\cos x\sin^{n-1} x + (n-1)\int \sin^{n-2} xdx$$

$$\Rightarrow \quad I = -\frac{\cos x\sin^{n-1} x}{n} + \frac{(n-1)}{n}\int \sin^{n-2} xdx.$$

$$\int \sin^n x\,dx = -\frac{\cos x \sin^{n-1} x}{n} + \frac{(n-1)}{n}\int \sin^{n-2} x dx. \tag{9.1}$$

Reduction formula for $\int \cos^n x dx; n > 0$.

Put $x = \frac{\pi}{2} - y, dx = -dy$ in (9.1), we have

$$\int \cos^n y dy = \frac{\sin y \cos^{n-1} y}{n} + \frac{(n-1)}{n}\int \cos^{n-2} y dy. \tag{9.2}$$

Remark: Instead of reduction formulae (9.1) and (9.2), we can also use the substitution method for evaluating the value of integrals $\int \sin^n x\,dx$ and $\int \cos^n x\,dx$.

(a) For $\int \sin^n x\,dx$; if n is odd, put $\cos x = t$ and integrate the integral in t. Then replace $t = \cos x$.

(b) For $\int \cos^n x\,dx$; if n is odd, put $\sin x = t$ and integrate the integral in t. Then replace $t = \sin x$.

(c) For both the integrals $\int \sin^n x\,dx$ and $\int \cos^n x\,dx$; if n is an even positive integer and small, then replace express powers of $\sin x$ and $\cos x$ in terms of multiple angles using trigonometric formulas like $\sin^2 x = \frac{1-\cos 2x}{2}$, $\cos^2 x = \frac{1+\cos 2x}{2}$, etc. If n is a large positive integer, use De Moivre's theorem to express the given integral in terms of multiple angle.

Reduction formula for $\int \tan^n x dx$.

Let $I = \int \tan^n x\,dx = \int \tan^{n-2} x \tan^2 x dx$

$$= \int \tan^{n-2} x\,(\sec^2 x - 1)dx$$

$$= \int \tan^{n-2} x \sec^2 x\,dx - \int \tan^{n-2} x\,dx$$

$$= \int \tan^{n-2} x d(\tan x) - \int \tan^{n-2} x\,dx$$

$$= \frac{\tan^{n-1} x}{n-1} - \int \tan^{n-2} x\,dx.$$

$$\int \tan^n x\,dx = \frac{\tan^{n-1} x}{n-1} - \int \tan^{n-2} x\,dx. \tag{9.3}$$

This reduction formula reduces power of $\tan x$ by 2. After repeated application of reduction formula, we have $\int \tan x\, dx = \log \sec x$ if n is odd and $\int \tan^2 x dx = (\tan x) - x$, if n is even.

Reduction formula for $\int \cot^n x\, dx$.

Let $I = \int \cot^n x\, dx = \int \cot^{n-2} x(\csc^2 x - 1)dx$

$$= \int \cot^{n-2} x \csc^2 x\, dx - \int \cot^{n-2} xdx$$

$$\int \cot^n x\, dx = \frac{\cot^{n-1} x}{n-1} - \int \cot^{n-2} x\, dx. \tag{9.4}$$

If n is odd, then $\int \cot x\, dx = \log \sin x$

and

If n is even, then $\int \cot^2 x\, dx = -\cot x - x$.

Reduction formula for $\int \sec^n x\, dx$.

Let $I = \int \sec^n x\, dx = \int \sec^{n-2} x \sec^2 x\, dx = \int \sec^{n-2} x\, d(\tan x)\, dx$.

Now, using integration by parts, we get

$$I = \tan x \sec^{n-2} x - \int \tan x (n-2) \sec^{n-3} x \sec x \tan x\, dx$$

$$= \tan x \sec^{n-2} x - (n-2) \int \sec^{n-2} x \tan^2 x\, dx$$

$$= \tan x \sec^{n-2} x - (n-2) \int \sec^{n-2} x \ (\sec^2 x - 1) dx$$

$$= \tan x \sec^{n-2} x - (n-2) \int \sec^n x\, dx - (n-2) \int \sec^{n-2} x\, dx.$$

$$\Rightarrow I(1+n-2) = \tan x \sec^{n-2} x + (n-2) \int \sec^{n-2} x\, dx.$$

$$\Rightarrow \qquad I = \frac{\tan x \ \sec^{n-2} x}{n-1} + \frac{n-2}{n-1} \int \sec^{n-2} x\, dx.$$

$$\int \sec^n x dx = \frac{\tan x \ \sec^{n-2} x}{n-1} + \frac{n-2}{n-1} \int \sec^{n-2} x\, dx. \tag{9.5}$$

If n is odd, then $I = \int \sec x\, dx = \log(\sec x + \tan x)$

and

If n is even, then $I = \int \sec^2 x\, dx = \tan x$.

Reduction formula for $\int \operatorname{cosec}^n x\,dx$.

Let $I = \int \operatorname{cosec}^n x\, dx = \int \operatorname{cosec}^{n-2} x \operatorname{cosec}^2 x\, dx = \int \operatorname{cosec}^{n-2} x\, d(-\cot x)\, dx.$

Now, using integration by parts, we get

$$I = -\cot x \operatorname{cosec}^{n-2} x - \int \cot x\,(n-2) \operatorname{cosec}^{n-3} x \operatorname{cosec} x \cot x\, dx$$

$$= -\cot x \operatorname{cosec}^{n-2} x - (n-2)\int \operatorname{cosec}^{n-2} x \cot^2 x\, dx$$

$$= -\cot x \operatorname{cosec}^{n-2} x - (n-2)\int \operatorname{cosec}^{n-2} x\ (\operatorname{cosec}^2 x - 1)\, dx$$

$$= -\cot x \operatorname{cosec}^{n-2} x - (n-2)\int \operatorname{cosec}^{n} x\, dx + (n-2)\int \operatorname{cosec}^{n-2} x\, dx.$$

$$\Rightarrow \qquad I(1+n-2) = -\cot x \operatorname{cosec}^{n-2} x + (n-2)\int \operatorname{cosec}^{n-2} x\, dx$$

$$\Rightarrow \qquad I = -\frac{\cot x\ \operatorname{cosec}^{n-2} x}{n-1} + \frac{n-2}{n-1}\int \operatorname{cosec}^{n-2} x\, dx.$$

$$\int \operatorname{cosec}^n x\, dx = -\frac{\cot x\ \operatorname{cosec}^{n-2} x}{n-1} + \frac{n-2}{n-1}\int \operatorname{cosec}^{n-2} x\, dx. \qquad (9.6)$$

If n is odd, then $I = \int \operatorname{cosec} x\, dx = \log \tan \frac{x}{2}$

and

If n is even, then $I = \int \operatorname{cosec}^2 x\, dx = -\cot x.$

Reduction formula for $\int \sin^m x \cos^n x\, dx$; $m, n > 0$.

We can write,

$$I = \int \sin^m x \cos^n x\, dx = \int \sin^{m-1} x \left(\sin x \cos^n x\right) dx.$$

Now, using integration by parts, we get

$$I = \int \sin^m x \cos^n x\, dx = -\frac{\cos^{n+1} x}{n+1} \sin^{m-1} x + \frac{m-1}{n+1}\int \cos^{n+1} x \sin^{m-2} x \cos x\, dx$$

$$= -\frac{\cos^{n+1} x}{n+1} \sin^{m-1} x + \frac{m-1}{n+1}\int \sin^{m-2} x \cos^n x \left(1-\sin^2 x\right) dx$$

$$= \frac{\cos^{n+1} x}{n+1} \sin^{m-1} x + \frac{m-1}{n+1}\int \sin^{m-2} x \cos^n x\, dx - \frac{m-1}{n+1}\int \sin^m x \cos^n x\, dx.$$

$$\Rightarrow \qquad I\left(1+\frac{m-1}{n+1}\right)=\frac{\cos^{n+1}x}{n+1}\sin^{m-1}x+\frac{m-1}{n+1}\int \sin^{m-2}x\cos^{n}x\,dx.$$

$$\Rightarrow \qquad I=\frac{\cos^{n+1}x}{m+n}\sin^{m-1}x+\frac{m-1}{m+n}\int \sin^{m-2}x\cos^{n}x\,dx.$$

$$\int \sin^{m}x\cos^{n}x\,dx=\frac{\cos^{n+1}x}{m+n}\sin^{m-1}x+\frac{m-1}{m+n}\int \sin^{m-2}x\cos^{n}x\,dx. \tag{9.7}$$

Similarly, we can find reduction formulae in reduced power of cos x by writing

$$\int \sin^{m}x\cos^{n}x\,dx=\int \sin^{m}x\cos^{n-1}x\cos x\,dx,$$

we get another reduction formulae as

$$\int \sin^{m}x\cos^{n}x\,dx=\frac{\sin^{m+1}x\cos^{n-1}x}{m+n}+\frac{n-1}{m+n}\int \sin^{m}x\cos^{n-2}x\,dx. \tag{9.8}$$

EXAMPLE 1 Evaluate $\int_0^{\pi/2}\sin^{m}x\cos^{n}x\,dx$, where m and n are positive integers.

Solution: Let $I_{m,n}=\int_0^{\pi/2}\sin^{m}x\cos^{n}x\,dx.$

Using the formula (9.7), we obtain

$$I_{m,n}=\left|\frac{\cos^{n+1}x}{m+n}\sin^{m-1}x\right|_0^{\frac{\pi}{2}}+\frac{m-1}{m+n}\int_0^{\frac{\pi}{2}}\sin^{m-2}x\cos^{n}x\,dx$$

$$=\frac{m-1}{m+n}\int_0^{\frac{\pi}{2}}\sin^{m-2}x\cos^{n}x\,dx.$$

Then $I_{m,n}=\dfrac{m-1}{m+n}I_{m-2,\,n}$. If we apply this recursive formula and replace by $(m-2)$, $(m-4)$, $(m-6)$, $4,3,2$, we get

$$I_{m-2,n}=\frac{m-3}{m+n-2}I_{m-4,\,n}.$$

$$I_{m-4,n}=\frac{m-5}{m+n-4}I_{m-6,\,n}.$$

$$\vdots$$

$$I_{3,n} = \frac{2}{3+n} I_{1,n}, \text{ if } m \text{ is odd.}$$

$$I_{2,n} = \frac{1}{2+n} I_{0,n}, \text{ if } m \text{ is even.}$$

where,

$$I_{1,n} = \int_0^{\frac{\pi}{2}} \sin^m x \cos^n x \, dx = -\left| \frac{\cos^{n+1} x}{n+1} \right|_0^{\frac{\pi}{2}} = \frac{1}{n+1}.$$

and

$$I_{0,n} = \int_0^{\frac{\pi}{2}} \cos^n x \, dx = \begin{cases} \dfrac{n-1}{n} \cdot \dfrac{n-3}{n-2} \cdot \dfrac{n-5}{n-4} \cdots \dfrac{2}{3}, & \text{if } n \text{ is odd} \\ \dfrac{n-1}{n} \cdot \dfrac{n-3}{n-2} \cdot \dfrac{n-5}{n-4} \cdots \dfrac{1}{2} \cdot \dfrac{\pi}{2}, & \text{if } n \text{ is even} \end{cases}.$$

Combining the above relations, we obtain

(a) If m is odd and n may be even or odd, then

$$I_{m,n} = \frac{m-1}{m+n} \cdot \frac{m-3}{m+n-2} \cdot \frac{m-5}{m+n-4} \cdots\cdots\cdots \frac{2}{3+n} \cdot \frac{1}{1+n}.$$

(b) If m is even and n is odd, then

$$I_{m,n} = \frac{m-1}{m+n} \cdot \frac{m-3}{m+n-2} \cdot \frac{m-5}{m+n-4} \cdots\cdots\cdots \frac{1}{2+n} \cdot \frac{n-1}{n} \cdot \frac{n-3}{n-2} \cdot \frac{n-5}{n-4} \cdots\cdots \frac{2}{3}.$$

(c) If m is even and n is even, then

$$I_{m,n} = \frac{m-1}{m+n} \cdot \frac{m-3}{m+n-2} \cdot \frac{m-5}{m+n-4} \cdots\cdots\cdots \frac{1}{2+n} \cdot \frac{n-1}{n} \cdot \frac{n-3}{n-2} \cdot \frac{n-5}{n-4} \cdots\cdots \frac{1}{2} \cdot \frac{\pi}{2}.$$

Integration of $x^n \sin mx$ and $x^n \cos mx$, where m and n are positive integers.

EXAMPLE 2 Evaluate $\int \sin^3 x dx$

Solution: Using reduction formulae (1), we obtain

$$\int \sin^3 x dx = -\frac{\cos x \sin^2 x}{3} - \frac{2}{3} \cos x + c$$

$$= -\frac{\cos x(1-\cos^2 x)}{3} - \frac{2}{3} \cos x + c$$

$$= -\frac{\cos x}{3} + \frac{\cos^3 x}{3} - \frac{2}{3}\cos x + c$$

$$= -\cos x + \frac{\cos^3 x}{3} + c.$$

9.2 AREA BETWEEN TWO CURVES

Theorem: Let $f(x)$ and $g(x)$ be the two continuous functions on the interval $[a,b]$ and $f(x) \geq g(x)$ for all x in $[a,b]$ (Figure 9.1). Then the area of the region bounded by curves $y = f(x)$ and $y = g(x)$ and ordinates $x = a,\ x = b$ is

$$A = \int_a^b [f(x) - g(x)]\,dx \tag{9.9}$$

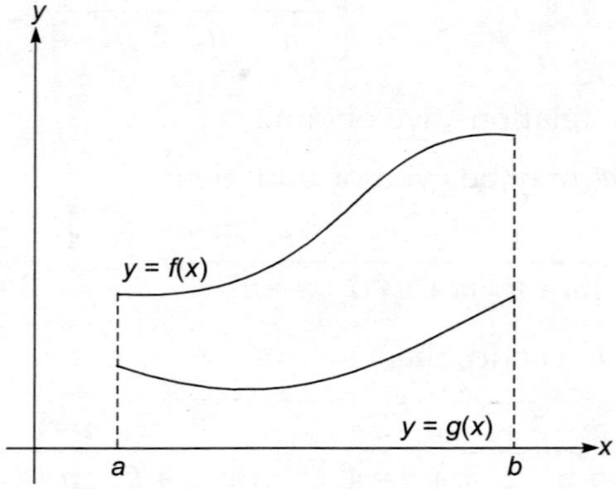

Figure 9.1

EXAMPLE 3 Find the area of the region enclosed by the curves $y = x^2$ and $y = 4x - 3$.

Solution: We know that the lower boundary is $y = x^2$ and the upper boundary is $y = 4x - 3$.

Now to find the value of ordinates, we equate

$$x^2 = 4x - 3 \text{or } x^2 - 4x + 3 = 0.$$

$\Rightarrow (x-1)(x-3) = 0 \Rightarrow x = 1, x = 3.$

Area enclosed between the curve, $A = \int_1^3 \left[(4x-3) - x^2\right]dx$

$$= \int_1^3 \left[-x^2 + 4x - 3\right]dx$$

$$= \left[-\frac{x^3}{3} + 2x^2 - 3x \right]_1^3 = \frac{4}{3}.$$

EXAMPLE 4 Find the area of the region that is enclosed between the parabolas $y = x^2$ and $y = 2x - x^2$.

Solution: First, we find the point of intersection of parabolas by solving the following equation,

$$x^2 = 2x - x^2 \Rightarrow x^2 - x = 0 \Rightarrow x = 0 \text{ or } x = 1.$$

Therefore, the points of intersection are $(0,0)$ and $(1,1)$.

The area enclosed by the parabolas, $A = \int_0^1 \left[\left(2x - x^2\right) - x^2 \right] dx$

$$= 2\int_0^1 \left(x - x^2\right) dx$$

$$= 2\left[\frac{x^2}{2} - \frac{x^3}{3} \right]_0^1 = 2\left[\frac{1}{2} - \frac{1}{3} \right] = \frac{1}{3}.$$

EXAMPLE 5 Find the area of the region that is enclosed between the curve $y^2(2a - x) = x^3$ and its asymptote.

Solution: From the equation of curve, we see that the curve is symmetrical about the x axis and passes through the origin. The equation $x = 2a$ is the asymptote of the curve.

So, the area enclosed by the curve and its asymptote,

$$A = 2\int_0^{2a} y dx = 2\int_0^{2a} \frac{x^{\frac{3}{2}}}{\sqrt{2a - x}} dx.$$

To solve the above integral, we put $x = 2a\sin^2\theta \Rightarrow dx = 4a\sin\theta\cos\theta$, we obtain

$$A = 2\int_0^{\frac{\pi}{2}} \frac{(2a)^{3/2} \sin^3\theta \, 4a\sin\theta\cos\theta}{(2a)^{\frac{1}{2}} \cos\theta} d\theta$$

$$= 16a^2 \int_0^{\frac{\pi}{2}} \sin^4\theta \, d\theta = 16a^2 \frac{3}{4}\frac{1}{2}\frac{\pi}{2} = 3\pi a^2.$$

EXAMPLE 6 Find the area of the region that is enclosed between the curve $xy^2 = 4a^2(2a - x)$ and its asymptote.

Solution: From the equation of curve, we see that the curve is symmetrical about the x axis and lies between $x = 0$ and $x = 2a$. The curve passes through the point $(2a, 0)$ and has the asymptote $x = 0$ (as shown in Figure 9.2).

So, the area enclosed by the curve and its asymptote,

$$A = 2\int_0^{2a} y\,dx = 2\int_0^{2a} 2a\sqrt{\frac{2a - x}{x}}\,dx.$$

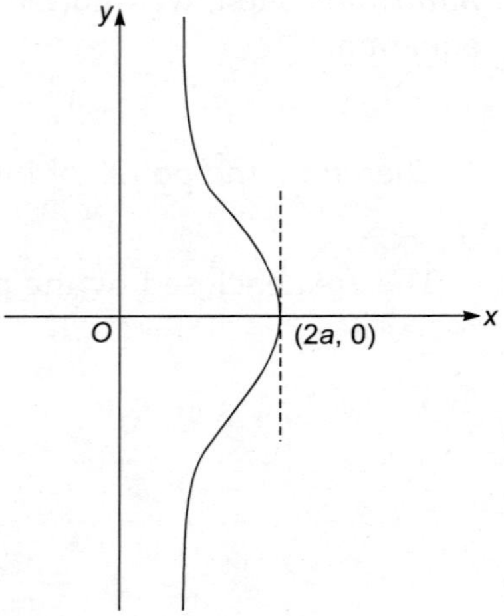

Figure 9.2

To solve the above integral, we put

$x = 2a\sin^2\theta \Rightarrow dx = 4a\sin\theta\cos\theta$, we obtain

$$A = 4a\int_0^{\frac{\pi}{2}} \sqrt{\frac{2a - 2a\sin^2\theta}{2a\sin^2\theta}}\, 4a\sin\theta\cos\theta\, d\theta$$

$$= 16a^2\int_0^{\frac{\pi}{2}} \cos^2\theta\, d\theta$$

$$= 16a^2\frac{1}{2}\frac{\pi}{2}$$

$$= 4\pi a^2.$$

9.3 VOLUMES BY SLICING: DISKS AND WASHERS

9.3.1 Volume by Slice

Let S be a solid bounded by two parallel planes perpendicular to the x-axis at $x = a$ and $= b$. If, for each $x \in [a, b]$, the cross-sectional area of S perpendicular to the x-axis is $A(x)$, then the **volume of the solid** is

$$V = \int_a^b A(x)\,dx, \text{ provided } A(x) \text{ is integrable.} \tag{9.10}$$

Let S be a solid bounded by two parallel planes perpendicular to the y-axis at $y = c$ and $= d$. If, for each $y \in [c, d]$, the cross-sectional area of S perpendicular to the y-axis is $A(y)$, then the **volume of the solid** is

$$V = \int_c^d A(y)\,dy, \text{ provided } A(y) \text{ is integrable.} \tag{9.11}$$

EXAMPLE 7 Find the volume of a right pyramid whose altitude is h and whose base is a square with sides of length a.

Solution: In Figure 9.3, we see that a rectangular coordinate system in which y-axis passes through the apex of pyramid and is perpendicular to the base and x-axis passes through the base and is parallel to a side of the base.

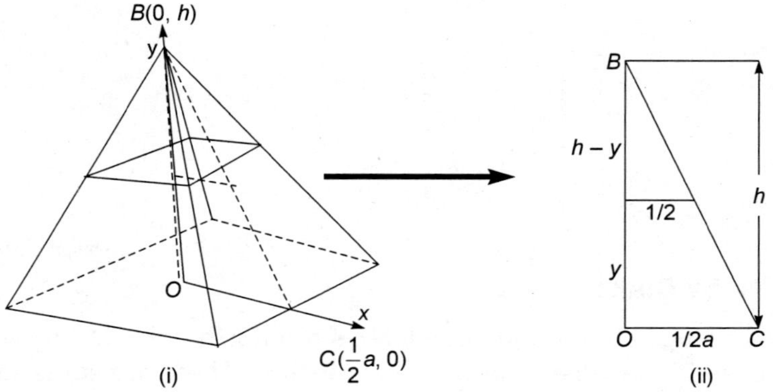

Figure 9.3

At any point y in the interval $[0,h]$ on the y-axis, the cross-sectional perpendicular to the y-axis is square. If s is the length of a side of triangle. Then by similar triangle theorem, we have

$$\frac{\frac{1}{2}s}{\frac{1}{2}a} = \frac{h-y}{h} \text{ or } s = \frac{a}{h}(h-y).$$

Thus, the area of cross section at y is

$$A(y) = s^2 = \left(\frac{a}{h}(h-y)\right)^2.$$

Using the volume formula, we obtain

$$V = \int_0^h A(y)dy$$

$$= \int_0^h \left(\frac{a}{h}(h-y)\right)^2 dy$$

$$= \frac{a^2}{h^2}\int_0^h (h-y)^2 dy$$

$$= \frac{a^2}{h^2}\int_0^h (h-y)^2\, dy$$

$$= \frac{a^2}{h^2}\left[-\frac{1}{3}(h-y)^3\right]_0^h$$

$$= \frac{1}{3}\frac{a^2}{h^2}h^3$$

$$= \frac{1}{3}a^2 h.$$

9.3.2 Volume by Disks

If we revolve a plane about a line, it generates a solid that lies in the same plane as the region and the solid is called a **solid of revolution.** There are some familiar solids of this type (Figure 9.4).

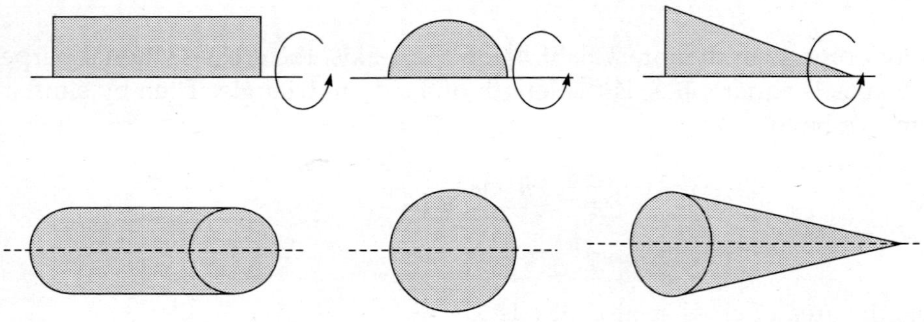

Figure 9.4

Let f be a continuous and non-negative on $[a,b]$ and R be the region that is bounded above by $y = f(x)$ and the lines $x = a$ and $x = b$ (Figure 9.5(a)). Then the volume of solid of revolution that is generated by revolving the region R about the x-axis is

$$V = \int_a^b \pi\left[f(x)\right]^2 dx. \tag{9.12}$$

Let g be a continuous and non-negative on $[c,d]$ and R be the region that is bounded above by $x = g(y)$ and the lines $y = c$ and $y = d$ (Figure 9.5(b)). Then the volume of solid of revolution that is generated by revolving the region R about the y-axis is

$$V = \int_c^d \pi\left[g(y)\right]^2 dy. \tag{9.13}$$

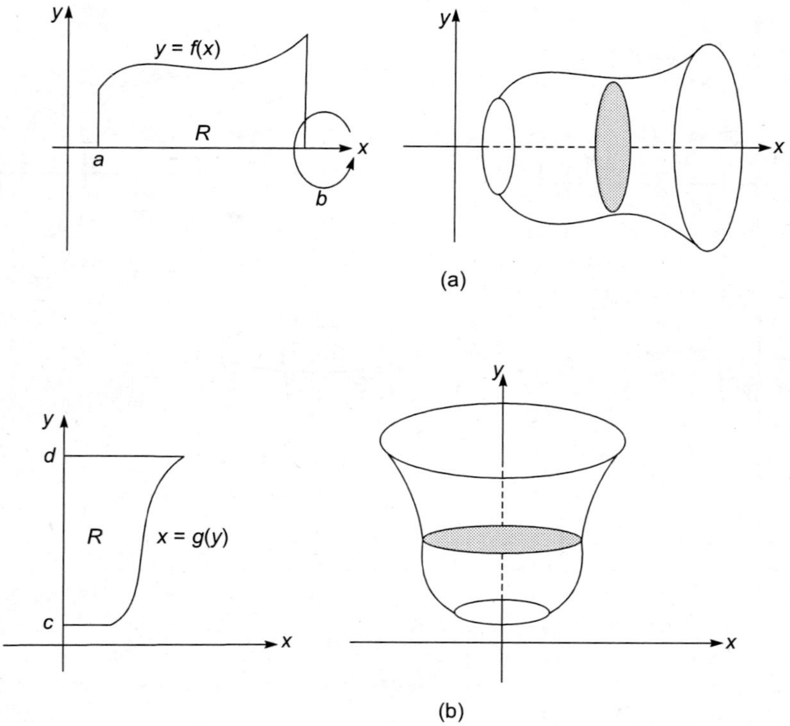

Figure 9.5

9.3.3 Volume by Washers

Let f and g be the two continuous and non-negative on $[a,b]$ and $f(x) \geq g(x)$ for all $x \in [a,b]$. Let R be the region that is bounded above by $y = f(x)$ and below by $y = g(x)$ and on the sides by the lines $x = a$ and $x = b$ (Figure 9.6(a)). Then the volume of solid of revolution that is generated by revolving the region R about the x-axis is

$$V = \int_a^b \pi\left(\left[f(x)\right]^2 - \left[g(x)\right]^2\right) dx. \tag{9.14}$$

Similarly, let w and u be the two continuous and non-negative on $[a,b]$ and $w(y) \geq u(y)$ for all $x \in [a,b]$. Let R be the region that is bounded above by $x = w(y)$ and below by $x = u(y)$ and on the sides by the lines $y = c$ and $y = d$ (Figure 9.6(b)). Then the volume of solid of revolution that is generated by revolving the region R about the y-axis is

$$V = \int_c^d \pi\left(\left[w(y)\right]^2 - \left[u(y)\right]^2\right) dy. \tag{9.15}$$

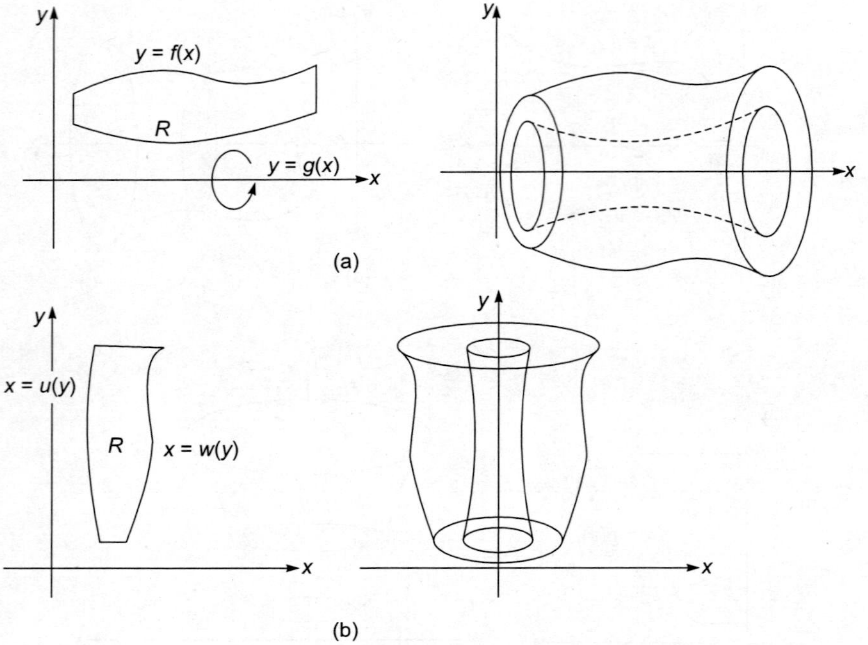

Figure 9.6

EXAMPLE 8 Find the volume of the solid generated when the region enclosed by the loop of the curve $2a\,y^2 = x(x-a)^2$ is revolved about the x-axis.

Solution: Use formula (9.12), because cross sections revolved about the x-axis. Then the volume of solid of revolution is given by

$$V = \int_c^d \pi\left[f(x)\right]^2 dy = \frac{1}{2a}\int_0^a \pi y^2 dx$$

$$= \frac{\pi}{2a}\int_0^a x(x-a)^2\, dx$$

$$= \frac{\pi}{2a}\int_0^a \left(x^3 - 2ax^2 + a^2x\right)dx$$

$$= \frac{\pi}{2a}\left[\frac{x^4}{4} - \frac{2a}{3}x^3 + \frac{a^2x^2}{2}\right]_0^a$$

$$= \frac{\pi}{2a}\left[\frac{a^4}{4} - \frac{2a^4}{3} + \frac{a^4}{2}\right] = \frac{\pi a^3}{24}.$$

EXAMPLE 9 Find the volume of the solid generated when the region enclosed by $y=\sqrt{x}$, $y=1$ and $x=0$ is revolved about the y-axis.

Solution: Use formula (9.13), because cross sections revolved about the y-axis. We can rewrite $y=\sqrt{x}$ as $x=y^2$. Then the volume of solid of revolution is given by

$$V=\int_c^d \pi\left[f(y)\right]^2 dy=\int_0^1 \pi y^4 dx=\left[\frac{\pi y^5}{5}\right]_0^1=\frac{\pi}{5}.$$

9.4 VOLUME OF CYLINDRICAL SHELLS

Let f be the continuous and non-negative on $[a,b]$, and R be the region that is bounded above by $y=f(x)$, below by x-axis, and on the sides by the lines $x=a$ and $x=b$ (Figure 9.7(a)). Then the volume of solid of revolution that is generated by revolving the region R about the y-axis is

$$V=\int_{x=a}^{x=b} 2\pi x\, f(x)\, dx \tag{9.16}$$

Similarly, let g be the continuous and non-negative on $[c,d]$, and R be the region bounded above by $x=g(y)$, below by y-axis, and on the sides by the lines $y=c$ and $x=d$ (Figure 9.7(b)). Then the volume of solid of revolution that is generated by revolving the region R about the x-axis is

$$V=\int_{y=c}^{y=d} 2\pi y\, g(y)\, dy \tag{9.17}$$

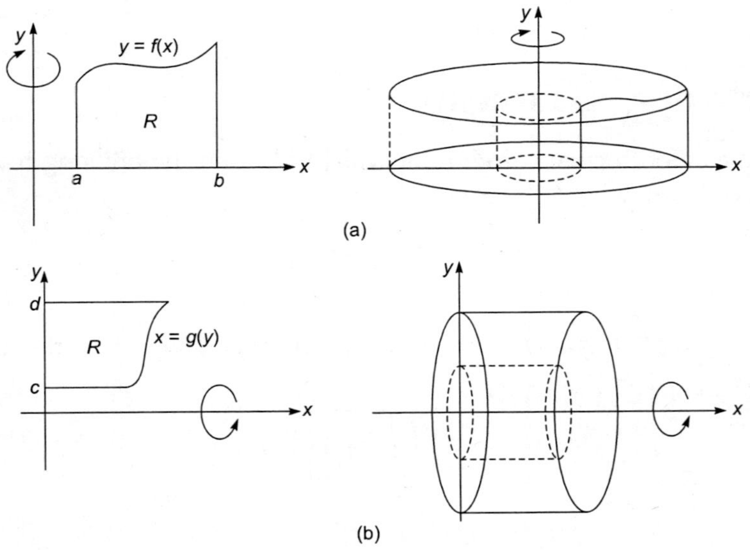

Figure 9.7

EXAMPLE 10 Use cylindrical shells to find the volume of the solid generated when the region enclosed between $y=\sqrt{x}$, $x=1$, $x=4$ and the x-axis is revolved about the y-axis.

Solution: Here $f(x)=\sqrt{x}$, $a=1$ and $b=4$.

By using formula (9.16), we get

$$V=\int_1^4 2\pi x\sqrt{x}\,dx=2\pi\int_1^4 x^{\frac{3}{2}}\,dx=\frac{4}{5}\pi\left[x^{\frac{5}{2}}\right]_1^4=\frac{4\pi}{5}[32-1]=\frac{124\pi}{5}.$$

EXAMPLE 11 Use the cylindrical shells method to find the volume of the solid generated when the region enclosed between $y=x^2$, $x=0$, $x=3$ and the x-axis is revolved about the y-axis.

Solution: Here $f(x)=x^2$, $a=0$ and $b=3$.

By using formula (9.16), we get

$$V=\int_0^3 2\pi x x^2\,dx=2\pi\int_0^3 x^3\,dx=\frac{2\pi}{4}\left[x^4\right]_0^3=\frac{81\pi}{2}.$$

EXAMPLE 12 Use the cylindrical shells method to find the volume of the solid generated when the region enclosed between $y=x^2$, $x=0$, $x=2$ and the x-axis is revolved about the x-axis.

Solution: Here we revolve about the x-axis. We must rewrite the $y=x^2$ as $=\sqrt{y}$.

By using formula (9.17), we get

$$V=\int_0^4 2\pi y\left(2-\sqrt{y}\right)dy=2\pi\int_0^4\left(2y-y^{\frac{3}{2}}\right)dy=2\pi\left[y^2-\frac{2}{5}y^{\frac{5}{2}}\right]_0^4=\frac{32\pi}{2}.$$

9.5 LENGTH OF A PLANE CURVE

If $y=f(x)$ is a smooth curve on the interval $[a,b]$, then the arc length L of this curve over $[a,b]$ is defined as

$$L=\int_a^b\sqrt{1+\left(\frac{dy}{dx}\right)^2}\,dx. \tag{9.18}$$

Similarly, if $x=g(y)$ is a smooth curve on interval $[c,d]$, then the arc length L of this curve over $[c,d]$ is

$$L=\int_c^d\sqrt{1+\left(\frac{dx}{dy}\right)^2}\,dy. \tag{9.19}$$

Next, when the equation of curve is in parametric form $x = x(t), y = y(t)$ with parameter $t \in [a,b]$, then the arc length of parametric curves is defined as

$$L = \int_a^b \sqrt{\left(\left(\frac{dx}{dt}\right)^2 + \left(\frac{dy}{dt}\right)^2\right)} dt. \tag{9.20}$$

Moreover, arc length of the curve $r = f(\theta)$ between the points $\theta = \alpha$ and $\theta = \beta$, is given by

$$L = \int_{\theta=\alpha}^{\theta=\beta} \sqrt{r^2 + \left(\frac{dr}{d\theta}\right)^2} d\theta.$$

If equation of the curve is of the form $\theta = f(r)$, then the arc length of the curve between the points $r = r_1$ and $r = r_2$ is given by

$$L = \int_{\theta=\alpha}^{\theta=\beta} \sqrt{r^2 + \left(\frac{dr}{d\theta}\right)^2} d\theta.$$

EXAMPLE 13 Find the arc length of the parabola $x^2 = 4ay$ intersected by its latus rectum.

Solution: From Figure 9.8, we see that latus rectum intersects the parabola at the points $(2a, a)$ and $(-2a, a)$.

Here, $x^2 = 4ay \Rightarrow y = \dfrac{x^2}{4a}$, then $\dfrac{dy}{dx} = \dfrac{x}{2a}$.

Use equation (9.18), the arc length of the parabola is

$$L = \int_{-2a}^{2a} \sqrt{1 + \left(\frac{dy}{dx}\right)^2} dx$$

$$= 2\int_0^{2a} \sqrt{1 + \left(\frac{x}{2a}\right)^2} dx$$

$$= \frac{2}{2a}\int_0^{2a} \sqrt{x^2 + (2a)^2} dx$$

$$= \frac{1}{a}\left[\frac{x\sqrt{x^2+4a^2}}{2} + \frac{4a^2}{2}\log\left(x + \sqrt{x^2+4a^2}\right)\right]_0^{2a}$$

$$= 2a\sqrt{2} + 2a\left(\log\left(2a + 2\sqrt{2}\,a\right) - \log 2a\right)$$

$$= 2a\,[\sqrt{2} + \log\left(1+\sqrt{2}\right)].$$

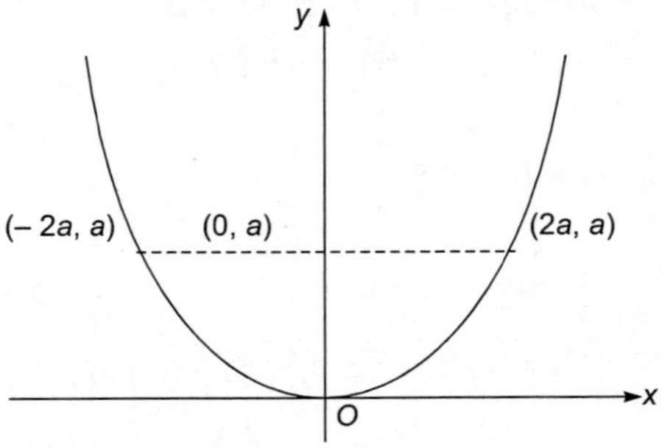

Figure 9.8

EXAMPLE 14 Find the arc length of the curve $y = 3x^{\frac{3}{2}} - 1$ from the $x = 0$ to $x = 1$.

Solution: Here, $y = 3x^{\frac{3}{2}} - 1 \Rightarrow \dfrac{dy}{dx} = \dfrac{9}{2}\sqrt{x}.$

Use equation (9.18), the arc length of the curve is

$$L = \int_0^1 \sqrt{1 + \left(\frac{9}{2}\sqrt{x}\right)^2}\, dx$$

$$= \int_0^1 \sqrt{\frac{81x + 4}{4}x}\, dx$$

$$= \frac{1}{2}\int_0^1 \sqrt{81x + 4}\, dx$$

$$= \frac{1}{2} \cdot \frac{1}{81}\left[\frac{(81x + 4)^{\frac{3}{2}}}{\frac{3}{2}}\right]_0^1$$

$$= \frac{1}{243}\left(85\sqrt{85} - 8\right).$$

EXAMPLE 15 Find the arc length of the loop of curve $3ay^2 = x(x - a)^2$.

Solution: From the figure of the curve $3ay^2 = x(x - a)^2$, we see that the loop intersects at points $(0, 0)$ and $(a, 0)$.

Here, $y = \dfrac{1}{\sqrt{3a}}\left[\sqrt{x}(x - a)\right] \Rightarrow \dfrac{dy}{dx} = \dfrac{3x - a}{2\sqrt{3ax}}.$

Use equation (9.18), the arc length of loop of the curve is

$$L = 2\int_0^a \sqrt{1 + \left(\frac{3x - a}{2\sqrt{3ax}}\right)^2}\, dx$$

$$= 2\int_0^a \sqrt{\frac{(3x + a)^2}{12ax}}\, dx$$

$$= 2\int_0^a \frac{3x + a}{2\sqrt{3ax}}\, dx$$

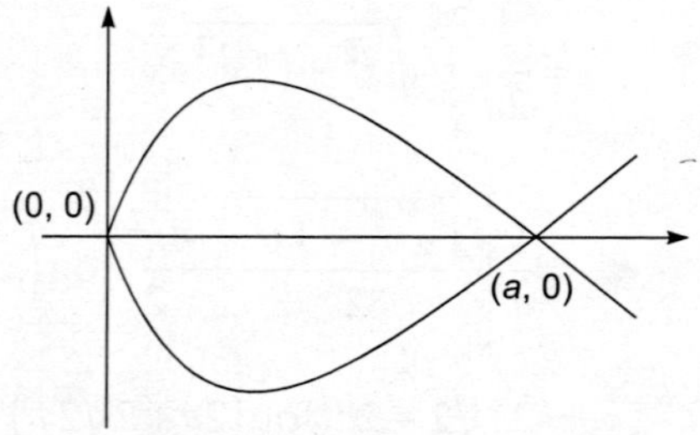

Figure 9.9

$$= \frac{1}{\sqrt{3a}} \int_0^a \left[3\sqrt{x} + \frac{a}{\sqrt{x}} \right] dx$$

$$= \frac{1}{\sqrt{3a}} \left[2x^{\frac{3}{2}} + 2ax^{\frac{1}{2}} \right]_0^a$$

$$= \frac{2}{\sqrt{3a}} \left[a^{\frac{3}{2}} + a a^{\frac{1}{2}} \right]$$

$$= \frac{4a^{\frac{3}{2}}}{\sqrt{3a}} = \frac{4a}{\sqrt{3}}.$$

EXAMPLE 16 Find the arc length of the four cusped hypocycloids whose parametric equations are $x = a\cos^3 \phi, y = a\sin^3 \phi$.

Solution: Here $x = a\cos^3 \phi, y = a\sin^3 \phi$, we get

$$\frac{dx}{d\phi} = -3a \cos^2 \phi \sin\phi \text{ and } \frac{dy}{d\phi} = 3a \sin^2\phi \cos\phi.$$

Then, the arc length of loop of the curve is

$$L = 4 \int_0^{\frac{\pi}{2}} \sqrt{\left(\frac{dx}{d\phi}\right)^2 + \left(\frac{dy}{d\phi}\right)^2} dx$$

$$= 4 \int_0^{\frac{\pi}{2}} \sqrt{\left(-3a \cos^2\phi \sin\phi\right)^2 + \left(3a \sin^2\phi \cos\phi\right)^2} dx$$

$$= 12a \int_0^{\frac{\pi}{2}} \sqrt{\cos^4\phi \sin^2\phi + \sin^4\phi \cos^2\phi} \, dx$$

$$= 6a \int_0^{\frac{\pi}{2}} 2 \cos\phi \sin\phi \, dx = 6a \int_0^{\frac{\pi}{2}} \sin 2\phi \, dx$$

$$= 6a \left[-\frac{1}{2} \cos 2\phi \right]_0^{\frac{\pi}{2}}$$

$$= 3a[-\cos\pi + \cos 0]$$

$$= 6a.$$

9.6 AREA OF SURFACE OF REVOLUTION

If we revolve a plane curve about an axis, then it generates a surface that lies in the same plane as the curve. This surface is called a surface of revolution. For example, if we revolve a circle about its diameter, it generates a surface of sphere and lateral surface of a right circular cylinder can be generated by revolving a line segment about an axis that is parallel to it (Figure 9.10).

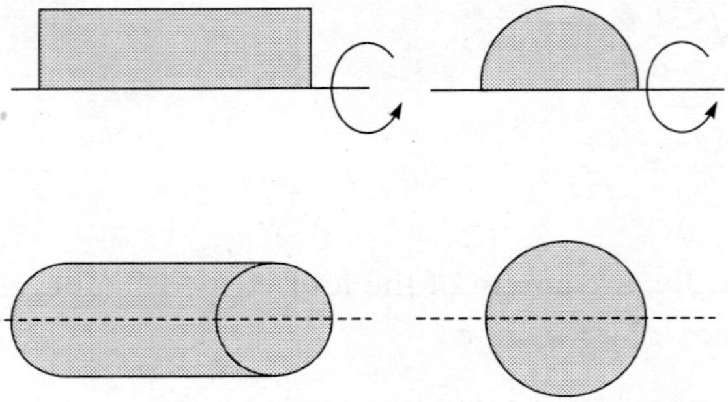

Figure 9.10

Definition 9.6.1

Let $f(x)$ be a smooth, non-negative function on $[a, b]$. The area of surface S of the surface of revolution that is generated by revolving the portion of the curve $y = f(x)$ between $x = a$ and $x = b$ about the x-axis is defined as

$$S = \int_{x=a}^{x=b} 2\pi y \sqrt{1+\left(\frac{dy}{dx}\right)^2}\, dx \tag{9.21}$$

Definition 9.6.2

Let $g(y)$ be a smooth, non-negative function on $[c, d]$. The area of surface S of the surface of revolution that is generated by revolving the portion of the curve $x = g(y)$ between $y = c$ and $y = d$ about the y-axis is defined as

$$S = \int_{y=c}^{y=d} 2\pi x \sqrt{1+\left(\frac{dx}{dy}\right)^2}\, dy \tag{9.22}$$

Moreover, when the equation of the curve is in parametric form $x = x(t)$, $y = y(t)$ with parameter $t \in [a, b]$, then the surface of revolution of parametric curves about x-axis defined as

$$S = \int_a^b 2\pi y \sqrt{\left(\left(\frac{dx}{dt}\right)^2 + \left(\frac{dy}{dt}\right)^2\right)}\, dt. \tag{9.23}$$

Further, when the equation of the curve is in polar form $r = f(\theta)$ with parameter $\theta \in [a, b]$, then the surface of revolution of the polar curves about x-axis defined as

$$S = \int_a^b 2\pi r \sin\theta \sqrt{\left(r^2 + \left(\frac{dr}{d\theta}\right)^2\right)}\, d\theta. \tag{9.24}$$

EXAMPLE 17 Find the area of the surface of revolution that is generated by revolving the upper half of the circle centred at the origin with radius a about x-axis.

Solution: The surface of a sphere is generated by revolving the upper half of the circle $x^2 + y^2 = a^2$ about the x-axis.

Since $x^2 + y^2 = a^2$, we have $2x + 2y\dfrac{dy}{dx} = 0 \Rightarrow \dfrac{dy}{dx} = -\dfrac{x}{y}$.

From equation (9.21), the area of surface S is

$$S = \int_{x=-a}^{x=a} 2\pi\, y \sqrt{1+\left(\frac{dy}{dx}\right)^2}\, dx$$

$$= \int_{x=-a}^{x=a} 2\pi\, y \sqrt{1+\left(-\frac{x}{y}\right)^2}\, dx$$

$$= 2\int_{x=0}^{x=a} 2\pi\, y \sqrt{\frac{x^2+y^2}{y^2}}\, dx$$

$$= 4\pi \int_{x=0}^{x=a} y\,\frac{a}{y}\, dx = 4\pi a^2.$$

EXAMPLE 18 Find the area of the surface that is generated by revolving the arc of the parabola $y^2 = 4x$ bounded by its latus rectum about x-axis.

Solution: Here, $y^2 = 4x \Rightarrow \dfrac{dy}{dx} = 1/\sqrt{x}$.

From equation (9.21), the area of surface S is

$$S = \int_{x=0}^{x=1} 2\pi\, y \sqrt{1+\left(\frac{dy}{dx}\right)^2}\, dx$$

$$= \int_{x=0}^{x=1} 2\pi\, y \sqrt{1+\left(1/\sqrt{x}\right)^2}\, dx$$

$$= 2\pi \int_{x=0}^{x=1} 2\sqrt{x}\,\frac{\sqrt{1+x}}{\sqrt{x}}\,dx$$

$$= 4\pi \int_{x=0}^{x=1} \sqrt{1+x}\,dx$$

$$= \frac{8\pi}{3}\left[(1+x)^{\frac{3}{2}}\right]_0^1 = \frac{8\pi}{3}\left[2\sqrt{2}-1\right].$$

EXAMPLE 19 Find the area of the surface that is generated by revolving the portion of the curve $y = x^2$ between $x = 1$ and $x = 3$ about y-axis.

Solution: The surface of revolution is generated by revolving the portion of curve $y = x^2$ between $x = 1$ and $x = 3$ about the y-axis (as shown in Figure 9.11).

Since $y = x^2$, we have $x = \sqrt{y} \Rightarrow \dfrac{dx}{dy} = \dfrac{1}{2\sqrt{y}}$.

From equation (9.22), the area of surface S is

$$S = \int_{y=1}^{y=9} 2\pi x\sqrt{1+\left(\frac{1}{2\sqrt{y}}\right)^2}\,dy$$

$$= \int_{y=1}^{y=9} 2\pi\sqrt{y}\sqrt{\frac{4y+1}{4y}}\,dy$$

$$= \int_{y=1}^{y=9} 2\pi\sqrt{y}\,\frac{1}{2\sqrt{y}}\sqrt{4y+1}\,dy$$

$$= \pi\int_{y=1}^{y=9} \sqrt{4y+1}\,dy = \frac{\pi}{4}\cdot\frac{2}{3}\cdot\left[(4y+1)^{\frac{3}{2}}\right]_1^9$$

$$= \frac{\pi}{6}\left[(37)^{\frac{3}{2}} - (5)^{\frac{3}{2}}\right] \approx 112.$$

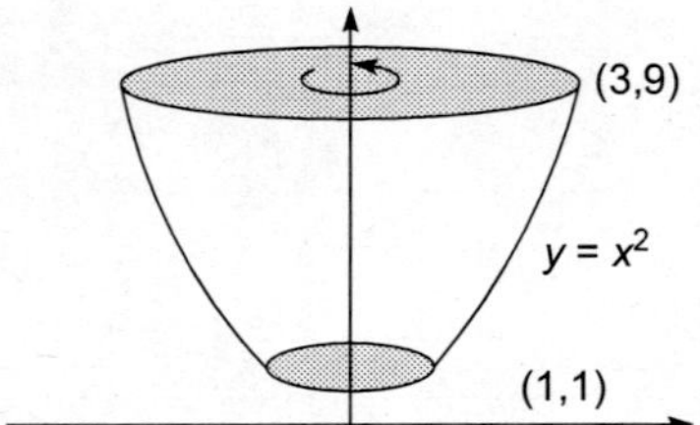

Figure 9.11

EXAMPLE 20 Find the area of the surface that is generated by revolving the portion of the curve $x = 9y + 1$ between $y = 0$ and $y = 2$ about y-axis.

Solution: From equation (9.22), the area of surface S is

$$S = \int_{y=0}^{y=2} 2\pi x \sqrt{1+(9)^2}\, dy\, S$$

$$= 2\pi \sqrt{82} \int_{y=0}^{y=2} (9y+1)\, dy$$

$$= 2\pi \sqrt{82} \left[\frac{9}{2}y^2 + y\right]_0^2$$

$$= 2\pi \sqrt{82}\,[20-0] = 40\pi \sqrt{82}.$$

EXAMPLE 21 Find the area of the surface of revolution that is generated by revolving the arc of the cycloid $x = a\phi + a\sin\phi, y = a + a\cos\phi$ about x-axis.

Solution: If the equation of curve is in parametric form $x = a\phi + a\sin\phi, y = a + a\cos\phi$ with parameter $\phi \in [-\pi, \pi]$(Figure 9.12), then the surface area of revolution of parametric curves about x-axis is defined as

$$S = \int_{-\pi}^{\pi} 2\pi y \sqrt{\left(\left(\frac{dx}{d\phi}\right)^2 + \left(\frac{dy}{d\phi}\right)^2\right)}\, d\phi.$$

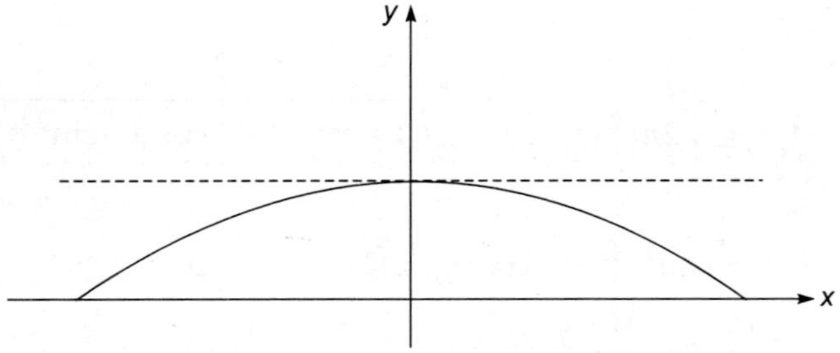

Cycloid: $x = a(\phi + \sin\phi)$, $y = a(1 + \cos\phi)$, $-\pi \le \phi \le \pi$

Figure 9.12

Here, $x = a\phi + a\sin\phi, y = a + a\cos\phi$, and $\dfrac{dx}{d\phi} = a + a\cos\phi, \dfrac{dy}{d\phi} = -a\sin\phi$.

So, we obtain

$$S = 2\int_{0}^{\pi} 2\pi (a + a\cos\phi) \sqrt{\left((a+a\cos\phi)^2 + (-a\sin\phi)^2\right)}\, d\phi$$

$$= 2\int_{0}^{\pi} 2\pi a^2 (1+\cos\phi) \sqrt{\left(1+\cos^2\phi + 2\cos\phi + \sin^2\phi\right)}\, d\phi$$

$$= 4\pi a^2 \int_0^{\pi} (1+\cos\phi)\sqrt{2(1+\cos\phi)}\, d\phi$$

$$= 4\pi a^2 \int_0^{\pi} \left(2\cos^2\frac{\phi}{2}\right)\sqrt{4\cos^2\frac{\phi}{2}}\, d\phi$$

$$= 16\pi a^2 \int_0^{\pi} \cos^3\frac{\phi}{2}\, d\phi = 32\pi a^2 \int_0^{\frac{\pi}{2}} \cos^3 t\, dt = \frac{64}{3}\pi a^2.$$

EXAMPLE 22 The equations $x = a\phi - a\sin\phi$ and $y = a - a\cos\phi$ $(0 \le \phi \le 2\pi)$ represent one arc of a cycloid. Show that the area of the surface of revolution that is generated by revolving the arc of about x-axis is $\frac{64}{3}\pi a^2$.

Solution: The area of surface of revolution of parametric curves about x-axis is defined as

$$S = \int_0^{2\pi} 2\pi y \sqrt{\left(\left(\frac{dx}{d\phi}\right)^2 + \left(\frac{dy}{d\phi}\right)^2\right)}\, d\phi.$$

Here, $x = a\phi - a\sin\phi, y = a - a\cos\phi$, and $\frac{dx}{d\phi} = a - a\cos\phi, \frac{dy}{d\phi} = a\sin\phi$.
So, we obtain

$$S = \int_0^{2\pi} 2\pi a^2 (1-\cos\phi)\sqrt{\left((1-\cos\phi)^2 + (\sin\phi)^2\right)}\, d\phi$$

$$= \int_0^{2\pi} 2\pi a^2 (1+\cos\phi)\sqrt{\left(1+\cos^2\phi - 2\cos\phi + \sin^2\phi\right)}\, d\phi$$

$$= 2\pi a^2 \int_0^{2\pi} (1-\cos\phi)\sqrt{2(1-\cos\phi)}\, d\phi$$

$$= 2\pi a^2 \int_0^{2\pi} \left(2\sin^2\frac{\phi}{2}\right)\sqrt{4\sin^2\frac{\phi}{2}}\, d\phi$$

$$= 8\pi a^2 \int_0^{2\pi} \sin^3\frac{\phi}{2}\, d\phi = 16\pi a^2 \int_0^{\pi} \sin^3 t dt = \frac{64}{3}\pi a^2.$$

EXAMPLE 23 Show that the area of the surface generated by revolving the curve $r = 2a\cos\theta$, about the initial line is $4\pi a^2$.

Solution: If we put $x = r\cos\theta, y = r\sin\theta$, then we have an equation of circle $x^2 + y^2 = 2ax$ centred at point $(a, 0)$ with radius a. Thus, the value of θ varies from 0 to $\frac{\pi}{2}$ in the upper half.

Then the area of surface about the initial line is

$$S = \int_{\theta=0}^{\frac{\pi}{2}} 2\pi r \sin\theta \sqrt{\left(r^2 + \left(\frac{dr}{d\theta}\right)^2\right)}\, d\theta.$$

Here, $$r = 2a\cos\theta \Rightarrow \frac{dr}{d\theta} = -2a\sin\theta.$$

So, we obtain

$$S = \int_{\theta=0}^{\frac{\pi}{2}} 2\pi\,(2a\cos\theta)\sin\theta \sqrt{\left((2a\cos\theta)^2 + (-2a\sin\theta)^2\right)}\, d\theta$$

$$= 4\pi a^2 \int_{\theta=0}^{\frac{\pi}{2}} 2\sin\theta\cos\theta\, d\theta$$

$$= 4\pi a^2 \int_{\theta=0}^{\frac{\pi}{2}} \sin 2\theta\, d\theta = 4\pi a^2 \left[-\frac{\cos 2\theta}{2}\right]_0^{\frac{\pi}{2}} = 4\pi a^2.$$

Exercise

1. If $I_{m,n} = \int_0^{\pi/2} \sin^m x \cos^n x\, dx$, where m and n are positive integers, then prove that $I_{m,n} = \frac{n-1}{m+n} I_{m,n-2}$.

2. Evaluate $\int \sin^5 x\, dx$ and also find $\int_0^{\frac{\pi}{2}} \sin^5 x\, dx$

3. Evaluate $\int \sin^6 x\, dx$ and also find $\int_0^{\frac{\pi}{2}} \sin^6 x\, dx$

4. If $u_n = \int_0^{\frac{\pi}{2}} x^n \sin x dx\ (n > 1)$, then show that $u_n + n(n-1)u_{n-2} = n\left(\frac{\pi}{2}\right)^{n-1}$. Also obtain u_5.

5. Evaluate $\int_0^{\frac{\pi}{2}} \sin^4 x \cos^5 x\, dx$

6. Find the area of the region enclosed by the curves $y = x^2$ and $y = 4x$ by integrating with respect to x and y.
7. Sketch the region enclosed by the following curves and find its area:

   (a) $y = x^2, y = \sqrt{x}, x = \frac{1}{4}, x = 1.$

   (b) $y = \cos 2x, y = 0, x = \frac{\pi}{4}, x = \frac{\pi}{2}.$

   (c) $x = \sin y, x = 0, y = \frac{\pi}{4}, y = \frac{3\pi}{4}.$

   (d) $y = 2 + |x - 1|, y = -\frac{1}{5}x + 7.$

8. Find the area between the curve $y = \sin x$ and the line segment joining the points (0, 0) and $\left(\frac{5\pi}{6}, \frac{1}{2}\right)$.
9. Find the area of the region enclosed by the parabola $y = 2x - x^2$ and x-axis.
10. Find the area of region enclosed between the cubic $y = x^3$ and the parabola $y = x^2$.
11. Show that the area of the region enclosed between the curve $y^2(a+x) = (a-x)^3$ and its asymptotes is $3\pi a^2$.
12. Show that common area included between the parabolas $y^2 = 4a(x+a)$ and $y^2 = 4b(b-x)$ is $\frac{8}{3}\sqrt{ab}\,(a+b)$ units.
13. Find the area of the region enclosed between the curve $x^{\frac{1}{2}} + y^{\frac{1}{2}} = a^{\frac{1}{2}}$ and the coordinate axis.
14. Show that the area enclosed by the ellipse $\frac{x^2}{a^2} + \frac{y^2}{b^2} = 1$ is πab.
15. Find the volume of solid that is obtained when the region under the following curve is revolved about the x-axis:

   (a) $y = x^2, x = 0, x = 2, y = 0.$

   (b) $y = \sqrt{\cos x}, x = \frac{\pi}{4}, x = \frac{\pi}{2}, y = 0.$

   (b) $y = \sqrt{25 - x^2}, y = 3.$

   (c) $x = \sqrt{y}, x = \frac{y}{4}.$

16. Find the volume of solid that is obtained when the region under the following curve is revolved about the y-axis:

(a) $y = x^3, x = 0, y = 1$.

(b) $x = \sqrt{1+y}, x = 0, y = 3$.

(c) $x = \text{cosec}\, y,\ y = \dfrac{\pi}{4},\ y = \dfrac{3\pi}{4},\ x = 0$.

(d) $x = y^2, x = y + 2$.

17. Find the volume of solid that is obtained when the region under the curve $y = \sqrt{x}$ over the interval [1, 4] is revolved about the x-axis.
18. Find the volume of solid that is obtained when the region under the curve $x = y^2$ and $x = y$ is revolved about the line $y = -1$.
19. Show that the volume of solid that is obtained when the region under the curve $y^2(2a - x) = x^2$ revolving about its asymptote is $2\pi^2 a^3$.
20. Find the volume of the solid generated when the region between the curves $y = \dfrac{1}{2} + x^2$ and $y = x$ over the interval [0, 2] is revolved about the x-axis.
21. Find the volume of the solid generated when the region enclosed by $y = \sqrt{x}$, $y = 2$ and $x = 0$ is revolved about the y-axis.
22. Find the volume of the solid generated when the region enclosed by the loop of the curve $y^2(a + x) = x^2(3a - x)$ is revolved about the x-axis.
23. Show that the volume of the solid generated when the region enclosed by the loop of the curve $y^4 = x(4 - x)$ revolving about the x-axis is $2\pi^2$.
24. Find the volume of the solid generated when the region above the x-axis and below the ellipse $\dfrac{x^2}{a^2} + \dfrac{y^2}{a^2} = 1\ (a > 0, b > 0)$ is revolved about the x-axis.
25. Use cylindrical shells to find the volume of the solid generated when the region in the first quadrant enclosed between $y = x$ and $y = x^2$ is revolved about the y-axis (figure below).

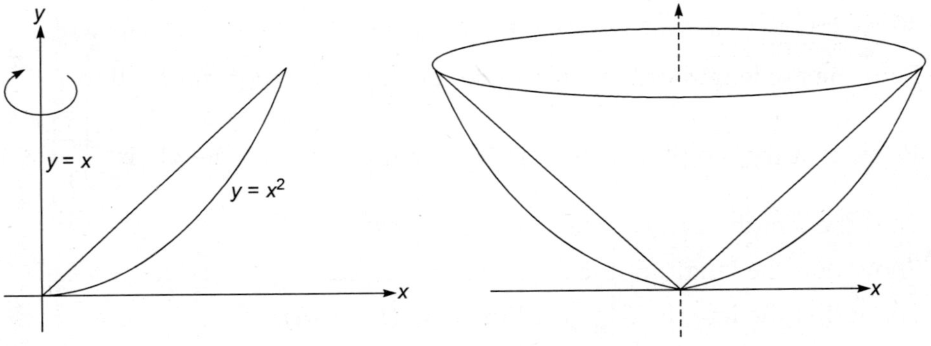

Figure

26. Use cylindrical shells to find the volume of the solid generated when the region enclosed by the following given curves is revolved about the y-axis:
    (a) $y = x^3$, $x = 1$, $y = 0$.
    (b) $y = \sqrt{x}$, $x = 4$, $x = 9$, $y = 0$.
    (c) $y = 2x - 1$, $y = -2x + 3$, $x = 2$.
    (d) $y = 2x - x^2$, $y = 0$.
27. Use cylindrical shells to find the volume of the solid generated when the region enclosed by the following given curves is revolved about the x-axis:
    (a) $y^2 = x$, $y = 1$, $x = 0$.
    (b) $y = x^2$, $x = 1$, $y = 0$.
    (c) $x = 2y$, $y = 2$, $y = 3$, $x = 0$.
28. Find the length of arc of $y = \sin x$ from $x = 0$ to $x = \pi$.
29. Find the arc length of the following curves over the stated interval:
    (a) $y = 3x^{\frac{3}{2}} - 1$, $0 \le x \le 1$.
    (b) $y = x^{\frac{2}{3}}$, $1 \le x \le 8$.
    (c) $y = \dfrac{x^6 + 8}{16x^2}$, $2 \le x \le 3$.
    (d) $24xy = y^4 + 48$, $2 \le y \le 4$.
30. Find the arc length of the following parametric curves over the stated interval:
    (a) $x = \dfrac{1}{3}t^3$, $y = \dfrac{1}{2}t^2$, $0 \le t \le 1$.
    (b) $x = \cos 2t$, $y = \sin 2t$; $0 \le t \le \dfrac{\pi}{2}$.
    (c) $x = \cos t + t\sin t$, $y = \sin t - t\cos t$; $0 \le t \le \pi$.
    (d) $x = (1+t)^2$, $y = (1+t)^3$; $0 \le t \le 1$.
31. Find the arc length of the curve $x = e^t \sin t$, $y = e^t \cos t$ form $t = 0$ to $t = \dfrac{\pi}{2}$.
32. Prove that the length of the loop of the curve $3ay^2 = x(x-a)^2$ is $\dfrac{4a}{\sqrt{3}}$.
33. Find the length of the curve $x^2\left(a^2 - x^2\right) = 8a^2y^2$.
34. Show that the length of the cycloid $x = a(1+\cos t)$, $y = a(1-\cos t)$ is $8a$.
35. Show that the length of the cardioid $r = a(1+\cos\theta)$ is $8a$.

36. Show that the length of the loop of the curve $9ay^2 = (x-2a)(x-5a)^2$ is $4\sqrt{3}a$.
37. Find the area of the surface generated by revolving the following curves about the x-axis:
    (a) $y = 7x,\ 0 \le x \le 1$.
    (b) $y = \sqrt{4-x^2},\ -1 \le x \le 1$.
    (c) $y = \sqrt{x} - \frac{1}{3}x^{\frac{3}{2}},\ 1 \le x \le 3$.
    (d) $y = \sin x,\ 0 \le x \le \pi$.
38. Find the area of the surface generated by revolving the following curves about the y-axis:
    (a) $x = 9y+1,\ 0 \le y \le 2$.
    (b) $x = \sqrt{9-y^2},\ -2 \le y \le 2$.
    (c) $8xy^2 = 2y^6 + 1,\ 1 \le y \le 2$.
39. Find the area of the surface generated by revolving the curve $x^2 + 4y^2 = 16$ about the x-axis.
40. Find the area of the surface generated by revolving the parametric curve $x(t) = t^2,\ y(t) = t - \frac{1}{3}t^3$ about the x-axis.
41. Find the area of the surface generated by revolving the parametric curve $x(t) = \cos^2 t,\ y(t) = 5\sin t$ about the x-axis.
42. Show that the area of the surface generated by revolving the curve $r = a(1+\cos\theta)$ (cardiode) about the initial line is $\frac{32}{5}\pi a^2$.

10 CHAPTER

Vector-Valued Functions

The real-valued functions have been studied so far. In this chapter, we will study the vector-valued functions whose domain is the set of real numbers but the range is as set of vectors. Such functions are basic tools for studying the motion of particles along the curve and to derive the Kepler's laws of planetary motion.

10.1 VECTOR-VALUED FUNCTIONS AND THEIR GRAPHS

In section 5.1 (chapter 5), we have described the parametric equations of a plane curves as

$$x = f(t) \text{ and } y = g(t), \tag{10.1}$$

where f and g are functions of a parameter t. The parametric equation (10.1) traces a curve in two-dimensional space, i.e., in $\mathbb{R}^2$. So, a curve in $\mathbb{R}^3$ is generated by the sets of all ordered triples $(f(t), g(t), h(t))$ satisfying the parametric equations $x = f(t)$, $y = g(t)$, $z = h(t)$, where f, g and h are functions of t on some domain D, which is a set of real numbers. The following definition of vector-valued functions is given below:

Definition: A vector-valued function (or a vector function) $\boldsymbol{F}$ is a function whose domain D is the set of real numbers and it assigns to each element t in D, a unique vector $\boldsymbol{F}(t)$. The range of $\boldsymbol{v} = \boldsymbol{F}(t)$ for $t \in D$, is the set of vectors $\boldsymbol{v}$. This means that for every real number $t \in D$, there is a unique vector $\boldsymbol{F}(t)$.

Thus, $\boldsymbol{F}(t) = f(t)\boldsymbol{i} + g(t)\boldsymbol{j}$ in two-dimensional space $\left(\text{in } \mathbb{R}^2\right)$

and $\boldsymbol{F}(t) = f(t)\boldsymbol{i} + g(t)\boldsymbol{j} + h(t)\boldsymbol{k}$ in three-dimensional space $\left(\text{in } \mathbb{R}^3\right)$

where f, g and h are real valued (or scalar-valued) functions of the real variable t in the domain D. These functions f, g and h are known as *components of a vector* $\boldsymbol{F}$.

The vector-valued function $\boldsymbol{F}(t)$ can also be written as $\boldsymbol{F}(t) = \langle f(t), g(t) \rangle$ in $\mathbb{R}^2$ and $\boldsymbol{F}(t) = f(t), g(t), h(t)$ in $\mathbb{R}^3$.

EXAMPLE 1 Find the domain of vector function $\boldsymbol{F}(t) = \langle t^2, \ln(2-t), \sqrt{t} \rangle$.

Solution: The components of the vector function $F(t)$ are:

$$f(t) = t^2,\ g(t) = \ln(2-t) \text{ and } h(t) = \sqrt{t}.$$

Now, $f(t) = t^2$ is defined for all values of t. But, $g(t) = \ln(2-t)$ and $h(t) = \sqrt{t}$ are defined for the values t when $2 - t > 0$ and $t \geq 0$, respectively. Therefore, the domain of the vector function $\boldsymbol{F}(t)$ is $(0, 2)$.

EXAMPLE 2 Find the domain of vector function $\boldsymbol{F}(t) = \langle \ln|t-1|, \sqrt{t}, e^t \rangle$.

Solution: The components of the vector function $F(t)$ are

$$f(t) = \ln|t-1|,\ g(t) = \sqrt{t} \text{ and } h(t) = e^t.$$

Now, $h(t) = e^t$ is defined for all values of t. But, $f(t) = \ln|t-1|$ and $g(t) = \sqrt{t}$ are defined for the values t when $t \in (-\infty, 1) \cup (1, \infty)$ and $t \geq 0$, respectively. Therefore, the domain of the vector function $\boldsymbol{F}(t)$ is $[0, 1) \cup (1, \infty)$.

10.1.1 Graph of Vector Functions

Let $\boldsymbol{F}(t) = f(t)\boldsymbol{i} + g(t)\boldsymbol{j} + h(t)\boldsymbol{k}$ be a vector function and suppose that its initial position is at the origin. Then the graph of $\boldsymbol{F}(t)$ is the curve C which is traced out by the terminal point of the vector $\boldsymbol{F}(t)$ as t varies over the domain D (as shown in Figure 10.1).

In this context, the vector $\boldsymbol{F}(t)$ is known as the *radius vector* or *position vector* of the point $P(f(t), g(t), h(t))$ on the curve C.

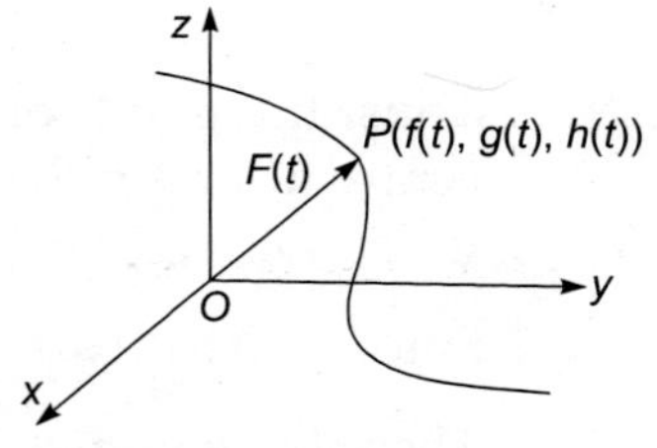

Figure 10.1

Note: For simplicity, we can use the notation $\boldsymbol{F}$ instead $\boldsymbol{F}(t)$ for the position vector as it is understood that the function $\boldsymbol{F}$ is dependent on t.

EXAMPLE 3 Sketch the graph of the following vector functions:

(a) $\boldsymbol{F}(t) = (2-t)\boldsymbol{i} + (2t)\boldsymbol{j} + (3t-6)\boldsymbol{k}$ for all t.

(b) $\boldsymbol{F}(t) = \cos t\boldsymbol{i} + \sin t\boldsymbol{j},\ t \in [0, 2\pi]$.

(c) $\boldsymbol{F}(t) = \cos t\boldsymbol{i} + \sin t\boldsymbol{j} + 3\boldsymbol{k}, t \in [0, 2\pi]$.

Solution:

(a) Since, $\boldsymbol{F}(t) = (2-t)\boldsymbol{i} + (2t)\boldsymbol{j} + (3t-6)\boldsymbol{k}$. Then, the parametric equations of $\boldsymbol{F}$ are $x = 2 - t$, $y = 2t$, $z = 3t - 6$, for all t.

$$v = -\boldsymbol{i} + 2\boldsymbol{j} + 3\boldsymbol{k}$$

Thus, the graph of the vector $\boldsymbol{F}$ is the line in $\mathbb{R}^3$ which passes through the point $P(2, 0, -6)$ and parallel to the vector $\boldsymbol{v} = -\boldsymbol{i} + 2\boldsymbol{j} + 3\boldsymbol{k}$ (as shown in Figure 10.2(a)).

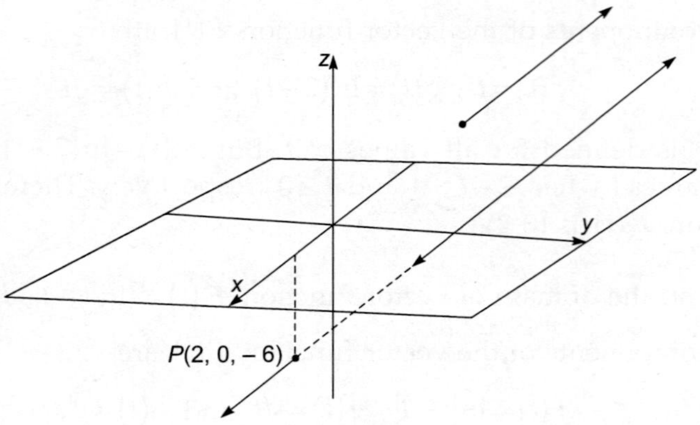

Figure 10.2(a)

(b) Since, $\boldsymbol{F}(t)=\cos t\boldsymbol{i}+\sin t\boldsymbol{j}$, $t\in[0,2\pi]$.

Then, the parametric equations of $\boldsymbol{F}$ are

$$x=\cos t,\ y=\sin t,\ 0\le t\le 2\pi.$$

Thus, the graph of the vector $\boldsymbol{F}$ is the circle in $\mathbb{R}^2$ with centre at origin and radius 1. The orientation of the circle is counterclockwise (as shown in Figure 10.2(b)).

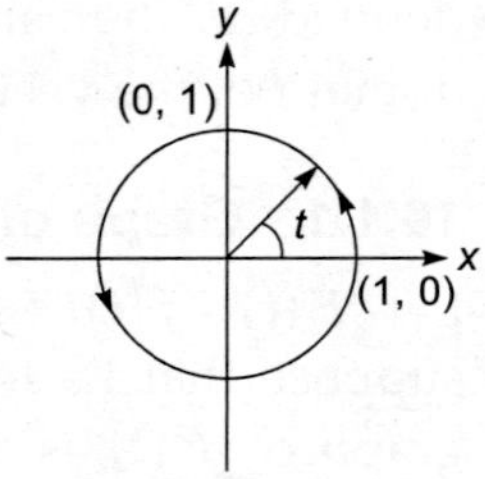

Figure 10.2(b)

(c) Since, $\boldsymbol{F}(t)=\cos t\boldsymbol{i}+\sin t\boldsymbol{j}+3\boldsymbol{k}, t\in[0,2\pi]$.

Then, the parametric equations of $\boldsymbol{F}$ are

$x=\cos t$, $y=\sin t$, $z=3$, $0\le t\le 2\pi$.

Thus, the graph of the vector $\boldsymbol{F}$ is the cylinder in $\mathbb{R}^3$ which is traced out by the curve in a plane $z=3$ and the circle centred at $(0,0,3)$ and radius 1. The orientation of the graph of $\boldsymbol{F}$ is counterclockwise looking down the z-axis (as shown in Figure 10.2(c)).

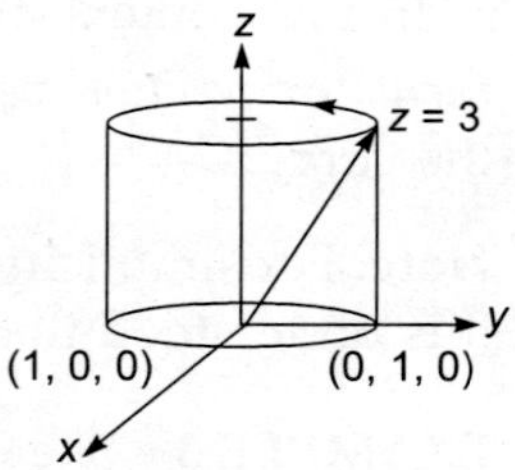

Figure 10.2(c)

EXAMPLE 4 Sketch the graph of the vector function:

$$\boldsymbol{F}(t)=\cos t\boldsymbol{i}+\sin t\boldsymbol{j}+2t\boldsymbol{k}.$$

Solution: Since, $\boldsymbol{F}(t)=\cos t\boldsymbol{i}+\sin t\boldsymbol{j}+2t\boldsymbol{k}$.

Then, the parametric equations of $\boldsymbol{F}$ are

$$x=\cos t,\ y=\sin t,\ z=2t, \text{ for all value of } t.$$

Since $x^2+y^2=\left(\cos^2 t+\sin^2 t\right)=1$, which means the graph of the vector function $\boldsymbol{F}$ must lies on the surface of right circular cylinder with radius 1 (as shown in Figure

10.3(a)). The axis of symmetry for the graph of $\boldsymbol{F}$ is z-axis. Since $z=2t$, the point (x,y,z) on the graph rises in a spiral upward counterclockwise around the cylinder $x^2+y^2=1$ as t increases (as shown in Figure 10.3(b)).

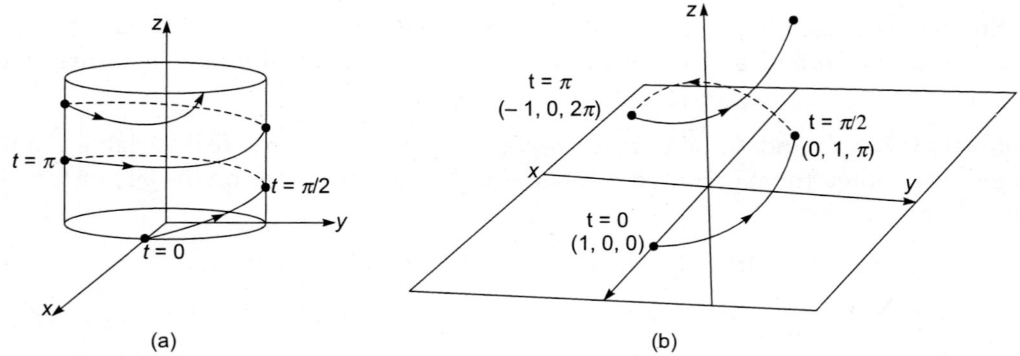

Figure 10.3

This graph is known as ***right circular helix***.

Remark: In Examples 3 and 4, the graph of vector function $\boldsymbol{F}$ is obtained by eliminating the parameter t. Now, we give some examples in which the graph of vector function $\boldsymbol{F}$ is the intersection of two curves.

EXAMPLE 5 Find the vector function $\boldsymbol{F}$ whose graph is the curve of intersection of the hemisphere $z=\sqrt{1-x^2-y^2}$ and the parabolic cylinder $y=x^2$.

Solution: Parameterizing the curve, by assuming $x=t$.

Since $y=x^2 \Rightarrow y=t^2$

and $\quad z=\sqrt{1-x^2-y^2}=\sqrt{1-t^2-t^4}$.

Therefore, the vector function is $\boldsymbol{F}(t)=t\boldsymbol{i}+t^2\boldsymbol{j}+\sqrt{1-t^2-t^4}\,\boldsymbol{k}$.

EXAMPLE 6 Find the vector function $\boldsymbol{F}$ whose graph is the curve in intersection of the cylinder $x^2+y^2=4$ and the plane $y+z=1$.

Solution: Since the projection of the curve C of intersection onto the xy-plane is $x^2+y^2=4$, $z=0$.

Then, the parametric equations are:

$$x=\cos t,\ y=\sin t,\ 0\le t\le 2\pi.$$

Since, $$y+z=1\Rightarrow z=1-y\Rightarrow z=1-\sin t.$$

Thus, the parametric equations of the curve C are:

$$x = \cos t,\ y = \sin t,\ z = 1 - \sin t,\ 0 \le t \le 2\pi.$$

Therefore, the vector function is $\boldsymbol{F}(t) = \cos t\boldsymbol{i} + \sin t\boldsymbol{j} + (1 - \sin t)\boldsymbol{k}$.

10.1.2 Operations with Vector Functions

In this section, we will study that the two vector functions $\boldsymbol{F}$ and $\boldsymbol{G}$ can be added, subtracted, multiplied and a vector function $\boldsymbol{F}$ can be multiplied by a real-valued function f.

Definition: Let $\boldsymbol{F}$ and $\boldsymbol{G}$ be two vector-valued functions of the real variable t and f be the real-valued function of the real variable t. The operations on the vector functions are defined as follows:

1. Addition: $(\boldsymbol{F} + \boldsymbol{G})(t) = \boldsymbol{F}(t) + \boldsymbol{G}(t)$, for all $t \in dom(\boldsymbol{F}) \cap dom(\boldsymbol{G})$.
2. Subtraction: $(\boldsymbol{F} - \boldsymbol{G})(t) = \boldsymbol{F}(t) - \boldsymbol{G}(t)$, for all $t \in dom(\boldsymbol{F}) \cap dom(\boldsymbol{G})$.
3. Scalar Multiplication: $(f\boldsymbol{F})(t) = f(t)\boldsymbol{F}(t)$, for all $t \in dom(\boldsymbol{F}) \cap dom(f)$.
4. Cross Product: $(\boldsymbol{F} \times \boldsymbol{G})(t) = \boldsymbol{F}(t) \times \boldsymbol{G}(t)$, for all $t \in dom(\boldsymbol{F}) \cap dom(\boldsymbol{G})$.
5. Dot Product: $(\boldsymbol{F}.\boldsymbol{G})(t) = \boldsymbol{F}(t).\boldsymbol{G}(t)$, for all $t \in dom(\boldsymbol{F}) \cap dom(\boldsymbol{G})$.

The first four operations on vector functions which give again a vector function, but the last operation is on two vector functions which give a scalar function.

EXAMPLE 7 Let $\boldsymbol{F}(t) = 2t\boldsymbol{i} - 5\boldsymbol{j} + t^2\boldsymbol{k}$ and $\boldsymbol{G}(t) = (1 - t)\boldsymbol{i} + \frac{1}{t}\boldsymbol{k}$. Then, find

(a) $(2\boldsymbol{F} - 3\boldsymbol{G})(t)$.
(b) $(t^2\boldsymbol{F} - \boldsymbol{G})(t)$.
(c) $(\boldsymbol{F}.\boldsymbol{G})(t)$.
(d) $(\boldsymbol{F} \times \boldsymbol{G})(t)$.
(e) $(\boldsymbol{F} + \boldsymbol{G})(t)$.

Solution:

(a) $(2\boldsymbol{F} - 3\boldsymbol{G})(t) = 2\boldsymbol{F}(t) - 3\boldsymbol{G}(t)$

$$= 2\left(2t\boldsymbol{i} - 5\boldsymbol{j} + t^2\boldsymbol{k}\right) - 3\left((1 - t)\boldsymbol{i} + \frac{1}{t}\boldsymbol{k}\right)$$

$$= (4t - 3 + 3t)\boldsymbol{i} - 10\boldsymbol{j} + \left(2t^2 - \frac{1}{t}\right)\boldsymbol{k}$$

$$= (7t - 3)\boldsymbol{i} - 10\boldsymbol{j} + \left(2t^2 - \frac{1}{t}\right)\boldsymbol{k}.$$

(b) $\left(t^2\boldsymbol{F}-\boldsymbol{G}\right)(t)=t^2\boldsymbol{F}(t)-\boldsymbol{G}(t)=t^2\left(2t\boldsymbol{i}-5\boldsymbol{j}+t^2\boldsymbol{k}\right)-\left((1-t)\boldsymbol{i}+\frac{1}{t}\boldsymbol{k}\right)$

$$=\left(2t^3+t-1\right)\boldsymbol{i}-5t^2\boldsymbol{j}+\left(t^4-\frac{1}{t}\right)\boldsymbol{k}.$$

(c) $(\boldsymbol{F}.\boldsymbol{G})(t)=\boldsymbol{F}(t).\boldsymbol{G}(t)=\left(2t\boldsymbol{i}-5\boldsymbol{j}+t^2\boldsymbol{k}\right).\left((1-t)\boldsymbol{i}+\frac{1}{t}\boldsymbol{k}\right)$

$$=2t(1-t)+t^2\cdot\frac{1}{t}$$

$$=3t-2t^2.$$

(d) $(\boldsymbol{F}\times\boldsymbol{G})(t)=\boldsymbol{F}(t)\times\boldsymbol{G}(t)=\left(2t\boldsymbol{i}-5\boldsymbol{j}+t^2\boldsymbol{k}\right)\times\left((1-t)\boldsymbol{i}+\frac{1}{t}\boldsymbol{k}\right)$

$$=\begin{vmatrix}\boldsymbol{i} & \boldsymbol{j} & \boldsymbol{k}\\ 2t & -5 & t^2\\ 1-t & 0 & 1/t\end{vmatrix}$$

$$=\left(-\frac{5}{t}\right)\boldsymbol{i}-\left(2-t^2+t^3\right)\boldsymbol{j}+(5-5t)\boldsymbol{k}.$$

(e) $(\boldsymbol{F}+\boldsymbol{G})(t)=\boldsymbol{F}(t)+\boldsymbol{G}(t)=\left(2t\boldsymbol{i}-5\boldsymbol{j}+t^2\boldsymbol{k}\right)+\left((1-t)\boldsymbol{i}+\frac{1}{t}\boldsymbol{k}\right)$

$$=(t+1)\boldsymbol{i}-5\boldsymbol{j}+\left(t^2+\frac{1}{t}\right)\boldsymbol{k}.$$

10.2 LIMITS AND CONTINUITY OF VECTOR-VALUED FUNCTIONS

Definition: Let $\boldsymbol{F}(t)=f(t)\boldsymbol{i}+g(t)\boldsymbol{j}+h(t)\boldsymbol{k}$ be a vector-valued function. Then, the limit of $F(t)$ as $t\to t_0$ is defined as

$$\lim_{t\to t_0}\boldsymbol{F}(t)=\left\{\lim_{t\to t_0}f(t)\right\}\boldsymbol{i}+\left\{\lim_{t\to t_0}g(t)\right\}\boldsymbol{j}+\left\{\lim_{t\to t_0}h(t)\right\}\boldsymbol{k},$$

provided that, the components of $\boldsymbol{F}$, i.e., f, g and h all have finite value as $t\to t_0$, where t is any real number or $\pm\infty$.

EXAMPLE 8 Find the limit of the following vector functions:

(a) $\lim\limits_{t\to 1}\left\{2t\boldsymbol{i}-3\boldsymbol{j}+e^t\boldsymbol{k}\right\}.$

(b) $\lim\limits_{t\to 0}\left\{\dfrac{(\sin t)\boldsymbol{i}-t\boldsymbol{k}}{t^2+t-1}\right\}.$

(c) $\lim_{t\to 0^+}\left\{\frac{\sin 3t}{(\sin 2t)}\boldsymbol{i}+\frac{\ln(\sin t)}{\ln(\tan t)}\boldsymbol{j}+t(\ln t)\boldsymbol{k}\right\}$.

Solution:

(a) $\lim_{t\to 1}\left\{2t\boldsymbol{i}-3\boldsymbol{j}+e^t\boldsymbol{k}\right\}=\left\{\lim_{t\to 1}2t\right\}\boldsymbol{i}-\left\{\lim_{t\to 1}3\right\}\boldsymbol{j}+\left\{\lim_{t\to 1}e^t\right\}\boldsymbol{k}.$

$$=2\boldsymbol{i}-3\boldsymbol{j}+e^1\boldsymbol{k}.$$

(b) $\lim_{t\to 0}\left\{\frac{(\sin t)\boldsymbol{i}-t\boldsymbol{k}}{t^2+t-1}\right\}=\left\{\lim_{t\to 0}\frac{\sin t}{t^2+t-1}\right\}\boldsymbol{i}-\left\{\lim_{t\to 0}\frac{t}{t^2+t-1}\right\}\boldsymbol{k}$

$$=0\boldsymbol{i}-0\boldsymbol{k}=\boldsymbol{0}.$$

(c) $\lim_{t\to 0^+}\left\{\frac{\sin 3t}{(\sin 2t)}\boldsymbol{i}+\frac{\ln(\sin t)}{\ln(\tan t)}\boldsymbol{j}+t(\ln t)\boldsymbol{k}\right\}$

$$=\left\{\lim_{t\to 0^+}\frac{\sin 3t}{(\sin 2t)}\right\}\boldsymbol{i}+\left\{\lim_{t\to 0^+}\frac{\ln(\sin t)}{\ln(\tan t)}\right\}\boldsymbol{j}+\left\{\lim_{t\to 0^+}t(\ln t)\right\}\boldsymbol{k}$$

(all terms are indeterminate forms except third term)

$$=\left\{\lim_{t\to 0^+}\frac{3\cos 3t}{2\cos 2t}\right\}\boldsymbol{i}+\left\{\lim_{t\to 0^+}\cos^2 t\right\}\boldsymbol{j}$$

(by using L' Hospital's rule)

$$=\frac{3}{2}\boldsymbol{i}+\boldsymbol{j}.$$

Theorem 10.1 (Algebra of Limits on Vector Functions)

Let $\boldsymbol{F}$ and $\boldsymbol{G}$ be two vector-valued functions and h be a scalar function of real variable t such that all the three functions have finite limits as $t\to t_0$.

Then, the following statements are hold:

1. Limit of a Sum: $\lim_{t\to t_0}\{\boldsymbol{F}(t)+\boldsymbol{G}(t)\}=\lim_{t\to t_0}\boldsymbol{F}(t)+\lim_{t\to t_0}\boldsymbol{G}(t)$.
2. Limit of a Difference: $\lim_{t\to t_0}\{\boldsymbol{F}(t)-\boldsymbol{G}(t)\}=\lim_{t\to t_0}\boldsymbol{F}(t)-\lim_{t\to t_0}\boldsymbol{G}(t)$.
3. Limit of Scalar Multiple: $\lim_{t\to t_0}\{h(t)\boldsymbol{F}(t)\}=\left\{\lim_{t\to t_0}h(t)\right\}\left\{\lim_{t\to t_0}\boldsymbol{F}(t)\right\}$.
4. Limit of a Dot Product: $\lim_{t\to t_0}\{\boldsymbol{F}(t).\boldsymbol{G}(t)\}=\left\{\lim_{t\to t_0}\boldsymbol{F}(t)\right\}.\left\{\lim_{t\to t_0}\boldsymbol{G}(t)\right\}$.
5. Limit of a Cross Product: $\lim_{t\to t_0}\{\boldsymbol{F}(t)\times\boldsymbol{G}(t)\}=\left\{\lim_{t\to t_0}\boldsymbol{F}(t)\right\}\times\left\{\lim_{t\to t_0}\boldsymbol{G}(t)\right\}$.

Proof: Let $F(t)=f_1(t)i+f_2(t)j+f_3(t)k$ and $G(t)=g_1(t)i+g_2(t)j+g_3(t)k$ be two vector functions such that all the components of F and G have finite values as: $t\to t_0$.

1. $F(t)+G(t)=\{f_1(t)i+f_2(t)j+f_3(t)k\}+\{g_1(t)i+g_2(t)j+g_3(t)k\}$

$$=\{f_1(t)+g_1(t)\}i+\{f_2(t)+g_2(t)\}j+\{f_3(t)+g_3(t)\}k.$$

$$\Rightarrow \lim_{t\to t_0}\{F(t)+G(t)\}=\lim_{t\to t_0}\{f_1(t)+g_1(t)\}i+\lim_{t\to t_0}\{f_2(t)+g_2(t)\}j+\lim_{t\to t_0}\{f_3(t)+g_3(t)\}k$$

$$=\left(\left\{\lim_{t\to t_0}f_1(t)\right\}+\left\{\lim_{t\to t_0}g_1(t)\right\}\right)i+\left(\left\{\lim_{t\to t_0}f_2(t)\right\}+\left\{\lim_{t\to t_0}g_2(t)\right\}\right)j+\left(\left\{\lim_{t\to t_0}f_3(t)\right\}+\left\{\lim_{t\to t_0}g_3(t)\right\}\right)k$$

$$=\left(\left\{\lim_{t\to t_0}f_1(t)\right\}i+\left\{\lim_{t\to t_0}f_2(t)\right\}j+\left\{\lim_{t\to t_0}f_3(t)\right\}k\right)+\left(\left\{\lim_{t\to t_0}g_1(t)\right\}i+\left\{\lim_{t\to t_0}g_2(t)\right\}j+\left\{\lim_{t\to t_0}g_3(t)\right\}k\right)$$

$$=\lim_{t\to t_0}F(t)+\lim_{t\to t_0}G(t)$$ (by using definition of limit).

2. Similarly, as we have done in (1) (proof is left for the reader).

3. $\lim_{t\to t_0}\{h(t)F(t)\}=\lim_{t\to t_0}\{h(t)(f_1(t)i+f_2(t)j+f_3(t)k)\}$

$$=\lim_{t\to t_0}\{h(t)f_1(t)i+h(t)f_2(t)j+h(t)f_3(t)k\}$$

$$=\lim_{t\to t_0}\{h(t)f_1(t)\}i+\lim_{t\to t_0}\{h(t)f_2(t)\}j+\lim_{t\to t_0}\{h(t)f_3(t)\}k$$

$$=\left(\left\{\lim_{t\to t_0}h(t)\right\}\left\{\lim_{t\to t_0}f_1(t)\right\}\right)i+\left(\left\{\lim_{t\to t_0}h(t)\right\}\left\{\lim_{t\to t_0}f_2(t)\right\}\right)j+\left(\left\{\lim_{t\to t_0}h(t)\right\}\left\{\lim_{t\to t_0}f_3(t)\right\}\right)k$$

$$= \left(\lim_{t \to t_0} h(t) \right) \left(\left\{ \lim_{t \to t_0} f_1(t) \right\} \boldsymbol{i} + \left\{ \lim_{t \to t_0} f_2(t) \right\} \boldsymbol{j} + \left\{ \lim_{t \to t_0} f_3(t) \right\} \boldsymbol{k} \right)$$

$$= \left(\lim_{t \to t_0} h(t) \right) \left(\lim_{t \to t_0} \boldsymbol{F}(t) \right).$$

4. $$\lim_{t \to t_0} \{ \boldsymbol{F}(t).\boldsymbol{G}(t) \} = \lim_{t \to t_0} \{ (f_1(t)\boldsymbol{i} + f_2(t)\boldsymbol{j} + f_3(t)\boldsymbol{k}).(g_1(t)\boldsymbol{i} + g_2(t)\boldsymbol{j} + g_3(t)\boldsymbol{k}) \}$$

$$= \lim_{t \to t_0} (f_1(t)g_1(t) + f_2(t)g_2(t) + f_3(t)g_3(t))$$

$$= \left(\left\{ \lim_{t \to t_0} f_1(t) \right\} \left\{ \lim_{t \to t_0} g_1(t) \right\} \right) + \left(\left\{ \lim_{t \to t_0} f_2(t) \right\} \left\{ \lim_{t \to t_0} g_2(t) \right\} \right)$$

$$+ \left(\left\{ \lim_{t \to t_0} f_3(t) \right\} \left\{ \lim_{t \to t_0} g_3(t) \right\} \right)$$

$$= \left(\left\{ \lim_{t \to t_0} f_1(t) \right\} \boldsymbol{i} + \left\{ \lim_{t \to t_0} f_2(t) \right\} \boldsymbol{j} + \left\{ \lim_{t \to t_0} f_3(t) \right\} \boldsymbol{k} \right)$$

$$. \left(\left\{ \lim_{t \to t_0} g_1(t) \right\} \boldsymbol{i} + \left\{ \lim_{t \to t_0} g_2(t) \right\} \boldsymbol{j} + \left\{ \lim_{t \to t_0} g_3(t) \right\} \boldsymbol{k} \right)$$

$$= \left(\lim_{t \to t_0} \boldsymbol{F}(t) \right) . \left(\lim_{t \to t_0} \boldsymbol{G}(t) \right).$$

5. $$\boldsymbol{F}(t) \times \boldsymbol{G}(t) = \begin{vmatrix} \boldsymbol{i} & \boldsymbol{j} & \boldsymbol{k} \\ f_1(t) & f_2(t) & f_3(t) \\ g_1(t) & g_2(t) & g_3(t) \end{vmatrix}$$

$$= (f_2(t)g_3(t) - f_3(t)g_2(t))\boldsymbol{i} + (f_3(t)g_1(t) - f_1(t)g_3(t))\boldsymbol{j}$$

$$+ (f_1(t)g_2(t) - f_2(t)g_1(t))\boldsymbol{k}.$$

$$\Rightarrow \lim_{t \to t_0} (\boldsymbol{F}(t) \times \boldsymbol{G}(t)) = \lim_{t \to t_0} \{ (f_2(t)g_3(t) - f_3(t)g_2(t))\boldsymbol{i} + (f_3(t)g_1(t) - f_1(t)g_3(t))\boldsymbol{j}$$

$$+ (f_1(t)g_2(t) - f_2(t)g_1(t))\boldsymbol{k} \}$$

$$=\left(\left\{\lim_{t\to t_0} f_2(t)\right\}\left\{\lim_{t\to t_0} g_3(t)\right\}-\left\{\lim_{t\to t_0} f_3(t)\right\}\left\{\lim_{t\to t_0} g_2(t)\right\}\right)\boldsymbol{i}$$

$$+\left(\left\{\lim_{t\to t_0} f_3(t)\right\}\left\{\lim_{t\to t_0} g_1(t)\right\}-\left\{\lim_{t\to t_0} f_1(t)\right\}\left\{\lim_{t\to t_0} g_3(t)\right\}\right)\boldsymbol{j}$$

$$+\left(\left\{\lim_{t\to t_0} f_1(t)\right\}\left\{\lim_{t\to t_0} g_2(t)\right\}-\left\{\lim_{t\to t_0} f_2(t)\right\}\left\{\lim_{t\to t_0} g_1(t)\right\}\right)\boldsymbol{k}. \tag{10.2}$$

Now, $\left(\lim_{t\to t_0}\boldsymbol{F}(t)\right)\times\left(\lim_{t\to t_0}\boldsymbol{G}(t)\right)$

$$=\begin{vmatrix}\boldsymbol{i} & \boldsymbol{j} & \boldsymbol{k}\\ \lim_{t\to t_0} f_1(t) & \lim_{t\to t_0} f_2(t) & \lim_{t\to t_0} f_3(t)\\ \lim_{t\to t_0} g_1(t) & \lim_{t\to t_0} g_2(t) & \lim_{t\to t_0} g_3(t)\end{vmatrix}$$

$$=\left(\left\{\lim_{t\to t_0} f_2(t)\right\}\left\{\lim_{t\to t_0} g_3(t)\right\}-\left\{\lim_{t\to t_0} f_3(t)\right\}\left\{\lim_{t\to t_0} g_2(t)\right\}\right)\boldsymbol{i}$$

$$+\left(\left\{\lim_{t\to t_0} f_3(t)\right\}\left\{\lim_{t\to t_0} g_1(t)\right\}-\left\{\lim_{t\to t_0} f_1(t)\right\}\left\{\lim_{t\to t_0} g_3(t)\right\}\right)\boldsymbol{j}$$

$$+\left(\left\{\lim_{t\to t_0} f_1(t)\right\}\left\{\lim_{t\to t_0} g_2(t)\right\}-\left\{\lim_{t\to t_0} f_2(t)\right\}\left\{\lim_{t\to t_0} g_1(t)\right\}\right)\boldsymbol{k}. \tag{10.3}$$

By (10.2) and (10.3), we have

$$\lim_{t\to t_0}\left(\boldsymbol{F}(t)\times\boldsymbol{G}(t)\right)=\left(\lim_{t\to t_0}\boldsymbol{F}(t)\right)\times\left(\lim_{t\to t_0}\boldsymbol{G}(t)\right).$$

EXAMPLE 9 Let $\boldsymbol{F}(t)=t\boldsymbol{i}+t^2\boldsymbol{j}+t^3\boldsymbol{k}$ and $\boldsymbol{G}(t)=\frac{1}{t}\boldsymbol{i}-e^t\boldsymbol{j}$ be two vector functions. Find the value of the following:

(a) $\lim_{t\to 0}\left(e^t\boldsymbol{F}(t)\right)$.

(b) $\lim_{t\to 1}\left(\boldsymbol{F}(t).\boldsymbol{G}(t)\right)$.

(c) $\lim_{t\to 1}\left(\boldsymbol{F}(t)\times\boldsymbol{G}(t)\right)$.

Solution:

(a) $\lim_{t\to 0}\left(e^t \boldsymbol{F}(t)\right) = \left(\lim_{t\to 0} e^t\right)\left(\lim_{t\to 0}\left\{t\boldsymbol{i} + t^2\boldsymbol{j} + t^3\boldsymbol{k}\right\}\right)$ (by using Theorem 10.1(3))

$$= \left(\lim_{t\to 0} e^t\right)\left(\left\{\lim_{t\to 0} t\right\}\boldsymbol{i} + \left\{\lim_{t\to 0} t^2\right\}\boldsymbol{j} + \left\{\lim_{t\to 0} t^3\right\}\boldsymbol{k}\right)$$

$$= (1)(0i + 0j + 0k) = 0.$$

(b) $\lim_{t\to 1}\left(\boldsymbol{F}(t).\boldsymbol{G}(t)\right) = \left(\lim_{t\to 1}\boldsymbol{F}(t)\right).\left(\lim_{t\to 1}\boldsymbol{G}(t)\right)$ (by using Theorem 10.1(4)).

$$= \left(\lim_{t\to 1}\left\{t\boldsymbol{i} + t^2\boldsymbol{j} + t^3\boldsymbol{k}\right\}\right).\left(\lim_{t\to 1}\left\{\frac{1}{t}\boldsymbol{i} - e^t\boldsymbol{j}\right\}\right)$$

$$= \left(\left\{\lim_{t\to 1} t\right\}\boldsymbol{i} + \left\{\lim_{t\to 1} t^2\right\}\boldsymbol{j} + \left\{\lim_{t\to 1} t^3\right\}\boldsymbol{k}\right).\left(\left\{\lim_{t\to 1}\left(\frac{1}{t}\right)\right\}\boldsymbol{i} - \left\{\lim_{t\to 1} e^t\right\}\boldsymbol{j}\right)$$

$$= (\boldsymbol{i} + \boldsymbol{j} + \boldsymbol{k}).\left(\boldsymbol{i} - e^1\boldsymbol{j}\right)$$

$$= 1 - e^1.$$

(c) $\lim_{t\to 1}\left(\boldsymbol{F}(t)\times\boldsymbol{G}(t)\right) = \left(\lim_{t\to 1}\boldsymbol{F}(t)\right)\times\left(\lim_{t\to 1}\boldsymbol{G}(t)\right)$ (by using Theorem 10.1(5))

$$= \left(\lim_{t\to 1}\left\{t\boldsymbol{i} + t^2\boldsymbol{j} + t^3\boldsymbol{k}\right\}\right)\times\left(\lim_{t\to 1}\left\{\frac{1}{t}\boldsymbol{i} - e^t\boldsymbol{j}\right\}\right)$$

$$= \left(\left\{\lim_{t\to 1} t\right\}\boldsymbol{i} + \left\{\lim_{t\to 1} t^2\right\}\boldsymbol{j} + \left\{\lim_{t\to 1} t^3\right\}\boldsymbol{k}\right)\times\left(\left\{\lim_{t\to 1}\left(\frac{1}{t}\right)\right\}\boldsymbol{i} - \left\{\lim_{t\to 1} e^t\right\}\boldsymbol{j}\right)$$

$$= (\boldsymbol{i} + \boldsymbol{j} + \boldsymbol{k})\times\left(\boldsymbol{i} - e^1\boldsymbol{j}\right)$$

$$= \begin{vmatrix} \boldsymbol{i} & \boldsymbol{j} & \boldsymbol{k} \\ 1 & 1 & 1 \\ 1 & -e^1 & 0 \end{vmatrix}$$

$$= e^1\boldsymbol{i} + \boldsymbol{j} - \left(e^1 + 1\right)\boldsymbol{k}.$$

10.2.1 Continuity of Vector-Valued Functions

Definition: Let $\boldsymbol{F}(t)$ be a vector function of real variable t. Then $\boldsymbol{F}$ is said to be *continuous* at t_0, if t_0 is in the domain of $\boldsymbol{F}$ and $\lim_{t\to t_0}\boldsymbol{F}(t) = \boldsymbol{F}(t_0)$.

or equivalently, a vector function $\boldsymbol{F}(t)=f(t)\boldsymbol{i}+g(t)\boldsymbol{j}+h(t)\boldsymbol{k}$ is *continuous* at t_0, if each of its components is continuous at t_0, i.e., a vector function $\boldsymbol{F}(t)=f(t)\boldsymbol{i}+g(t)\boldsymbol{j}+h(t)\boldsymbol{k}$ is *continuous* at t_0, if t_0 is in the domain of each of the components f, g and h, and

$$\lim_{t\to t_0} f(t)=f(t_0),\ \lim_{t\to t_0} g(t)=g(t_0) \text{ and } \lim_{t\to t_0} h(t)=h(t_0).$$

EXAMPLE 10 For what values of t, the following vector functions are continuous:

(a) $\boldsymbol{F}(t)=e^t\left\{t\boldsymbol{i}+\frac{1}{t}\boldsymbol{j}+3\boldsymbol{k}\right\}$

(b) $\boldsymbol{G}(t)=\frac{1}{t^2+t}\boldsymbol{i}+\frac{2}{t^2+t}\boldsymbol{j}$

Solution:

(a) The components of vector function $\boldsymbol{F}(t)=f(t)\boldsymbol{i}+g(t)\boldsymbol{j}+h(t)\boldsymbol{k}$ are

$$f(t)=te^t,\ g(t)=\frac{e^t}{t} \text{ and } h(t)=3e^t$$

The function f is continuous for all values of t; g is continuous for all values of $t\neq 0$ and h is continuous for all values of t. Therefore, the vector function F is continuous for all values of $t\neq 0$.

(b) The components of vector function $\boldsymbol{F}(t)=f(t)\boldsymbol{i}+g(t)\boldsymbol{j}+h(t)\boldsymbol{k}$ are

$$f(t)=\frac{1}{t^2+t},\ g(t)=\frac{2}{t^2+t} \text{ and } h(t)=0$$

The function f is continuous for all values of t, except at -1 and 0; g is also continuous for all values of t, except at -1 and 0; h is continuous function for all values of t, being a zero function. Therefore, the vector function F is continuous for all values of $t\neq 0,-1$.

10.3 DIFFERENTIATION AND INTEGRATION OF VECTOR-VALUED FUNCTIONS

In this section, we will study the derivatives of vector functions, tangent vectors, properties satisfied by the derivatives of vector functions, modeling the motion of an object in $\mathbb{R}^3$ and finally, the integration of vector functions.

10.3.1 Differentiation of Vector-Valued Functions

In chapter 3, we defined the derivative of real-valued function f w.r.t. x as

$$f'(x) = \lim_{h \to 0} \frac{f(x+h) - f(x)}{h}, \text{ provided the limit exists.}$$

Definition: Let $\boldsymbol{F}$ be a vector-valued function of real variable t, then the derivative of $\boldsymbol{F}$ w.r.t. t is defined as

$$\boldsymbol{F}'(t) = \lim_{\Delta t \to 0} \frac{\boldsymbol{F}(t+\Delta t) - \boldsymbol{F}(t)}{\Delta t}, \text{ provided the limit exists.}$$

The derivative of vector function $\boldsymbol{F}$ w.r.t t can be denoted as $\frac{d\boldsymbol{F}}{dt}$.

A vector function $\boldsymbol{F}$ is *differentiable* at $t = t_0$, if $\boldsymbol{F}'(t_0) = \lim_{t \to t_0} \frac{\boldsymbol{F}(t) - \boldsymbol{F}(t_0)}{t - t_0}$ exists finitely.

Theorem 10.2. Let $\boldsymbol{F}(t) = f(t)\boldsymbol{i} + g(t)\boldsymbol{j} + h(t)\boldsymbol{k}$ be a vector function. Then, $\boldsymbol{F}$ is differentiable w.r.t. t, if each of its components f, g and h are differentiable w.r.t. t and

$$\boldsymbol{F}'(t) = f'(t)\boldsymbol{i} + g'(t)\boldsymbol{j} + h'(t)\boldsymbol{k}.$$

Proof: Let us suppose that the scalar functions f, g and h are differentiable w.r.t. t, i.e., f', g' and h' exist finitely.

Now, $\boldsymbol{F}'(t) = \lim_{\Delta t \to 0} \frac{\boldsymbol{F}(t+\Delta t) - \boldsymbol{F}(t)}{\Delta t}$

$$= \lim_{\Delta t \to 0} \frac{\left(\left(f(t+\Delta t)\boldsymbol{i} + g(t+\Delta t)\boldsymbol{j} + h(t+\Delta t)\boldsymbol{k}\right) - \left(f(t)\boldsymbol{i} + g(t)\boldsymbol{j} + h(t)\boldsymbol{k}\right)\right)}{\Delta t}$$

$$= \left\{\lim_{\Delta t \to 0} \frac{f(t+\Delta t) - f(t)}{\Delta t}\right\}\boldsymbol{i} + \left\{\lim_{\Delta t \to 0} \frac{g(t+\Delta t) - g(t)}{\Delta t}\right\}\boldsymbol{j}$$

$$+ \left\{\lim_{\Delta t \to 0} \frac{h(t+\Delta t) - h(t)}{\Delta t}\right\}\boldsymbol{k}.$$

Therefore, $\boldsymbol{F}'(t) = f'(t)\boldsymbol{i} + g'(t)\boldsymbol{j} + h'(t)\boldsymbol{k}$.

EXAMPLE 11 Find the derivative of vector function $\boldsymbol{F}(t) = (\ln t)\left(t\boldsymbol{i} + 5\boldsymbol{j} - e^t\boldsymbol{k}\right)$.

Solution: Since $\boldsymbol{F}(t) = (\ln t)\left(t\boldsymbol{i} + 5\boldsymbol{j} - e^t\boldsymbol{k}\right)$.

So, $f(t) = (\ln t)t$, $g(t) = 5(\ln t)$ and $h(t) = -e^t(\ln t)$.

$\Rightarrow f'(t) = 1 + \ln t$, $g'(t) = \frac{5}{t}$ and $h'(t) = -\left(\frac{e'}{t} + e^t(\ln t)\right)$.

Thus, $\boldsymbol{F}'(t)=(1+\ln t)\boldsymbol{i}+\left(\frac{5}{t}\right)\boldsymbol{j}-\left(\frac{e'}{t}+e^{t}(\ln t)\right)\boldsymbol{k}$.

EXAMPLE 12 Find the value of t for which the vector function $\boldsymbol{F}(t)=|t|\boldsymbol{i}+\sin t\boldsymbol{j}+(t-1)\boldsymbol{k}$ is differentiable.

Solution: Since $\boldsymbol{F}(t)=|t|\boldsymbol{i}+\sin t\boldsymbol{j}+(t-1)\boldsymbol{k}$.

So, $f(t)=|t|$, $g(t)=\sin t$ and $h(t)=(t-1)$.

Then $g(t)$ and $h(t)$ are differentiable for all values of t, but $f(t)$ is not differentiable at $t=0$.

Therefore, the vector function $\boldsymbol{F}(t)$ is differentiable for all values of $t\neq 0$.

10.3.2 Tangent Vectors

We know that $f'(x_0)$ represents a slope of the tangent line to the graph of a scalar function $f(x)$ at the point $x=x_0$ and that provides a measure of the direction of the graph of $f(x)$ at that point. So, in this section, we extend this concept to the vector function $\boldsymbol{F}$ in space and it will be shown that the derivatives of a vector function $\boldsymbol{F}$ is the tangent vector to the graph of $\boldsymbol{F}$ in space.

Let $\boldsymbol{F}(t)$ be the vector function of t. Let P and Q be two points on the graph of $\boldsymbol{F}$ corresponding to $t=t_0$ and $t=t_0+\Delta t$, respectively (as shown in Figure 10.4).

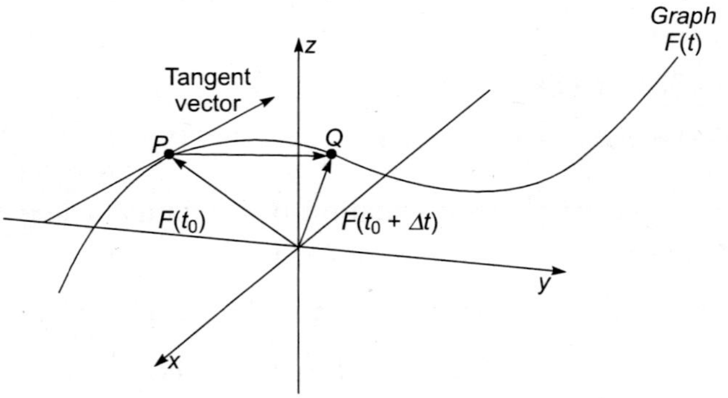

Figure 10.4

Then, both the vector $\boldsymbol{PQ}=\boldsymbol{F}(t_0+\Delta t)-\boldsymbol{F}(t_0)$ and, the difference quotient (which is also a vector)

$$\frac{\Delta \boldsymbol{F}}{\Delta t}=\frac{\boldsymbol{F}(t_0+\Delta t)-\boldsymbol{F}(t_0)}{\Delta t}$$

for any positive number Δt, have same direction.

Suppose that the difference quotient $\frac{\Delta \boldsymbol{F}}{\Delta t}$ has a limit $\Delta t\to 0$ such that $\lim_{\Delta t\to 0}\frac{\Delta \boldsymbol{F}}{\Delta t}\neq 0$.

Then, on taking limit $\Delta t \to 0$, then directions of the vector $\boldsymbol{PQ}$ and of the difference quotient $\frac{\Delta \boldsymbol{F}}{\Delta t}$, will tend to the direction of the tangent vector at P.

Therefore, the tangent vector at P is the limit vector $\lim_{\Delta t \to 0} \frac{\Delta \boldsymbol{F}}{\Delta t}$ and it is represented by $\frac{d\boldsymbol{F}}{dt}$ or $\boldsymbol{F}'(t_0)$.

So, from the above discussion, we have the following interpretation for the tangent vector to the graph of the vector function $\boldsymbol{F}$:

"Let $\boldsymbol{F}(t)$ be a differentiable vector function such that $\boldsymbol{F}'(t_0) \neq 0$. Then $\boldsymbol{F}'(t_0)$ is a tangent vector to the graph of $\boldsymbol{F}(t)$ at the point $t = t_0$ and is in the direction of increasing t".

EXAMPLE 13 Find the tangent vector to the graph of vector function

$$\boldsymbol{F}(t) = t^3\boldsymbol{i} + 2t\boldsymbol{j} + \left(t^3 + t^2\right)\boldsymbol{k}$$

at the point $t = 0, 1, -1$.

Solution: The derivative of a vector function $\boldsymbol{F}(t)$ is given by

$$\boldsymbol{F}'(t) = 3t^2\boldsymbol{i} + 2\boldsymbol{j} + \left(3t^2 + 2t\right)\boldsymbol{k}.$$

Thus, the tangent vector,

At $t = 0$, $\boldsymbol{F}'(0) = 2\boldsymbol{j}$.

At $t = 1$, $\boldsymbol{F}'(1) = 3\boldsymbol{i} + 2\boldsymbol{j} + 5\boldsymbol{k}$.

At $t = -1$, $\boldsymbol{F}'(-1) = 3\boldsymbol{i} + 2\boldsymbol{j} + \boldsymbol{k}$.

EXAMPLE 14 Find the parametric equation of the tangent line to the graph of the vector function $\boldsymbol{F}(t) = \sin t\boldsymbol{i} + \cos t\boldsymbol{j} + 3t\boldsymbol{k}$ at $t = 0$.

Solution: The derivative of the vector function $F(t)$ is given by

$$\boldsymbol{F}'(t) = \cos t\boldsymbol{i} - \sin t\boldsymbol{j} + 3\boldsymbol{k}.$$

Thus, the tangent vector at $t = 0$ is

$$\boldsymbol{F}'(0) = \boldsymbol{i} + 3\boldsymbol{k}.$$

The tangent line to the graph of $\boldsymbol{F}(t)$ at $t = 0$ is the line that passes through $t = 0$ and is parallel to the vector $\boldsymbol{F}(0)$. Since

$$\boldsymbol{F}(0) = j.$$

Thus, the tangent has parametric equation:

$$x = t,\ y = 1 \text{ and } z = 3t.$$

Smooth Curve: The graph of the vector function $\boldsymbol{F}(t)$ is said to be *smooth* on any interval of t, where $\boldsymbol{F}'$ is continuous and $\boldsymbol{F}'(t) \neq 0$. The graph is said to be *piecewise smooth* on an interval, if it can be subdivided into a finite number of subintervals on which $\boldsymbol{F}$ is smooth.

EXAMPLE 15 Determine whether the vector function $\boldsymbol{F}(t) = (e^t - t)\boldsymbol{i} + t^{\frac{3}{2}}\boldsymbol{j} + \cos t\boldsymbol{k}$ is smooth over the interval $[-1,2]$.

Solution: The derivative of a vector function $\boldsymbol{F}(t)$ is

$$\boldsymbol{F}'(t) = \left(e^t - 1\right)\boldsymbol{i} + \frac{3}{2}t^{1/2}\boldsymbol{j} - \sin t\boldsymbol{k}.$$

Thus, $\boldsymbol{F}'(t) \neq 0$, for all values of $0 \neq t \in [-1,2]$, but $\boldsymbol{F}'(0) = 0$. Therefore, the given vector function is not smooth over the interval $[-1,2]$. However, the given vector function is piecewise smooth as it consists of a finite number of smooth pieces.

10.3.3 Properties of Vector Derivatives

In chapter 4, we have studied higher order derivatives of real-valued functions. So, higher order of vector-valued function $\boldsymbol{F}(t)$ of real variable t can also be obtained by successively differentiating each of its components.

Let $\boldsymbol{F}(t) = f(t)\boldsymbol{i} + g(t)\boldsymbol{j} + h(t)\boldsymbol{k}$ be the vector function of real variable t. Then, the second order derivative of $\boldsymbol{F}(t)$ w.r.t. t is given as

$$\boldsymbol{F}''(t) = \{\boldsymbol{F}'(t)\}' = f''(t)\boldsymbol{i} + g''(t)\boldsymbol{j} + h''(t)\boldsymbol{k}.$$

The second order derivative of $\boldsymbol{F}(t)$ w.r.t. t is also denoted by $\dfrac{d^2\boldsymbol{F}}{dt^2}$.

Similarly, the third order derivative of $\boldsymbol{F}(t)$ w.r.t. t is given as

$$\boldsymbol{F}'''(t) = \{\boldsymbol{F}''(t)\}' = f'''(t)\boldsymbol{i} + g'''(t)\boldsymbol{j} + h'''(t)\boldsymbol{k}$$

and also denoted by $\dfrac{d^3\boldsymbol{F}}{dt^3}$.

EXAMPLE 16 Find the second and third order derivatives of the following vector functions:

(a) $\boldsymbol{F}(t) = \sin^2 t\boldsymbol{i} + \cos 2t\boldsymbol{j} + t^2\boldsymbol{k}$.

(b) $\boldsymbol{F}(t) = \left(1 - 2t^2\right)\boldsymbol{i} + (t\cos t)\boldsymbol{j} - t\boldsymbol{k}$.

Solution:

(a) $\boldsymbol{F}(t) = \sin^2 t\boldsymbol{i} + \cos 2t\boldsymbol{j} + t^2\boldsymbol{k}$.

$$\Rightarrow \boldsymbol{F}'(t) = \sin 2t\boldsymbol{i} - 2\sin 2t\boldsymbol{j} + 2t\boldsymbol{k}.$$

$$\Rightarrow \boldsymbol{F}''(t) = 2\cos 2t\boldsymbol{i} - 4\cos 2t\boldsymbol{j} + 2\boldsymbol{k}.$$

$$\Rightarrow \boldsymbol{F}'''(t) = -4\sin 2t\boldsymbol{i} + 8\sin 2t\boldsymbol{j} + 2\boldsymbol{k}.$$

(b) $\boldsymbol{F}(t) = \left(1 - 2t^2\right)\boldsymbol{i} + (t\cos t)\boldsymbol{j} - t\boldsymbol{k}.$

$$\Rightarrow \boldsymbol{F}'(t) = -4t\boldsymbol{i} + (\cos t - t\sin t)\boldsymbol{j} - \boldsymbol{k}.$$

$$\Rightarrow \boldsymbol{F}''(t) = -4\boldsymbol{i} + (-2\sin t - t\cos t)\boldsymbol{j}.$$

$$\Rightarrow \boldsymbol{F}'''(t) = (-2\cos t - (\cos t - t\sin t))\boldsymbol{j} = (-3\cos t + t\sin t)\boldsymbol{j}.$$

Theorem 10.3 (Algebra of Derivatives for Vector Functions)

Let $\boldsymbol{F}$ and $\boldsymbol{G}$ be two vector functions and h be a scalar function of real variable t. If $\boldsymbol{F},\boldsymbol{G}$ and h all are differentiable at t, then $a\boldsymbol{F} + b\boldsymbol{G}$, $h\boldsymbol{F}$, $\boldsymbol{F}.\boldsymbol{G}$ and $\boldsymbol{F}\times\boldsymbol{G}$ all are also differentiable at t, where a and b are constants and we have the following rules:

1. Linearity: $(a\boldsymbol{F} + b\boldsymbol{G})'(t) = a\boldsymbol{F}'(t) + b\boldsymbol{G}'(t).$
2. Scalar Multiple: $(h\boldsymbol{F})'(t) = h'(t)\boldsymbol{F}(t) + h(t)\boldsymbol{F}'(t).$
3. Dot Product: $(\boldsymbol{F}.\boldsymbol{G})'(t) = \boldsymbol{F}'(t).\boldsymbol{G}(t) + \boldsymbol{F}(t).\boldsymbol{G}'(t).$
4. Cross Product: $(\boldsymbol{F}\times\boldsymbol{G})'(t) = \boldsymbol{F}'(t)\times\boldsymbol{G}(t) + \boldsymbol{F}(t)\times\boldsymbol{G}'(t).$
5. Chain Rule: $\left(\boldsymbol{F}(h(t))\right)' = h'(t)\boldsymbol{F}'(h(t)).$

Proof: (1) By definition of derivative of vector function, we have

$$(a\boldsymbol{F} + b\boldsymbol{G})'(t) = \lim_{\Delta t\to 0}\frac{(a\boldsymbol{F} + b\boldsymbol{G})(t+\Delta t) - (a\boldsymbol{F} + b\boldsymbol{G})(t)}{\Delta t}$$

$$= \lim_{\Delta t\to 0}\frac{\left((a\boldsymbol{F}(t+\Delta t) + b\boldsymbol{G}(t+\Delta t)) - (a\boldsymbol{F}(t) + b\boldsymbol{G}(t))\right)}{\Delta t}$$

$$= \lim_{\Delta t\to 0}\frac{\left(a\left(\boldsymbol{F}(t+\Delta t) - \boldsymbol{F}(t)\right) + b\left(\boldsymbol{G}(t+\Delta t) - \boldsymbol{G}(t)\right)\right)}{\Delta t}$$

$$= \lim_{\Delta t\to 0}\left(a\left(\frac{\left(\boldsymbol{F}(t+\Delta t) - \boldsymbol{F}(t)\right)}{\Delta t}\right)\right) + \lim_{\Delta t\to 0}\left(b\left(\frac{\left(\boldsymbol{G}(t+\Delta t) - \boldsymbol{G}(t)\right)}{\Delta t}\right)\right).$$

Therefore, $(a\boldsymbol{F} + b\boldsymbol{G})'(t) = a\boldsymbol{F}'(t) + b\boldsymbol{G}'(t)$.

Proof of (2) to (4) are left for the reader as an exercise.

(5) Since, the vector function $\boldsymbol{F}$ is differentiable at $v = h(t)$ and h is differentiable at t. Then by definition of differentiation, we have

$$h'(t) = \lim_{\Delta t \to 0} \frac{\big(h(t+\Delta t) - h(t)\big)}{\Delta t}.$$

$$\Rightarrow \qquad \frac{\big(h(t+\Delta t) - h(t)\big)}{\Delta t} - h'(t) \to 0 \text{ as } \Delta t \to 0.$$

Let
$$u(t) = \frac{h(t+\Delta t) - h(t)}{\Delta t} - h'(t). \tag{10.4}$$

Then, $u(t) \to 0$ as $\Delta t \to 0$.

Since $\boldsymbol{F}$ is differentiable function at $v = h(t)$, then

$$\boldsymbol{F}'(v) = \lim_{k \to 0} \frac{\big(\boldsymbol{F}(v+k) - \boldsymbol{F}(v)\big)}{k}.$$

$\Rightarrow \dfrac{\boldsymbol{F}(v+k) - \boldsymbol{F}(v)}{k} - \boldsymbol{F}'(v) \to 0$ as $k \to 0$, here we are using the variable k which stands for Δt.

Let $\boldsymbol{W}(t) = \dfrac{\boldsymbol{F}(v+k) - \boldsymbol{F}(v)}{k} - \boldsymbol{F}'(v)$. (10.5)

Then $\boldsymbol{W}(t) \to 0$ as $k \to 0$.

From, (10.4) and (10.5), we have

$$h(t+\Delta t) = h(t) + \{u(t) + h'(t)\}\,\Delta t \tag{10.6}$$

and
$$\boldsymbol{F}(v+k) = \boldsymbol{F}(v) + \{\boldsymbol{W}(t) + \boldsymbol{F}'(v)\}k. \tag{10.7}$$

$$\Rightarrow \qquad \boldsymbol{F}\big(h(t+\Delta t)\big) = \boldsymbol{F}\big(h(t) + \{u(t) + h'(t)\}\,\Delta t\big) \text{ (by using (10.6))}$$

$$= \boldsymbol{F}\big(h(t)\big) + \{\boldsymbol{W}(t) + \boldsymbol{F}'\big(h(t)\big)\}\big(\{u(t) + h'(t)\}\,\Delta t\big). \tag{10.8}$$

(by using (10.7) and here, $k = \{u(t) + h'(t)\}\,\Delta t)$.

Now, as $\Delta t \to 0, k = \{u(t) + h'(t)\}\,\Delta t \to 0$.

From (10.8), we have

$$\frac{\boldsymbol{F}(h(t+\Delta t)) - \boldsymbol{F}(h(t))}{\Delta t} = \big(\boldsymbol{F}'(h(t)) + \boldsymbol{W}(t)\big)\big(u(t) + h'(t)\big).$$

$$\Rightarrow \qquad \lim_{\Delta t \to 0} \frac{\boldsymbol{F}(h(t+\Delta t)) - \boldsymbol{F}(h(t))}{\Delta t} = \lim_{\Delta t \to 0} \big(\boldsymbol{F}'(h(t)) + \boldsymbol{W}(t)\big)\big(u(t) + h'(t)\big)$$

$$= h'(t)(\boldsymbol{F}'\big(h(t)\big) \text{ (as } \boldsymbol{W}(t) \text{ and } u(t) \to 0 \text{ as } \Delta t \to 0).$$

Therefore, $\boldsymbol{F}'\big(h(t)\big) = h'(t)(\boldsymbol{F}'\big(h(t)\big)$.

EXAMPLE 17 Let $F(t)=(3+t^2)i-(\cos 3t)j+t^{-1}k$ and $G(t)=\sin(2-t)i-e^{2t}k$ be two vector functions. Then, prove the following:

(a) $(3F-2G)'(t)=3F'(t)-G'(t)$.

(b) $(F.G)'(t)=F'(t).G(t)+F(t).G'(t)$.

Solution:

(a) $(3F-2G)(t)=3F(t)-2G(t)$

$$=3\left((3+t^2)i-(\cos 3t)j+t^{-1}k\right)-2\left(\sin(2-t)i-e^{2t}k\right)$$

$$=\left(9+3t^2-2\sin(2-t)\right)i-(\cos 3t)j+\left(t^{-1}+2e^{2t}\right)k$$

$$\Rightarrow \qquad (3F-2G)'(t)=\left(6t+2\cos(2-t)\right)i+3\sin 3t\,j+\left(-t^{-2}+4e^{2t}\right)k. \qquad (10.9)$$

Now,

$$3F'(t)-2G'(t)=3\left(2ti+3\sin 3t\,j-t^{-2}k\right)-2\left(-\cos(2-t)i-2e^{2t}k\right)$$

$$=\left(6t+2\cos(2-t)\right)i+3\sin 3t\,j+\left(-t^{-2}+4e^{2t}\right)k. \qquad (10.10)$$

From (10.9) and (10.10), we get

$(3F-2G)'(t)=3F'(t)-G'(t)$.

(b) $(F.G)(t)=F(t).G(t)=\left((3+t^2)i-(\cos 3t)j+t^{-1}k\right).\left(\sin(2-t)i-e^{2t}k\right)$

$$=\left(3+t^2\right)\sin(2-t)-t^{-1}e^{2t}.$$

$$\Rightarrow (F.G)'(t)=\left(3+t^2\right)\left(-\cos(2-t)\right)+2t\sin(2-t)-2t^{-1}e^{2t}+t^{-2}e^{2t}. \qquad (10.11)$$

Now,

$$F'(t).G(t)+F(t).G'(t)=\left(2ti+3\sin 3t\,j-t^{-2}k\right).\left(\sin(2-t)i-e^{2t}k\right)$$
$$+\left((3+t^2)i-(\cos 3t)j+t^{-1}k\right)$$
$$.\left(-\cos(2-t)i-2e^{2t}k\right)$$
$$=\left(2t\sin(2-t)+t^{-2}e^{2t}\right)$$
$$+\left(3+t^2\right)\left(-\cos(2-t)\right)-2t^{-1}. \qquad (10.12)$$

From (10.11) and (10.12), we get

$$(F.G)'(t)=F'(t).G(t)+F(t).G'(t).$$

In the following theorem, we will discuss the geometric property of derivatives of vector-valued function.

Theorem 10.4 (Orthogonality of Vector Function of Constant Length and Its Derivative)
Let $\boldsymbol{F}(t)$ be a non-zero vector function of real variable t. So, if $\boldsymbol{F}(t)$ is differentiable and has constant length, then $\boldsymbol{F}(t)$ is orthogonal to its derivative $\boldsymbol{F}'(t)$.

Proof: Let a non-zero vector function $\boldsymbol{F}(t)$ is differentiable and has constant length.
So, let $\|\boldsymbol{F}(t)\| = a$, for all t and where a is some constant.
We have to show that $\boldsymbol{F}(t)$ is orthogonal to its derivative $\boldsymbol{F}'(t)$.
For this, we will show that

$$\boldsymbol{F}(t)\cdot\boldsymbol{F}'(t) = 0, \text{ for all } t.$$

Now, $a^2 = \|\boldsymbol{F}(t)\|^2 = \boldsymbol{F}(t)\cdot\boldsymbol{F}(t)$, for all t.

Differentiate w.r.t. t, we get

$\boldsymbol{F}'(t)\cdot\boldsymbol{F}(t) + \boldsymbol{F}(t)\cdot\boldsymbol{F}'(t) = 0$ (as a is constant and by using Theorem 10.3(3)).

$\Rightarrow \quad 2\boldsymbol{F}'(t)\cdot\boldsymbol{F}(t) = 0$ (as $\boldsymbol{a}\cdot\boldsymbol{b} = \boldsymbol{b}\cdot\boldsymbol{a}$).

$\Rightarrow \quad \boldsymbol{F}'(t)\cdot\boldsymbol{F}(t) = 0.$

Hence, $\boldsymbol{F}(t)$ and $\boldsymbol{F}'(t)$ are orthogonal.

We know that in the plane geometry that a tangent line to the circle and the radial line from the centre are orthogonal at the point of tangency. We can extend this property to spheres in $\mathbb{R}^3$ by the help of Theorem 10.4.

Let $\boldsymbol{R}(t) = f(t)\boldsymbol{i} + g(t)\boldsymbol{j} + h(t)\boldsymbol{k}$ be a vector function whose graph completely lies on the sphere $x^2+y^2+z^2=r^2$ in $\mathbb{R}^3$. Let $P(t_0)$ be a point on the graph at $t = t_0$ (as shown in Figure 10.5). Then $\boldsymbol{R}(t_0)$ is the radial vector from the centre O to P and $\boldsymbol{R}'(t_0)$ is the tangent vector to the graph of $\boldsymbol{R}(t)$ at the point P. Thus, $\boldsymbol{R}'(t_0)$ is also tangent to the sphere at P.

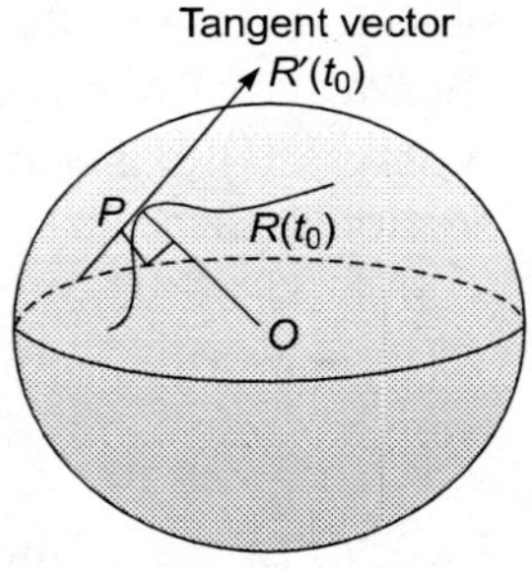

Figure 10.5

Now, $\boldsymbol{R}(t) = \sqrt{f^2(t)+g^2(t)+h^2(t)}$, for all t.

Since the graph of the $\boldsymbol{R}(t)$ completely lies on the sphere, so that it satisfies the equation of sphere.

i.e., $$f^2(t)+g^2(t)+h^2(t) = r^2, \text{ for all } t.$$

$$\Rightarrow \quad \|\boldsymbol{R}(t)\| = r, \quad \text{for all } t.$$

Therefore, the vector function $\boldsymbol{R}(t)$ is of constant length.

Hence, by Theorem 10.4, $\boldsymbol{R}(t)$ and its derivative $\boldsymbol{R}'(t)$ are orthogonal, i.e., the tangent vector $\boldsymbol{R}'(t_0)$ at the point P and the radial vector $\boldsymbol{R}(t_0)$ from the centre O of the sphere to the point P of tangency are orthogonal.

10.3.4 Motion of an Object in $\mathbb{R}^2$

In chapter 3, we discussed that the if x is displacement covered by an object at any time t, then the derivative of x w.r.t. t is the velocity of a moving object and the derivative of velocity w.r.t. t is an acceleration of an object. Conversely, by using integration, we can find the velocity from the acceleration and position from the velocity of an object.

Now, we will discuss the same concepts in terms of vector function.

Let $\boldsymbol{R}(t) = f(t)\boldsymbol{i} + g(t)\boldsymbol{j} + h(t)\boldsymbol{k}$ be the position of a moving object at time t.

Then the derivative of $\boldsymbol{R}(t)$ w.r.t. t is the velocity of an object and the derivative of velocity is an acceleration of an object.

So, the *velocity vector* is $\boldsymbol{V} = \dfrac{d\boldsymbol{R}}{dt}$ and the *acceleration vector* is $\boldsymbol{A} = \dfrac{d\boldsymbol{V}}{dt} = \dfrac{d^2\boldsymbol{R}}{dt^2}$.

The magnitude of velocity vector $\boldsymbol{V}$, i.e. $\|\boldsymbol{V}\|$ is known as the *speed* of an object and the *direction of motion* of an object is the unit vector $\dfrac{\boldsymbol{V}}{\|\boldsymbol{V}\|}$.

Geometrically, the graph of the position vector $\boldsymbol{R}(t)$ is known as the *trajectory* of the moving object. The velocity $\boldsymbol{V}(t) = \boldsymbol{R}'(t)$ is the tangent to the trajectory at any point (as shown in Figure 10.6), where $\boldsymbol{V}(t)$ exists and $\boldsymbol{V}(t) \neq 0$ (as we discussed in section 10.3.2).

If $\boldsymbol{V}(t) = 0$ at any point P on the graph of the $\boldsymbol{R}(t)$, then that point P is known as the *stationary point.*

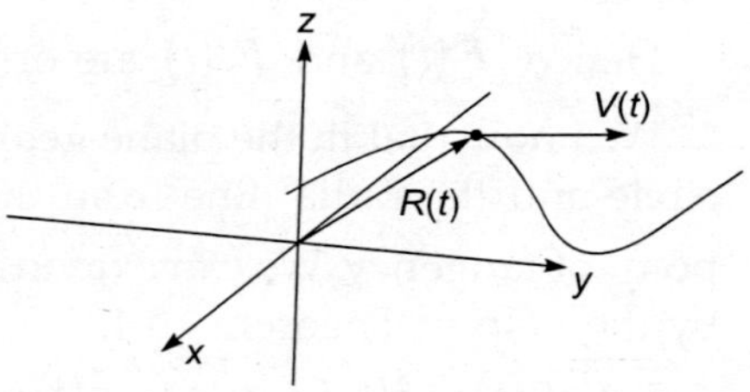

Figure 10.6

EXAMPLE 18 Find the particle's velocity and acceleration vectors and then find the speed and direction of motion for the following position vector $\boldsymbol{R}(t)$ of a particle at given time t.

(a) $\boldsymbol{R}(t) = (\cos t)\boldsymbol{i} + (\sin t)\boldsymbol{j} + 3t\boldsymbol{k}$ at $t = \dfrac{\pi}{4}$.

(b) $\boldsymbol{R}(t) = e^{t}\boldsymbol{i} + e^{-t}\boldsymbol{j} + e^{2t}\boldsymbol{k}$ at $t = \ln 2$.

(c) $\boldsymbol{R}(t) = (\ln t)\boldsymbol{i} + \dfrac{1}{2}t^3\boldsymbol{j} - t\boldsymbol{k}$ at $t = 1$.

Solution:

(a) $\boldsymbol{V} = \dfrac{d\boldsymbol{R}}{dt} = -\sin t\,\boldsymbol{i} + \cos t\,\boldsymbol{j} + 3k$

and $\boldsymbol{A} = \dfrac{d\boldsymbol{V}}{dt} = -\cos t\,\boldsymbol{i} - \sin t\,\boldsymbol{j}$.

At time $t = \dfrac{\pi}{4}$, $\boldsymbol{V}\left(\dfrac{\pi}{4}\right) = -\sin\dfrac{\pi}{4}\boldsymbol{i} + \cos\dfrac{\pi}{4}\boldsymbol{j} + 3\left(\dfrac{\pi}{4}\right)\boldsymbol{k} = \dfrac{1}{\sqrt{2}}(-\boldsymbol{i} + \boldsymbol{j}) + \dfrac{3\pi}{4}\boldsymbol{k}$

and $A\left(\frac{\pi}{4}\right) = -\cos\frac{\pi}{4}\,i - \sin\frac{\pi}{4}\,j = -\frac{1}{\sqrt{2}}(i + j)$.

The speed is $\|V\| = \sqrt{(-\sin t)^2 + (\cos t)^2 + 3^2} = \sqrt{10}$.

At time $t = \frac{\pi}{4}$, $\left\|V\left(\frac{\pi}{4}\right)\right\| = \sqrt{10}$.

The direction of motion is $\frac{V}{\|V\|} = \frac{1}{\sqrt{10}}(-\sin t\,i + \cos t j + 3k)$.

At time $t = \frac{\pi}{4}$, $\frac{V\left(\frac{\pi}{4}\right)}{\left\|V\left(\frac{\pi}{4}\right)\right\|} = \frac{1}{\sqrt{10}}\left(-\frac{1}{\sqrt{2}}i + \frac{1}{\sqrt{2}}j + 3k\right)$.

(b) $V = \frac{dR}{dt} = e^t i - e^{-t} j + 2e^{2t} k$

and $A = \frac{dV}{dt} = e^t i + e^{-t} j + 4e^{2t} k$.

At time $t = \ln 2$, $V(\ln 2) = 2i - \frac{1}{2}j + 8k$

and $A(\ln 2) = 2i + \frac{1}{2}j + 16k.$

The speed is $\|V\| = \sqrt{e^{2t} + e^{-2t} + 4e^{4t}}$.

At time $t = \ln 2$, $\|V(\ln 2)\| = \sqrt{4 + \frac{1}{4} + 64} = \frac{\sqrt{273}}{2}$.

The direction of motion is $\frac{V}{\|V\|} = \frac{2}{\sqrt{273}}\left(e^t i - e^{-t} j + 2e^{2t} k\right)$.

At time $t = \ln 2$, $\frac{V(\ln 2)}{\|V(\ln 2)\|} = \frac{2}{\sqrt{273}}\left(2i - \frac{1}{2}j + 8k\right)$.

(c) $V = \frac{dR}{dt} = \frac{1}{t}i + \frac{3}{2}t^2 j - k$

and $A = \frac{dV}{dt} = -\frac{1}{t^2}i + 3tj$.

At time $t=1$, $\boldsymbol{V}(1)=\boldsymbol{i}+\frac{3}{2}\boldsymbol{j}-\boldsymbol{k}$

and $\boldsymbol{A}(1)=-\boldsymbol{i}+3\boldsymbol{j}$.

The speed is $\|\boldsymbol{V}\|=\sqrt{\frac{1}{t^2}+\frac{9}{4}t^4+1}$.

At time $t=1$, $\|\boldsymbol{V}(1)\|=\sqrt{1+\frac{9}{4}+1}=\frac{\sqrt{17}}{2}$.

The direction of motion is $\frac{\boldsymbol{V}}{\|\boldsymbol{V}\|}=\frac{2}{\sqrt{17}}\left(\frac{1}{t}\boldsymbol{i}+\frac{3}{2}t^2\boldsymbol{j}-\boldsymbol{k}\right)$.

At time t = 1, $\frac{\boldsymbol{V}(1)}{\|\boldsymbol{V}(1)\|}=\frac{2}{\sqrt{17}}\left(\boldsymbol{i}+\frac{3}{2}\boldsymbol{j}-\boldsymbol{k}\right)$.

10.3.5 Integration of Vector-Valued Functions

Definition: Let $\boldsymbol{F}(t)=f(t)\boldsymbol{i}+g(t)\boldsymbol{j}+h(t)\boldsymbol{k}$ be any vector function such that its components f, g and h are continuous on the closed interval $a \le t \le b$. Then the *indefinite integral* of $\boldsymbol{F}(t)$ is defines as

$$\int \boldsymbol{F}(t)\,dt=\left(\int f(t)\,dt\right)\boldsymbol{i}+\left(\int g(t)\,dt\right)\boldsymbol{j}+\left(\int h(t)\,dt\right)\boldsymbol{k}+\boldsymbol{C},$$

where $\boldsymbol{C}=C_1 i+C_2 j+C_3 k$ is any constant vector.

And the *definite integral* of $\boldsymbol{F}(t)$ on $a \le t \le b$ is defined as

$$\int_a^b \boldsymbol{F}(t)\,dt=\left(\int_a^b f(t)\,dt\right)\boldsymbol{i}+\left(\int_a^b g(t)\,dt\right)\boldsymbol{j}+\left(\int_a^b h(t)\,dt\right)\boldsymbol{k}.$$

Remark: The integral of a vector function (definite or indefinite) is again a vector function.

EXAMPLE 19 Find the value of $\int_0^{2a}(\cos t\boldsymbol{i}+\sin t\boldsymbol{j}+\sin t\cos t\boldsymbol{k})\,dt$.

Solution: $\int_0^{2a}(\cos t\boldsymbol{i}+\sin t\boldsymbol{j}+\sin t\cos t\boldsymbol{k})\,dt$

$$=\left(\int_0^{2a}\cos t\,dt\right)i+\left(\int_0^{2a}\sin t\,dt\right)j+\left(\int_0^{2a}\sin t\cos t\,dt\right)k$$

$$=\sin 2a i+(1-\cos 2a)j+\frac{1}{2}(1-\cos 4a)k.$$

EXAMPLE 20 Find the position vector $\boldsymbol{R}(t)$ and the velocity vector $\boldsymbol{V}(t)$, when the acceleration vector is $\boldsymbol{A}(t)=(\cos t)\boldsymbol{i}-(t\sin t)\boldsymbol{k}$ and the initial conditions are $\boldsymbol{R}(0)=\boldsymbol{i}-2\boldsymbol{j}+\boldsymbol{k}$ and $\boldsymbol{V}(0)=2\boldsymbol{i}+3\boldsymbol{k}$.

Solution: $\boldsymbol{V}(t)=\int \boldsymbol{A}(t)dt=\int((\cos t)\boldsymbol{i}-(t\sin t)\boldsymbol{k})dt$

$$=\sin t\boldsymbol{i}+(t\cos t-\sin t)\boldsymbol{k}+\boldsymbol{C}.$$

At initial condition, $\boldsymbol{V}(0)=2\boldsymbol{i}+3\boldsymbol{k}$.

$\Rightarrow$ $$\boldsymbol{C}=2\boldsymbol{i}+3\boldsymbol{k}.$$

Thus, $\boldsymbol{V}(t)=(2+\sin t)\boldsymbol{i}+(t\cos t-\sin t+3)\boldsymbol{k}$.

Again, $\boldsymbol{R}(t)=\int \boldsymbol{V}(t)dt=\int((2+\sin t)\boldsymbol{i}+(t\cos t-\sin t+3)\boldsymbol{k})dt$

$=(2t-\cos t)\boldsymbol{i}+(t\sin t+2\cos t+3t)\boldsymbol{k}+D$, where $\boldsymbol{D}$ is a constant vector.

At initial condition, $\boldsymbol{R}(0)=\boldsymbol{i}-2\boldsymbol{j}+\boldsymbol{k}$.

$\Rightarrow$ $$-\boldsymbol{i}+2\boldsymbol{k}+D=\boldsymbol{i}-2\boldsymbol{j}+\boldsymbol{k}.$$

$\Rightarrow$ $$\boldsymbol{D}=2\boldsymbol{i}-2\boldsymbol{j}-\boldsymbol{k}.$$

Thus, $\boldsymbol{R}(t)=(2t-\cos t+2)\boldsymbol{i}-2\boldsymbol{j}+(t\sin t+2\cos t+3t-1)\boldsymbol{k}$.

10.4 MODELING BALLISTICS AND PLANETARY MOTION

In this section, we will study the mathematical modeling of the motion of a projectile in a vacuum and Kepler's second law of motion.

Definition: A *projectile* is any object that has been thrown, shot, or launched, and *ballistics* is the study of projectile motion. For instance: a golf ball hit by a player, a baseball thrown by a pitcher, a rocket fired into space.

The flight paths of all projectiles are affected by two factors: gravity and air resistance. So, if the projectile is launched with these two factors, then the projectile become too heavy and travels with relatively low speed and stays near to the earth surface. Thus, it is quite difficult to analyze the motion of a projectile with these two factors. But, if we assume that the acceleration due to gravity is constant and the projectile travels in vacuum, then we can analyze the motion of a projectile reasonably.

Let a projectile be launched from the point P and travel in a vertical plane, so that the point P is right above the origin O and it hit the ground at the point Q, which is assumed that on the x-axis. Let s_0 be the height of P from O, i.e., $OP=s_0$, v_0 be the initial speed of the projectile, which is known as muzzle speed and α be the angle of elevation (i.e., α is the angle at which the projectile is launched) as shown in Figure 10.7.

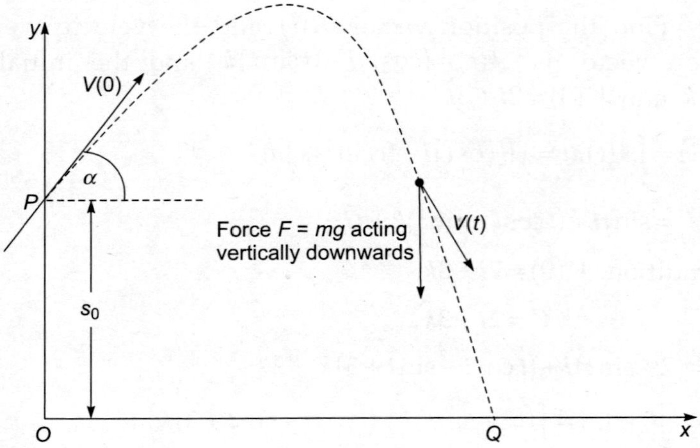

Figure 10.7

Since the projectile is travelling in vacuum, so the only force is action on the projectile at any time is the force due to gravity. This force $\boldsymbol{F}$ is acting downward.

Let m be the mass of a projectile and g be the acceleration due to gravity ("free-fall") which is approximately 9.8 m/s^2 or 32 ft/s^2. Then, the force

$$\boldsymbol{F} = -mg\boldsymbol{j}$$

(as it is acting downward in the negative direction of y-axis).

Also, according to Newton's second law of motion, the total force (sum of all forces) action on the projectile is equal to $m\boldsymbol{A}(t)$, where $A(t)$ is the acceleration of the projectile at any time t. Thus, we get

$$-mg\boldsymbol{j} = m\boldsymbol{A}(t).$$

$$\Rightarrow \qquad \boldsymbol{A}(t) = -g\boldsymbol{j}, \text{ for all } t \geq 0.$$

Let $\boldsymbol{V}(t)$ be the velocity of the projectile at any time t. Then, we have

$$\begin{aligned} \boldsymbol{V}(t) &= \int \boldsymbol{A}(t)\,dt \\ &= \int (-g\boldsymbol{j})\,dt \\ &= -gt\boldsymbol{j} + \boldsymbol{C}_1, \end{aligned} \tag{10.13}$$

where $\boldsymbol{C}_1 = \boldsymbol{V}(0)$ is the initial velocity.

Since the projectile is fired with the initial speed v_0 at an angle of elevation α. So, that

$$v_0 = \| \boldsymbol{V}(0) \|$$

and components of the initial velocity $\boldsymbol{V}(0)$ are given by (as shown in the Figure 10.8)

$$\boldsymbol{V}(0) = (v_0 \cos\alpha)\boldsymbol{i} + (v_0 \sin\alpha)\boldsymbol{j}.$$

Substitute in (10.13), we get

$$\Rightarrow \boldsymbol{V}(t) = -gt\boldsymbol{j} + (v_0 \cos\alpha)\boldsymbol{i} + (v_0 \sin\alpha)\boldsymbol{j}$$

$$= (v_0 \cos\alpha)\boldsymbol{i} + (v_0 \sin\alpha - gt)\boldsymbol{j}.$$

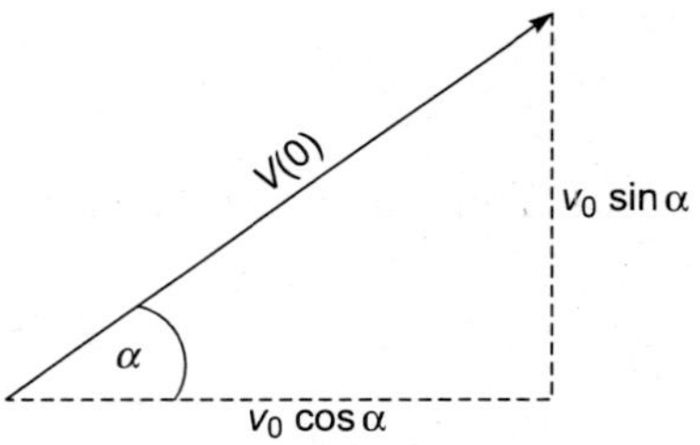

Figure 10.8

Let $\boldsymbol{R}(t)$ be the position of projectile at any time t. Then, we have

$$\boldsymbol{R}(t) = \int \boldsymbol{V}(t)\,dt$$

$$= \int \big((v_0 \cos\alpha)\boldsymbol{i} + (v_0 \sin\alpha - gt)\boldsymbol{j}\big)dt$$

$$= (v_0 \cos\alpha)t\boldsymbol{i} + \left((v_0 \sin\alpha)t - \frac{1}{2}gt^2\right)\boldsymbol{j} + \boldsymbol{C}_2, \tag{10.14}$$

where $\boldsymbol{C}_2 = \boldsymbol{R}(0)$ is the position vector at time $t = 0$.

Since, initially the projectile is fired from the point P. So, that

$$\boldsymbol{R}(0) = s\boldsymbol{j}.$$

Substitute in (10.14), we get

$$\boldsymbol{R}(t) = (v_0 \cos\alpha)t\boldsymbol{i} + \left((v_0 \sin\alpha)t - \frac{1}{2}gt^2 + s_0\right)\boldsymbol{j}, \tag{10.15}$$

which gives the position of the projectile at any time t (after it is fired and before it hits the ground).

Therefore, if $x(t)$ and $y(t)$ is the horizontal and vertical components of the position vector $\boldsymbol{R}(t)$, respectively. Then,

$$x(t) = (v_0 \cos\alpha)t \text{ and } y(t) = (v_0 \sin\alpha)t - \frac{1}{2}gt^2 + s_0. \tag{10.16}$$

Regarding all the above discussion of motion of projectile, we give the following statement.

Motion of a Projectile in Vacuum

Suppose a projectile is travelling in vacuum with constant acceleration due to gravity g, in a xy-coordinate plane, where ground level is along the x-axis. So, if the projectile is fired from the height s_0, at the angle of elevation α with the initial speed v_0. Then at any time $t (t \geq 0)$, the position of the projectile is $(x(t), y(t))$, where

$$x(t) = (v_0 \cos\alpha)t \text{ and } y(t) = (v_0 \sin\alpha)t - \frac{1}{2}gt^2 + s_0.$$

Since, the equation (10.16) of motion of a projectile in vacuum is in parametric form, which provides useful information about the path of projectile as follows:

If the angle of elevation $\alpha \neq 90°$, then we can eliminate the parameter t by solving the $x(t)$ in (10.16), which is given by

$$t = \frac{x}{v_0 \cos\alpha}$$

and then, substitute in $y(t)$, we get

$$y = (v_0 \sin\alpha)\frac{x}{v_0 \cos\alpha} - \frac{1}{2}g\left(\frac{x}{v_0 \cos\alpha}\right)^2 + s_0.$$

$$\Rightarrow \qquad y = \left(\frac{-g}{2(v_0 \cos\alpha)^2}\right)x^2 + (\tan\alpha)\,x + s_0. \tag{10.17}$$

This is a Cartesian equation for the trajectory of a projectile.

Since it is in the form of $y = ax^2 + bx + c$ with $a < 0$. So, it represents a downward opening parabola.

Therefore, if the projectile is fired at the angle of elevation $\alpha \neq 90°$ (i.e., not vertically upward), then the trajectory of a projectile is a downward-opening parabolic arc (as shown in Figure 10.9).

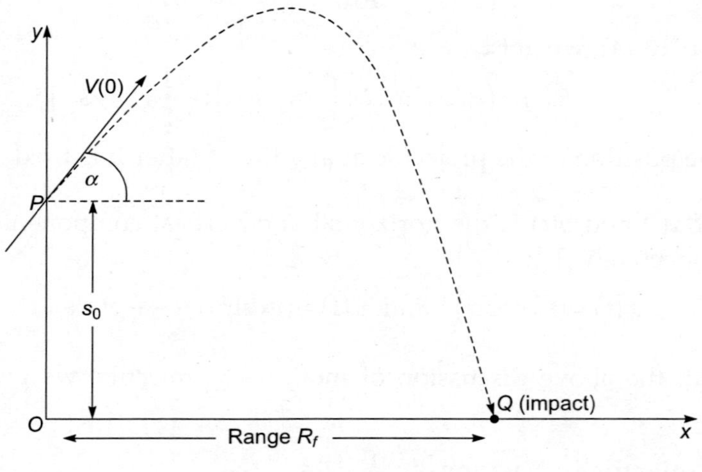

Figure 10.9

Let T_f be the time of a projectile which elapses between launching and hitting the ground and R_f is the total horizontal distance travelled by the projectile during its flight. Since $y = 0$ at the point Q, then T_f must satisfy the equation

$$\Rightarrow \qquad -\frac{1}{2}gT_f^2 + v_0 \sin\alpha T_f + s_0 = 0. \tag{10.18}$$

Since the flight starts from $x = 0$ at $t = 0$ and reaches the point Q, where $x = R_f$ at $t = T_f$, then R_f is given by $x(t)$ in (10.16) at time $t = T_f$, i.e.,

$$R_f = (v_0 \cos\alpha)T_f. \tag{10.19}$$

The time T_f is known as the *time of flight* and distance travelled by projectile R_f is known as the *range*.

In particular, if the projectile is fired from the ground, then

$$s_0 = 0.$$

From (10.18), we get

$$T_f = \frac{2}{g} v_0 \sin\alpha.$$

Now, since

$$R_f = (v_0 \cos\alpha) T_f$$

$$= (v_0 \cos\alpha) \frac{2}{g} v_0 \sin\alpha$$

$$= \frac{v_0^2}{g} \sin 2\alpha.$$

Time of Flight and Range of a Projectile Fired from the Ground Level

When a projectile is fired from the ground level, then the time of flight T_f and the range R_f are given by

$$T_f = \frac{2}{g} v_0 \sin\alpha \text{ and } R_f = \frac{v_0^2}{g} \sin 2\alpha.$$

Note: Since $R_f = \frac{v_0^2}{g} \sin 2\alpha$, then the range of a projectile is maximum when $2\alpha = 1$, i.e., $\alpha = 45°$ for a given initial speed v_0.

Hence, the projectile travelled maximum horizontal distance when it was fired at the angle of elevation $\alpha = 45°$ and so, the maximum range is given by

$$\boxed{R_{\max} = \frac{v_0^2}{g}}$$

EXAMPLE 21 A shell is fired from ground level with a muzzle speed of 280 ft/s and at an elevation of 45° from the ground level.

(a) Find the maximum height attained by the shell.
(b) Find the time of flight and the range of the shell.
(c) Find the velocity and speed of the shell at impact.

Solution:

We know that $g = 32$ ft/s^2 and we are given that $v_0 = 280$ ft/s, $\alpha = 45°$ and since the shell is fired from the ground, thus $s_0 = 0$.

So, the trajectory of the shell has parametric equations

$$x(t) = (v_0 \cos\alpha)t = 280\cos 45°t = \frac{280}{\sqrt{2}}t.$$

$$y(t) = -\frac{1}{2}gt^2 + (v_0 \sin\alpha)t + s_0$$

$$= -\frac{1}{2}(32)t^2 + (280\sin 45°)t + 0$$

$$= -\frac{1}{2}(32)t^2 + \frac{280}{\sqrt{2}}t.$$

(a) The shell attains maximum height, when the vertical component of its velocity $V(t)$ is zero, i.e.,

$$y'(t) = 0.$$

$$\Rightarrow \quad -32t + \frac{280}{\sqrt{2}} = 0.$$

$$\Rightarrow \quad t = 6.2.$$

Therefore, the maximum height attained by the shell is

$$y_{\max} = y(6.2) \approx 612.5.$$

So, the maximum height attained by the shell is 612.5 ft.

(b) The time of flight is given by

$$T_f = \frac{2}{g}v_0 \sin\alpha = \frac{2}{32}(280)\sin 45° = \frac{2}{32}(280)\frac{1}{\sqrt{2}} = 12.$$

Thus, time of flight of the shell is 12 seconds.

The range of the shell is given by

$$R_f = \frac{v_0^2}{g}\sin 2\alpha = \frac{280^2}{32}(\sin 90°) = 2450.$$

Therefore, the range of the shell is 2450 ft.

(c) Since $x(t) = \frac{280}{\sqrt{2}}t$ and $y(t) = -\frac{1}{2}(32)t^2 + \frac{280}{\sqrt{2}}t.$

$$\Rightarrow x'(t) = \frac{280}{\sqrt{2}} \text{ and } y'(t) = -32t + \frac{280}{\sqrt{2}}.$$

So, the velocity at time t is

$$\boldsymbol{V}(t)=\frac{280}{\sqrt{2}}\boldsymbol{i}+\left(-32t+\frac{280}{\sqrt{2}}\right)\boldsymbol{j}.$$

At impact $y(t)=0 \Rightarrow t=12.4$ (as if $t=0$, then $x(t)=0$, which is not possible).

$$\Rightarrow \qquad \boldsymbol{V}(12.4)=140\sqrt{2}\boldsymbol{i}-199\boldsymbol{j}$$

and its speed is given by

$$\|\boldsymbol{V}(12.4)\|=\sqrt{\left(140\sqrt{2}\right)^2+(-199)^2}\approx 280.7.$$

Hence, the speed at impact is about 280.7 ft/s.

EXAMPLE 22 If a shot-putter throws a shot from a height of 5 ft with an angle of 46° and initial speed of 25 ft/s. What is the horizontal distance of the throw (to the nearest unit)?

Solution: We have, $s_0 = 5$, $\alpha = 46°$, $v_0 = 25$ and $g = 32$ ft/s^2.

So, the shot's trajectory has parametric equations

$$x(t)=(v_0\cos\alpha)t=(25\cos 46°)t=17.4t.$$

$$\begin{aligned} y(t) &= -\frac{1}{2}gt^2+(v_0\sin\alpha)t+s_0 \\ &= -\frac{1}{2}(32)t^2+(25\sin 46°)t+5 \\ &= -16t^2+18t+5. \end{aligned}$$

The shot hits the ground, when $y = 0$,

i.e., $$-16t^2 + 18t + 5 = 0$$

$$\Rightarrow \qquad t = 0.5, 0.6.$$

So, when $t = 0.5$, then the horizontal distance covered by shot is

$$x(0.5) = 8.7.$$

and, when $t = 0.6$, then the horizontal distance covered by shot is

$$x(0.6) = 10.4.$$

Hence, the horizontal distance to the nearest unit is 8.7.

10.4.1 Kepler's Second Law

In 17th century, the German astronomer "Johannes Kepler" formulated three scientific laws describing the motion of planets around the sun.

Kepler's First Law (*Law of Orbits*)

The planets move about the sun in elliptical orbits with the sun is at one of the foci.

Kepler's Second Law (*Law of Areas*)

The radius vector joining the sun to the planets sweeps out equal areas in equal intervals of time, i.e., planets do not move at constant speed.

Kepler's Third Law (*Law of periods*)

The square of the time of one complete revolution of a planet about its orbit is proportional to the cube of the length of the semi-major axis of its elliptical path.

Now, we will prove the Kepler's second law for vector valued function. Before that, we will find the velocity and acceleration of a moving planet around the sun.

Let $P(r, \theta)$ be the position of the planet B in polar coordinate system in which the sun is at the pole. Let $\boldsymbol{u}_r$ and $\boldsymbol{u}_\theta$ be the unit vector along the radial axis and the orthogonal to that axis, respectively (as shown in Figure 10.10).

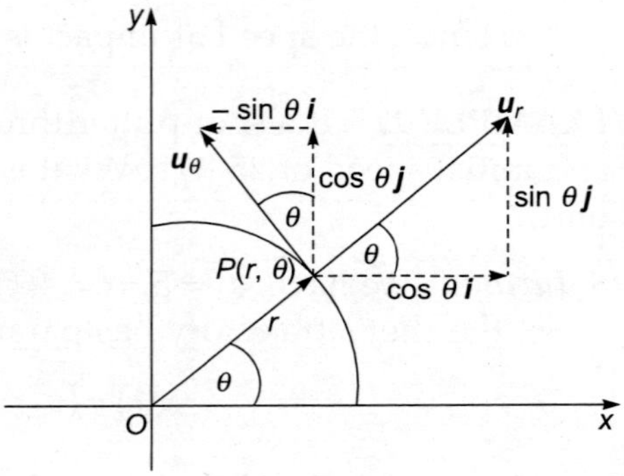

Figure 10.10

Then,

$$\boldsymbol{u}_r = \cos\theta\boldsymbol{i} + \sin\theta\boldsymbol{j}$$

and

$$\boldsymbol{u}_\theta = -\sin\theta\boldsymbol{i} + \cos\theta\boldsymbol{j}.$$

Differentiate w.r.t. θ, we get

$$\frac{d\boldsymbol{u}_r}{d\theta} = -\sin\theta\boldsymbol{i} + \cos\theta\boldsymbol{j} = \boldsymbol{u}_\theta \tag{10.20}$$

and

$$\frac{d\boldsymbol{u}_\theta}{d\theta} = -(\cos\theta\boldsymbol{i} + \sin\theta\boldsymbol{j}) = -\boldsymbol{u}_r. \tag{10.21}$$

Let $\boldsymbol{R}$ be the radial vector from the sun to the planet B, then

$$\boldsymbol{R} = r\boldsymbol{u}_r = (\cos\theta\,\boldsymbol{i} + \sin\theta\,\boldsymbol{j}),$$

where $r = \|\boldsymbol{R}\|$.

Let V be the velocity of a moving planet B, then

$$V = \frac{d\mathbf{R}}{dt} = \frac{dr}{dt}\mathbf{u}_r + r\frac{d\mathbf{u}_r}{dt}$$

$$= \frac{dr}{dt}\mathbf{u}_r + r\frac{d\mathbf{u}_r}{d\theta}\frac{d\theta}{dt}.$$

So, from (10.20), we get

$$V = \frac{dr}{dt}\mathbf{u}_r + r\mathbf{u}_\theta\frac{d\theta}{dt}.$$

Let A be the acceleration of a moving planet B, then

$$A = \frac{dV}{dt} = \frac{d^2r}{dt^2}\mathbf{u}_r + \frac{dr}{dt}\frac{d\mathbf{u}_r}{dt} + \frac{dr}{dt}\mathbf{u}_\theta\frac{d\theta}{dt} + r\frac{d\mathbf{u}_\theta}{dt}\frac{d\theta}{dt} + r\mathbf{u}_\theta\frac{d^2\theta}{dt^2}.$$

By using (10.20) and (10.21), we get

$$A = \frac{d^2r}{dt^2}\mathbf{u}_r + 2\frac{dr}{dt}\mathbf{u}_\theta\frac{d\theta}{dt} + r(-\mathbf{u}_r)\left(\frac{d\theta}{dt}\right)^2 + r\mathbf{u}_\theta\frac{d^2\theta}{dt^2}$$

$$= \left\{\frac{d^2r}{dt^2} - r\left(\frac{d\theta}{dt}\right)^2\right\}\mathbf{u}_r + \left\{r\frac{d^2\theta}{dt^2} + 2\frac{dr}{dt}\frac{d\theta}{dt}\right\}\mathbf{u}_\theta.$$

Hence, we obtained formulae of velocity and acceleration of a planet B in polar form, which are as follows:

Polar Formula for Velocity

$$V = \frac{dr}{dt}\mathbf{u}_r + r\mathbf{u}_\theta\frac{d\theta}{dt}. \tag{10.22}$$

Polar Formula for Acceleration

$$A = \left\{\frac{d^2r}{dt^2} - r\left(\frac{d\theta}{dt}\right)^2\right\}\mathbf{u}_r + \left\{r\frac{d^2\theta}{dt^2} + 2\frac{dr}{dt}\frac{d\theta}{dt}\right\}\mathbf{u}_\theta. \tag{10.23}$$

Theorem 10.5 (Kepler's Second Law)

The radius vector from the sun to the planet in its orbit sweeps out equal areas in equal intervals of time.

Proof: The geometrical interpretation of the statement of theorem is illustrated in Figure 10.11.

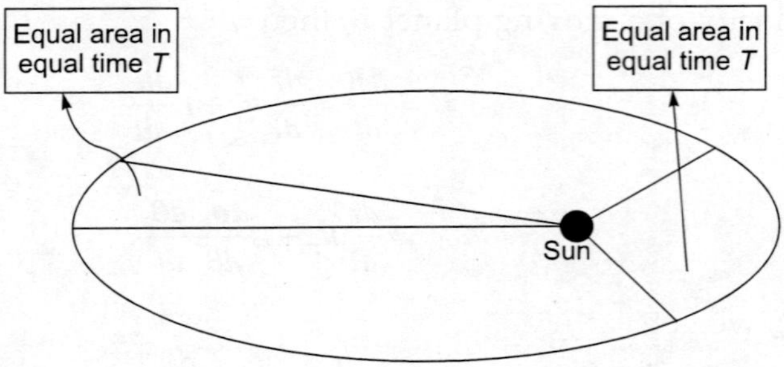

Figure 10.11

Let us suppose that the only force acting on a planet is the gravitational force of attraction of the sun. So, according to the law of gravitation, the force of attraction $\boldsymbol{F}$ is given by

$$\boldsymbol{F} = -G\frac{mM}{r^2}\boldsymbol{u}_r,$$

where G is a gravitational constant, m and M are the masses of the planet and the sun, respectively and $\boldsymbol{u}_r$ is the unit vector along the radial axis from the sun to the planet.

Now, according of Newton's second law of motion, the total force $\boldsymbol{F} = m\boldsymbol{A}$, where $\boldsymbol{A}$ is the acceleration of the planet in its orbit.

Thus, we have

$$m\boldsymbol{A} = -G\frac{mM}{r^2}\boldsymbol{u}_r.$$

$$\Rightarrow \qquad G(t) = (1-t)\boldsymbol{i} + \frac{1}{t}\boldsymbol{k}.$$

It means that the acceleration of a planet in its orbit has only one radial component. Thus, the $\boldsymbol{u}_\theta$ component of acceleration of a planet in its orbit is zero.

Then, by (10.23), it is equivalent to differential equation

$$r\frac{d^2\theta}{dt^2} + 2\frac{dr}{dt}\frac{d\theta}{dt} = 0.$$

Take $\omega = d\theta/dt$, for the sake of convenience. Then

$$\frac{rd\omega}{dt} + 2\omega\frac{d\theta}{dt} = 0.$$

By separation of variables, we get

$$\frac{1}{\omega}\frac{d\omega}{dt} = -\frac{2}{r}\frac{dr}{dt}.$$

On integration, we get

$$\int \frac{1}{\omega}\, d\omega = \int -\frac{2}{r}\, dr.$$

$$\Rightarrow \quad \ln|\omega| = -2\ln|r| + \ln C.$$

$$\Rightarrow \quad \ln \omega = \ln (Cr^{-2}) \text{ (as } \omega > 0 \text{ and } r > 0).$$

$$\Rightarrow \quad \omega = Cr^{-2}.$$

$$\Rightarrow \quad \frac{d\theta}{dt} = Cr^{-2}.$$

$$\Rightarrow \quad r^2 \frac{d\theta}{dt} = C. \tag{10.24}$$

Now, let $[t_1, t_2]$ and $[t_3, t_4]$ be two time intervals of equal length,

i.e., $$(t_2 - t_1) = (t_4 - t_3). \tag{10.25}$$

Then, the area S_1 swept out in the time interval $[t_1, t_2]$ is given by

$$S_1 = \int_{t_1}^{t_2} \frac{1}{2} r^2 d\theta$$

$$= \int_{t_1}^{t_2} \left(\frac{1}{2} r^2 \frac{d\theta}{dt}\right) dt$$

$$= \int_{t_1}^{t_2} C dt \qquad \text{(by using (10.24))}$$

$$= C(t_2 - t_1). \tag{10.26}$$

Now, the area S_2 swept out in the time interval $[t_3, t_4]$ is given by

$$S_2 = \int_{t_3}^{t_4} \frac{1}{2} r^2 d\theta$$

$$= \int_{t_2}^{t_4} \left(\frac{1}{2} r^2 \frac{d\theta}{dt}\right) dt$$

$$= \int_{t_3}^{t_4} C dt \qquad \text{(by using (10.24))}$$

$$= C(t_3 - t_4). \tag{10.27}$$

From (10.25), (10.26) and (10.27), we get

$$S_1 = S_2.$$

Hence, the radius vector from the sun to the planet in its orbit sweeps out equal areas in equal intervals of time.

Central Force Field

A central force on an object is a force whose magnitude depends only on the distance $\boldsymbol{r}$ of the object from the origin and is directed along the line joining them.

Thus, an object is said to move in a central force field, if only a single force acting on it and which is directed towards a particular point.

Thus, Kepler's second law of motion is based on the assumption that the motion of a planet occurs in central force field.

In particular, the formula

$$\boldsymbol{A} = -G\frac{M}{r^2}\boldsymbol{u}_r,$$

shows that the acceleration of a planet has only a radial component and so, in general this turn out be true for an object which is moving a plane in any central force field.

EXAMPLE 23 An object moves along the curve whose parametric equations are $r = \sin\theta$ and $\theta = 2t$. Find the velocity and acceleration of the object in terms of $\boldsymbol{u}_r$ and $\boldsymbol{u}_\theta$.

Solution: Since the parametric equations of the curve are (in polar form)

$$r = \sin\theta \text{ and } \theta = 2t.$$

$$\Rightarrow \qquad \frac{dr}{d\theta} = \cos\theta \text{ and } \frac{d\theta}{dt} = 2.$$

$$\Rightarrow \qquad \frac{dr}{dt} = \frac{dr}{d\theta}\frac{d\theta}{dt} = 2\cos\theta.$$

Also, $\dfrac{d^2r}{dt^2} = \dfrac{d}{dt}\left(\dfrac{dr}{dt}\right) = \dfrac{d}{dt}(2\cos\theta) = \dfrac{d}{d\theta}(2\cos\theta)\dfrac{d\theta}{dt} = -4\sin\theta$ and $\dfrac{d^2\theta}{dt^2} = 0$.

Now, the velocity of the object is given by

$$\boldsymbol{V}(t) = \frac{dr}{dt}\boldsymbol{u}_r + r\frac{d\theta}{dt}\boldsymbol{u}_\theta.$$

$$\Rightarrow \qquad \boldsymbol{V}(t) = (2\cos\theta)\boldsymbol{u}_r + 2r\boldsymbol{u}_\theta.$$

$$\Rightarrow \qquad \boldsymbol{V}(t) = (2\cos 2t)\boldsymbol{u}_r + (2\sin 2t)\boldsymbol{u}_\theta.$$

The acceleration of the object is given by

$$\boldsymbol{A} = \left\{\frac{d^2r}{dt^2} - r\left(\frac{d\theta}{dt}\right)^2\right\}\boldsymbol{u}_r + \left\{r\frac{d^2\theta}{dt^2} + 2\frac{dr}{dt}\frac{d\theta}{dt}\right\}\boldsymbol{u}_\theta.$$

$$\Rightarrow \qquad A = \left\{-4\sin\theta - r(2)^2\right\}\boldsymbol{u}_r + \left\{r(0) + 2(2\cos\theta)(2)\right\}\boldsymbol{u}_\theta.$$

$$\Rightarrow \qquad A = -8(\sin 2t)\,\boldsymbol{u}_r + 8(\cos 2t)\,\boldsymbol{u}_\theta.$$

10.5 CURVATURE

In chapter 6, we have studied the curvature in terms of real valued functions. In this section we will study the curvature in terms of vector valued functions, i.e., we will study the problems based on the bending of curves in 2-space or 3-space. These types of problems have applications in geometry and in the study of motion along the curved path.

Definition (Curvature): Let C be a smooth curve in 2-space or 3-space and parametrized by an arc length $\boldsymbol{r}(s)$, then the curvature of C, is denoted by $k = k(s)$ and defined by

$$k(s) = \left\| \frac{d\boldsymbol{T}}{ds} \right\| = \|\boldsymbol{r}''(s)\|, \tag{10.28}$$

where $\boldsymbol{T}$ is the unit tangent vector.

Remarks:

1. Since $k(s)$ is the length if $\dfrac{d\boldsymbol{T}}{ds}$, so that the curvature $k(s)$ is a real valued function of s.
2. If C is a straight line (no bending), then the direction of $\boldsymbol{T}$ remains constant (Figure 10.12(a)).
3. If C bends slightly, then the rate of change in $\boldsymbol{T}$ with respect to s is relatively low (Figure 10.12(b)).
4. If C bends sharply, then the rate of change in $\boldsymbol{T}$ with respect to s is relatively high (Figure 10.12(c)).

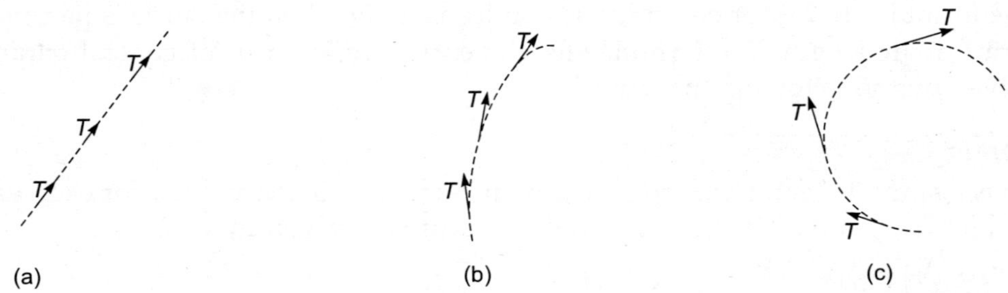

Figure 10.12

Therefore, in general the curvature will vary from point to point along the curve.

EXAMPLE 24 Show that the circle C with radius a, centred at the origin has constant curvature.

Solution: The parametric equation of the circle in terms of an arc length s is

$$\boldsymbol{r}(s) = a\cos\left(\frac{s}{a}\right)\boldsymbol{i} + a\sin\left(\frac{s}{a}\right)\boldsymbol{j} \quad (0 \le s \le 2\pi a).$$

$$\Rightarrow \qquad \boldsymbol{r}''(s) = -\frac{1}{a}\cos\left(\frac{s}{a}\right)\boldsymbol{i} - \frac{1}{a}\sin\left(\frac{s}{a}\right)\boldsymbol{j}.$$

Then, by (10.28), we have

$$k(s) = \|\boldsymbol{r}''(s)\| = \sqrt{\left(-\frac{1}{a}\cos\left(\frac{s}{a}\right)\right)^2 + \left(-\frac{1}{a}\sin\left(\frac{s}{a}\right)\right)^2}$$

$$= \frac{1}{a}.$$

Hence, the curvature of a circle is constant.

EXAMPLE 25 Show that the line in 2-space has zero curvature.

Solution: The parametric form of the line in terms of an arc length s is

$$\boldsymbol{r}(s) = \boldsymbol{r}_0 + s\boldsymbol{u},$$

where the terminal point $\boldsymbol{r}_0$ is a point on the line and vector $\boldsymbol{u}$ is a unit vector parallel to the line.

Now, differentiate w.r.t. s, we get

$$\boldsymbol{r}'(s) = \boldsymbol{u} \text{ (as } \boldsymbol{r}_0 \text{ is constant)}$$

$$\Rightarrow \qquad \boldsymbol{r}''(s) = 0 \text{ (as } \boldsymbol{u} \text{ is constant).}$$

Thus,

$$k(s) = \|\boldsymbol{r}''(s)\| = 0.$$

Hence, the curvature of the line is zero.

The formula (10.28) for curvature is applicable only when the curve is parametrized in terms of arc length. The formulae for the curvature in terms of general parameter t are given by the following theorem:

Theorem 10.6

If $r(t)$ is a smooth vector valued function in 2-space or 3-space, then for each value of t at which $T'(t)$ and $r''(t)$ exist, then the curvature k is given by

1. $k(t) = \dfrac{\|\boldsymbol{T}'(t)\|}{\|\boldsymbol{r}'(t)\|}$

2. $k(t) = \dfrac{\|\boldsymbol{r}'(t) \times \boldsymbol{r}''(t)\|}{\|\boldsymbol{r}'(t)\|^3}$

Proof:

1. By formulae of curvature (10.28), we have

$$k(s)=\left\|\frac{d\boldsymbol{T}}{ds}\right\|=\left\|\frac{d\boldsymbol{T}/dt}{ds/dt}\right\|=\left\|\frac{d\boldsymbol{T}/dt}{dr/dt}\right\|=\left\|\frac{\boldsymbol{T}'(t)}{\boldsymbol{r}'(t)}\right\|.$$

2. Since the unit tangent vector $\boldsymbol{T}(t)$ to the graph of smooth vector-valued function $\boldsymbol{r}(t)$ is given by

$$\boldsymbol{T}(t)=\frac{\boldsymbol{r}'(t)}{\|\boldsymbol{r}'(t)\|}.$$

$$\Rightarrow \qquad \boldsymbol{r}'(t)=\|\boldsymbol{r}'(t)\|\,\boldsymbol{T}(t). \qquad (10.29)$$

$$\Rightarrow \qquad \boldsymbol{r}''(t)=\|\boldsymbol{r}'(t)\|'\,\boldsymbol{T}(t)+\|\boldsymbol{r}'(t)\|'\,\boldsymbol{T}'(t) \qquad (10.30)$$

and since the unit normal vector $\boldsymbol{N}(t)$ to the graph of smooth vector-valued function $\boldsymbol{r}(t)$ at t is given by

$$\boldsymbol{N}(t)=\frac{\boldsymbol{T}'(t)}{\|\boldsymbol{T}'(t)\|}.$$

$$\Rightarrow \qquad \boldsymbol{T}'(t)=\|\boldsymbol{T}'(t)\|\,\boldsymbol{N}(t)$$

and by (1), we have $\|\boldsymbol{T}'(t)\|=k(t)\|\boldsymbol{r}'(t)\|$. Thus, we get

$$\boldsymbol{T}'(t)=k(t)\|\boldsymbol{r}'(t)\|\,\boldsymbol{N}(t).$$

Substitute the values of $T'(t)$ in (10.30), we get

$$\boldsymbol{r}''(t)=\|\boldsymbol{r}'(t)\|'\,\boldsymbol{T}(t)+k(t)\|\boldsymbol{r}'(t)\|^2\,\boldsymbol{N}(t). \qquad (10.31)$$

Now, consider

$$\boldsymbol{r}'(t)\times\boldsymbol{r}''(t)=\left(\|\boldsymbol{r}'(t)\|\,\boldsymbol{T}(t)\right)\times\left(\|\boldsymbol{r}'(t)\|'\,\boldsymbol{T}(t)+k(t)\|\boldsymbol{r}'(t)\|^2\,\boldsymbol{N}(t)\right)$$

(by using (10.29) and (10.31))

$$=k(t)\|\boldsymbol{r}'(t)\|^3\left(\boldsymbol{T}(t)\times\boldsymbol{N}(t)\right) \quad (\text{as } \boldsymbol{T}(t)\times\boldsymbol{T}(t)=0).$$

Since $\left(\boldsymbol{T}(t)\times\boldsymbol{N}(t)\right)$ is a unit vector, then

$$\|\boldsymbol{r}'(t)\times\boldsymbol{r}''(t)\|=k(t)\|\boldsymbol{r}'(t)\|^3.$$

$$\Rightarrow \qquad k(t)=\frac{\|\boldsymbol{r}'(t)\times\boldsymbol{r}''(t)\|}{\|\boldsymbol{r}'(t)\|^3}.$$

EXAMPLE 26 Find the curvature of the circular helix whose parametric equations are $x = a\cos t$, $y = a\sin t$ and $z = ct$, where $a > 0$.

Solution: The radius vector of the circular helix is given by

$$\boldsymbol{r}(t) = a\cos t\boldsymbol{i} + a\sin t\boldsymbol{j} + ct\boldsymbol{k}.$$

Then,

$$\boldsymbol{r}'(t) = (-a\sin t)\boldsymbol{i} + (a\cos t)\boldsymbol{j} + c\boldsymbol{k}$$

and

$$\boldsymbol{r}''(t) = (-a\cos t)\boldsymbol{i} + (-a\sin t)\boldsymbol{j}.$$

Now, $\boldsymbol{r}'(t) \times \boldsymbol{r}''(t) = \begin{vmatrix} \boldsymbol{i} & \boldsymbol{j} & \boldsymbol{k} \\ -a\sin t & a\cos t & c \\ -a\cos t & -a\sin t & 0 \end{vmatrix}$

$$= (ac\sin t)\boldsymbol{i} - (ac\cos t)\boldsymbol{j} + a^2\boldsymbol{k}.$$

So, $\|\boldsymbol{r}'(t) \times \boldsymbol{r}''(t)\| = \sqrt{(ac\sin t)^2 + (ac\cos t)^2 + a^4}$

$$= a\sqrt{a^2 + c^2}.$$

Also, $\|\boldsymbol{r}'(t)\| = \sqrt{(-a\sin t)^2 + (a\cos t)^2 + c^2} = \sqrt{a^2 + c^2}$.

Therefore, by (2) of Theorem 10.6, we have

$$k(t) = \frac{\|\boldsymbol{r}'(t) \times \boldsymbol{r}''(t)\|}{\|\boldsymbol{r}'(t)\|^3} = \frac{a\sqrt{a^2 + c^2}}{\sqrt{a^2 + c^2}} = a.$$

Hence, the curvature of circular helix is constant a.

EXAMPLE 27 Find the curvature the ellipse at the end points of major and minor axes, where the position vector is given as $\boldsymbol{r}(t) = 2\cos t\boldsymbol{i} + 3\sin t\boldsymbol{j}$, $0 \le t \le 2\pi$.

Solution: Since $\boldsymbol{r}(t) = 2\cos ti + 3\sin tj$.

Then,

$$\boldsymbol{r}'(t) = -2\sin t\boldsymbol{i} + 3\cos t\boldsymbol{j}$$

and

$$\boldsymbol{r}''(t) = -2\cos t\boldsymbol{i} - 3\sin t\boldsymbol{j}.$$

Now, $\boldsymbol{r}'(t) \times \boldsymbol{r}''(t) = \begin{vmatrix} \boldsymbol{i} & \boldsymbol{j} & \boldsymbol{k} \\ -2\sin t & 3\cos t & 0 \\ -2\cos t & -3\sin t & 0 \end{vmatrix}$

$$= 6k.$$

So, $\|\boldsymbol{r}'(t) \times \boldsymbol{r}''(t)\| = 6.$

Also, $\| \boldsymbol{r}'(t) \| = \sqrt{(-2\sin t)^2 + (3\cos t)^2} = \sqrt{4\sin^2 t + 9\cos^2 t}$.

Therefore, by (2) of Theorem 10.6, we have

$$k(t) = \frac{\| \boldsymbol{r}'(t) \times \boldsymbol{r}''(t) \|}{\| \boldsymbol{r}'(t) \|^3} = \frac{6}{\left(\sqrt{4\sin^2 t + 9\cos^2 t}\right)^3}.$$

Now, since at the end of minor axis of the ellipse $t = 0$ and $t = \pi$ (as the major axis of the given ellipse is along the y-axis).

So, $$k(0) = k(\pi) = \frac{6}{(9)^{\frac{3}{2}}} = \frac{2}{9}$$

and at the end of major axis of the ellipse $t = \dfrac{\pi}{2}$ and $t = \dfrac{3\pi}{2}$.

So, $$k\left(\frac{\pi}{2}\right) = k\left(\frac{3\pi}{2}\right) = \frac{6}{4^{\frac{3}{2}}} = \frac{3}{4}.$$

Hence, the radius of curvature at the end points of the major axis is greater than at the end points of the minor axis.

10.5.1 Radius of Curvature

The radius of curvature of a curve at any point P on a curve C in 2-space which has non-zero curvature, is defined as the radius $\rho = 1/k$ of a circle C with centred O on the concave side of the curve which has the same tangent and the same curvature at P as the given curve.

The circle C is called the *circle of curvature* at P or *osculating circle* at P and the centre O is called the *centre of curvature* at P.

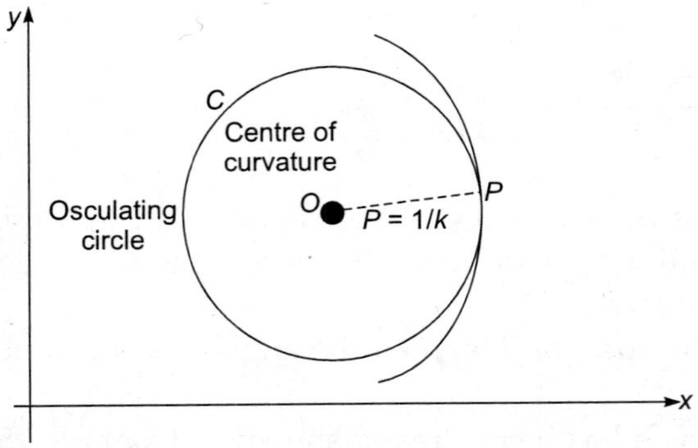

Figure 10.13

Remark: The osculating circle and the curve C not only touch at P, but they actually have the same curvature at P.

Geometrical Interpretation of Curvature in 2-Space

Let P be a point on the curve C in 2-space and the tangent vector T be drawn at P making an angle ϕ with the positive direction of x-axis (as shown in Figure 10.14).

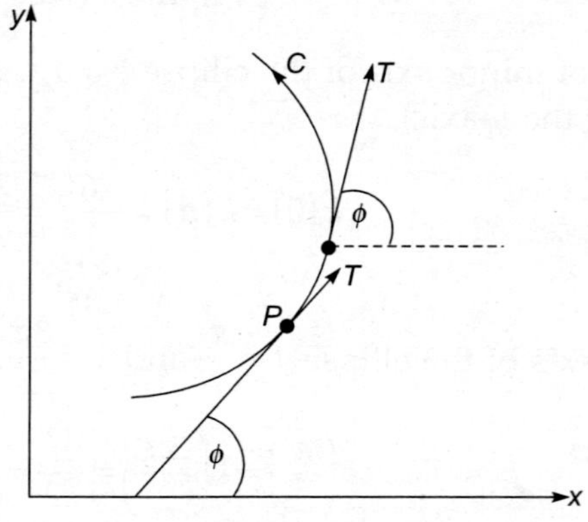

Figure 10.14

Then, the tangent vector $\boldsymbol{T}$ can be expressed in terms of ϕ as

$$\boldsymbol{T}(\phi) = \cos\phi\boldsymbol{i} + \sin\phi\boldsymbol{j}.$$

$$\Rightarrow \qquad \frac{d\boldsymbol{T}}{d\phi} = -\sin\phi\boldsymbol{i} + \cos\phi\boldsymbol{j}.$$

Also, $$\frac{d\boldsymbol{T}}{ds} = \frac{d\boldsymbol{T}}{d\phi}\frac{d\phi}{ds}.$$

Therefore, $$k(s) = \left\|\frac{d\boldsymbol{T}}{ds}\right\| = \left|\frac{d\phi}{ds}\right|\left\|\frac{d\boldsymbol{T}}{d\phi}\right\| = \left|\frac{d\phi}{ds}\right|\sqrt{(-\sin\phi)^2 + (\cos\phi)^2} = \left|\frac{d\phi}{ds}\right|.$$

Hence, the curvature is

$$k(s) = \left|\frac{d\phi}{ds}\right|.$$

This interprets that the curvature of the curve C in 2-space is the magnitude of the rate of change of ϕ with respect to s. So, the more rapidly change in ϕ with s, the greater the curvature of the curve C.

If the curve C is the straight line, then the angle ϕ is constant and so, the curvature $k(s) = \left|\frac{d\phi}{ds}\right| = 0$, which has been already shown in Example 25.

Formula Summary

The following formulae are for $\boldsymbol{T}$, $\boldsymbol{N}$ and $\boldsymbol{B}$:

$$\boldsymbol{T}(s)=\boldsymbol{r}'(s).$$

$$\boldsymbol{N}(s)=\frac{1}{k(s)}\frac{d\boldsymbol{T}}{ds}=\frac{\boldsymbol{r}''(s)}{\|\boldsymbol{r}''(s)\|}=\frac{\boldsymbol{r}''(s)}{k(s)}.$$

$$\boldsymbol{B}(s)=\frac{\boldsymbol{r}'(s)\times\boldsymbol{r}''(s)}{\|\boldsymbol{r}''(s)\|}=\frac{\boldsymbol{r}'(s)\times\boldsymbol{r}''(s)}{k(s)}.$$

$$\boldsymbol{T}(t)=\frac{\boldsymbol{r}'(t)}{\|\boldsymbol{r}'(t)\|}.$$

$$\boldsymbol{B}(t)=\frac{\boldsymbol{r}'(t)\times\boldsymbol{r}''(t)}{\|\boldsymbol{r}'(t)\times\boldsymbol{r}''(t)\|}.$$

$$\boldsymbol{N}(t)=\boldsymbol{B}(t)\times\boldsymbol{T}(t).$$

10.6 TANGENTIAL AND NORMAL COMPONENTS OF ACCELERATION AND APPLICATIONS

In this section, we will study the tangential and normal components of acceleration and its applications.

10.6.1 Tangential and Normal Components of Acceleration

When a body starts to accelerate or brakes are applied to stop it, then the acceleration acts in two directions; one is in the direction of the body's motion and other one in the direction of normal to the body. The following theorem enable us to know that how much acceleration acts in the direction of the body's motion which is indicated by the unit tangent vector T and how much in the direction of the principal unit normal N.

Theorem 10.7. An object is moving along a smooth curve such that $\boldsymbol{T}'\neq 0$, then the velocity and acceleration of the body are given by

$$\boldsymbol{V}=\left(\frac{ds}{dt}\right)\boldsymbol{T} \text{ and } \boldsymbol{A}=\left(\frac{d^2s}{dt^2}\right)\boldsymbol{T}+k\left(\frac{ds}{dt}\right)^2\boldsymbol{N}$$

respectively, and s is the arc length of the trajectory.

Proof: Since s is the arc length of the trajectory and the velocity of a moving object along a smooth curve is the rate of change of position vector w.r.t. t.

Thus, the formula of the velocity follows from the fact that the velocity has the magnitude $\|\boldsymbol{V}\| = \left(\frac{ds}{dt}\right)$ and it is in the direction of unit tangent vector.

To find the formula for A**:**

We know that

$$\frac{d\boldsymbol{T}}{ds} = \left\|\frac{d\boldsymbol{T}}{ds}\right\| \boldsymbol{N} = k\boldsymbol{N} \tag{10.32}$$

(as the derivative of $\boldsymbol{T}(s)$ is in the direction of unit normal vector to $\boldsymbol{T}$).

By, the definition of $\boldsymbol{A}$, we have

$$\begin{aligned}
\boldsymbol{A} &= \frac{d\boldsymbol{V}}{dt} \\
&= \frac{d}{dt}\left(\boldsymbol{T}\frac{ds}{dt}\right) (\text{by formula of } V) \\
&= \boldsymbol{T}\frac{d^2 S}{dt^2} + \frac{d\boldsymbol{T}}{dt}\frac{ds}{dt} \\
&= \boldsymbol{T}\frac{d^2 S}{dt^2} + \frac{ds}{dt}\left\{\frac{d\boldsymbol{T}}{ds}\frac{ds}{dt}\right\} \\
&= \boldsymbol{T}\frac{d^2 S}{dt^2} + \left(\frac{ds}{dt}\right)^2 \frac{d\boldsymbol{T}}{ds} \\
&= \boldsymbol{T}\frac{d^2 S}{dt^2} + \left(\frac{ds}{dt}\right)^2 (k\boldsymbol{N}) \qquad (\text{by using } (10.32)).
\end{aligned}$$

Thus, we conclude that the velocity of a moving object, at each point of the trajectory is towards in the direction of the unit tangent vector $\boldsymbol{T}$. But the acceleration $\boldsymbol{A}$ may have both tangential and normal components. The acceleration is always in the plane determined by $\boldsymbol{T}$ and the unit normal $\boldsymbol{N}$, even if the trajectory has turns and twists.

Therefore, the acceleration A of a moving object can be written as

$$\boldsymbol{A} = A_T\boldsymbol{T} + A_N\boldsymbol{N},$$

where $A_T = \frac{d^2 S}{dt^2}$ is, the tangential component and

$A_N = k\left(\frac{ds}{dt}\right)^2$ is the normal component.

These components of acceleration are useful to provide us the way of a moving object travels along its trajectory. But sometimes, they are not convenient computational

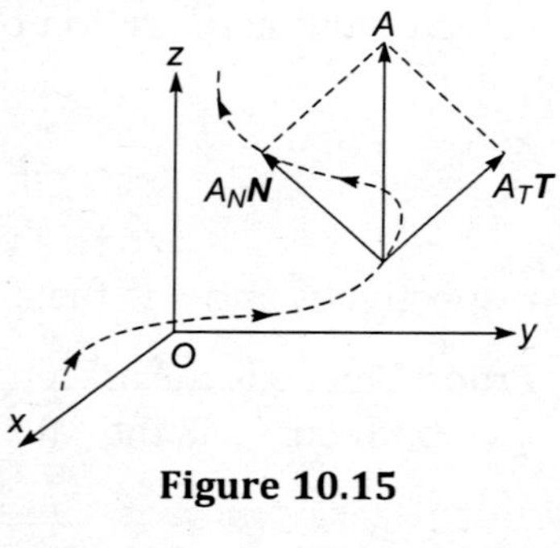

Figure 10.15

forms to find the acceleration. Thus, the following theorem gives an alternative formulae for finding acceleration in terms of position vector $\boldsymbol{R}(t)$.

Theorem 10.8 Let $\boldsymbol{R}(t)$ be the position vector of a moving object along a smooth curve C. Then the tangential and normal components of the acceleration of an object are given by

$$A_T \frac{\boldsymbol{R}' \cdot \boldsymbol{R}''}{\|\boldsymbol{R}'\|} \text{ and } A_N = \frac{\|\boldsymbol{R}' \times \boldsymbol{R}''\|}{\|\boldsymbol{R}'\|}.$$

Proof: Let θ be an angle between the vector $\boldsymbol{R}'$ and $\boldsymbol{R}''$. Since $\boldsymbol{R}''$ is an acceleration of the moving object, then

$$A_T = \| \boldsymbol{A} \| \cos\theta = \frac{\|\boldsymbol{R}'\| \, \|\boldsymbol{A}\| \cos\theta}{\|\boldsymbol{R}'\|} = \frac{\boldsymbol{R}' \cdot \boldsymbol{R}''}{\|\boldsymbol{R}'\|}$$

and

$$A_N = \| \boldsymbol{A} \| \sin\theta = \frac{\|\boldsymbol{R}'\| \, \|\boldsymbol{A}\| \sin\theta}{\|\boldsymbol{R}'\|} = \frac{\|\boldsymbol{R}' \times \boldsymbol{R}''\|}{\|\boldsymbol{R}'\|}$$

10.6.2 Applications

According to Newton's second law of motion, the total force acting on a moving body of mass m is $\boldsymbol{F} = m\boldsymbol{A}$, where $\boldsymbol{A}$ is the acceleration of the moving body.

Now, since $\boldsymbol{A} = A_T\boldsymbol{T} + A_N\boldsymbol{N}$.

$$\Rightarrow \quad \boldsymbol{F} = m(A_T\boldsymbol{T} + A_N\boldsymbol{N}).$$

$$\Rightarrow \quad \boldsymbol{F} = (mA_T)\boldsymbol{T} + (mA_N)\boldsymbol{N}.$$

Therefore, the force can be written as

$$\boldsymbol{F} = F_T\boldsymbol{T} + F_N\boldsymbol{N},$$

where $F_T = mA_T = m\dfrac{d^2s}{dt^2}$ and $F_N = mA_N = mk\left(\dfrac{ds}{dt}\right)^2$.

Now, suppose that a car is moving on the road and expected that the car is going to skid if the car takes sharp turn at moderate speed or even a gradual turn at high speed.

Mathematically, a sharp turn happens, when the radius of curvature $\rho = 1/k$ is small and high speed means that the ds/dt is large.

Either of the case occurs when $F_N = mk\left(\dfrac{ds}{dt}\right)^2$ will be relatively large.

And tyre the car will be running smoothly on the road, only if there is a correspondingly large fictional force between the tyre of the car and the surface of the road.

When forces, masses and accelerations are involved in the equation of motion of an object, then it is useful to express the mass m in slugs of an object of weight W (in pounds).

Note: A slug is the unit of mass that receives an acceleration of $\frac{1\text{ ft}}{s^2}$ and when a force of 1 lb is applied on it, i.e., $1\text{ slug} = \frac{1\text{ lb}}{\frac{1\text{ ft}}{s^2}} = \text{lb.}\frac{s^2}{\text{ft}}$.

Theorem 10.9 If an object is moving with constant speed, then its acceleration is always orthogonal to the direction of motion.

Proof: Let $\boldsymbol{R}(t)$ be the position vector of an object. Then, since the object is moving with constant speed $\|\boldsymbol{R}'(t)\|$ is constant.

Thus, by Theorem 10.4, $\boldsymbol{R}'(t)$ is orthogonal to the vector $\boldsymbol{R}''(t) = \boldsymbol{A}(t)$.

Now, since $\boldsymbol{R}'(t)$ is in the direction of the motion of an object along its trajectory, it implies that the acceleration $\boldsymbol{A}$ is orthogonal to the direction of motion.

As an application of Theorem 10.9, if an object is moving with constant speed v_0 along a circular path of radius $\rho\left(k = \frac{1}{\rho}\right)$, then the acceleration of the object is towards the centre of the path (as shown in Figure 10.16) and its magnitude is

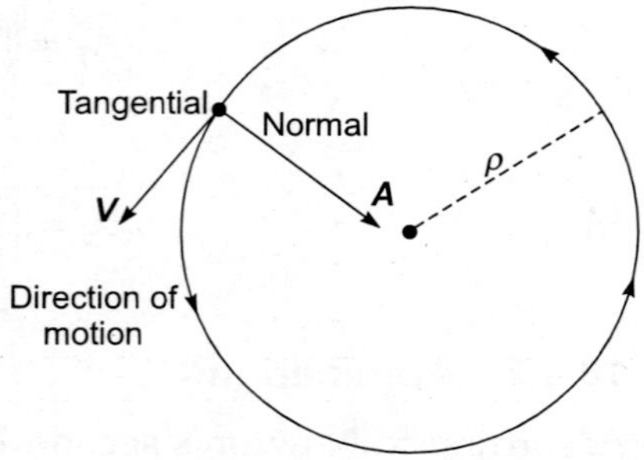

Figure 10.16

$$A_N = k\left(\frac{ds}{dt}\right)^2 = \frac{1}{\rho}v_0^2.$$

Now, we will find the period of a satellite which is based on the application of Theorem 10.9.

Problem. An artificial satellite travels at constant speed in a stable circular orbit 20,000 km above the earth's surface. How much time the satellite will take to make one complete path around the earth?

Solution: Let m be the mass of the satellite and v be the speed. Suppose the earth is a sphere of radius 6440 km, so the curvature of the path is $k = 1/R$, where

$$R = 6440 + 20000 = 26440 \text{ km (i.e., } \rho = \text{radius of the earth + height).}$$

is, the distance of the satellite from the centre of the earth.

The satellite maintains a stable orbit, because the magnitude of the force $\boldsymbol{F}_c = \frac{mv^2}{R}$ (produced by the rate of change of tangential velocity, called *centripetal acceleration*) is equal to the magnitude of the force due to gravity $\boldsymbol{F}_g$ in the opposite direction.

By, Newton's law of universal gravitation,

$$\|\boldsymbol{F}_g\| = GmM/R^2,$$

where M is the mass of the earth and G is the gravitational constant.

Thus, by stability, we have

$$\frac{mv^2}{R} = \frac{GmM}{R^2}.$$

$$\Rightarrow \qquad v = \sqrt{\frac{GM}{R}} = \sqrt{\frac{398,600}{26440}} \approx 3.88273653$$

(as GM = 398,600 km^3/s^2).

Therefore, the speed of the satellite is 3.883 km/s.

Let T be the time taken by the satellite to complete one path around the earth (called ***period of satellite***). In each period, the satellite travels a distance which is equal to the circumference of the earth of the radius R = 26,440 km and since, it travels at a speed v km/s, we have

$$vT = 2\pi R.$$

$$\Rightarrow \qquad T = \frac{2\pi R}{v} = \frac{2\pi(26,440)}{3.883} \approx 42,786.16852 \text{ seconds.}$$

Hence, the satellite completes one path around the earth in 713 minutes (i.e., 11 hr and 53 min).

EXAMPLE 28 Find the tangential and normal components of the object's acceleration, where the position vector $\boldsymbol{R}(t)$ of the object is given by $\boldsymbol{R}(t) = \langle t\sin t, t\cos t\rangle$.

Solution: Since $\boldsymbol{R}(t) = \langle t\sin t, t\cos t\rangle$.

$$\Rightarrow \qquad \boldsymbol{V}(t) = \boldsymbol{R}'(t) = \langle t\cos t + \sin t, -t\sin t + \cos t\rangle$$

$$\Rightarrow \qquad \boldsymbol{A}(t) = \boldsymbol{R}''(t) = \langle -t\sin t + 2\cos t, -t\cos t - 2\sin t\rangle.$$

Now, $\boldsymbol{R}' \cdot \boldsymbol{R}'' = (t\cos t + \sin t)(-t\sin t + 2\cos t)$

$$-(-t\sin t + \cos t)(t\cos t + 2\sin t).$$

$$\Rightarrow \qquad \boldsymbol{R}' \cdot \boldsymbol{R}'' = t.$$

and $\|\boldsymbol{R}'\| = \sqrt{(t\cos t + \sin t)^2 + (-t\sin t + \cos t)^2} = \sqrt{t^2 + 1}$.

Thus, the tangential component of an acceleration is given by

$$A_T = \frac{\boldsymbol{R}' \cdot \boldsymbol{R}''}{\|\boldsymbol{R}'\|} = \frac{t}{\sqrt{t^2 + 1}}.$$

$$\text{Now, } \boldsymbol{R}' \times \boldsymbol{R}'' = \begin{vmatrix} \boldsymbol{i} & \boldsymbol{j} & \boldsymbol{k} \\ t\cos t + \sin t & -t\sin t + \cos t & 0 \\ -t\sin t + 2\cos t & -t\cos t - 2\sin t & 0 \end{vmatrix}$$

$$= -(t^2 + 2)\boldsymbol{k}.$$

Therefore, the normal component of an acceleration is given by

$$A_N = \frac{\|\boldsymbol{R}' \times \boldsymbol{R}''\|}{\|\boldsymbol{R}'\|} = \frac{t^2 + 2}{\sqrt{t^2 + 1}}.$$

Hence, the acceleration of an object is given by

$$\boldsymbol{A} = \frac{t}{\sqrt{t^2 + 1}}\boldsymbol{i} + \frac{t^2 + 2}{\sqrt{t^2 + 1}}\boldsymbol{j}.$$

Note: From Figure 10.15, we see that $\|A\|$ is the hypotenuse of a right angled triangle and other two sides are $\|A_T\boldsymbol{T}\| = |A_T|$ and $\|A_N\boldsymbol{N}\| = |A_N|$. Therefore, by Pythagorean theorem, we have

$$\|\boldsymbol{A}\|^2 = (A_T)^2 + (A_N)^2$$

and so,

$$A_N = \sqrt{\|\boldsymbol{A}\|^2 - A_T^2}.$$

EXAMPLE 29 An object moves with the position vector $\boldsymbol{R}(t) = \sin t\,\boldsymbol{i} + \cos t\,\boldsymbol{j} + \sin t\,\boldsymbol{k}$. Find the tangential and normal components of an acceleration.

Solution: The velocity of an object is given by

$$\boldsymbol{V}(t) = \frac{d\boldsymbol{R}}{dt} = \cos t\boldsymbol{i} - \sin t\boldsymbol{j} + \cos t\boldsymbol{k}$$

and the acceleration of an object is given by

$$\boldsymbol{A}(t) = \frac{d\boldsymbol{V}}{dt} = -\sin t\boldsymbol{i} - \cos t\boldsymbol{j} - \sin t\boldsymbol{k}.$$

We know that,

$$\frac{ds}{dt} = \|\boldsymbol{V}\| = \sqrt{\cos^2 t + \sin^2 t + \cos^2 t} = \sqrt{1 + \cos^2 t}.$$

Thus, the tangential component of an acceleration is given by

$$A_T = \frac{d^2 s}{dt^2} = \frac{-\cos t \sin t}{\sqrt{1 + \cos^2 t}}.$$

and the normal component of an acceleration is given by

$$A_N = \sqrt{\|\boldsymbol{A}\|^2 - A_T^2}$$

$$= \sqrt{\left(1+\sin^2 t\right) - \frac{\cos^2 t \sin^2 t}{1+\cos^2 t}}$$

$$= \sqrt{\frac{2}{1+\cos^2 t}}.$$

Hence, the acceleration of an object is given by

$$A = \frac{-\cos t \sin t}{\sqrt{1+\cos^2 t}} i + \sqrt{\frac{2}{1+\cos^2 t}} j.$$

EXAMPLE 30 A car weighing 2000 lb makes a turn on a flat road while travelling at 50 ft/s. If the radius of the turn is 70 ft, how much frictional force is required to keep the car from skidding?

Solution: The force tending to push the car off the road is

$$F_N = mk\left(\frac{ds}{dt}\right)^2,$$

where m is the mass of the car and k is the curvature of the road, i.e., it is the force that is required so that the car does not skid.

We are given that,

$$\frac{ds}{dt} = 50 \text{ ft/s} \text{ and } W = 2000 \text{ lb}.$$

Thus, the mass of the car is $m = \frac{W}{g} = \frac{2000}{32} = 62.5$ slugs.

Since, the radius of the turn is 70 ft. So, the curvature is $k = \frac{1}{70}$.

Therefore,

$$F_N = 62.5\left(\frac{1}{70}\right)(50)^2 \approx 2232\, lb.$$

Exercise

1. Sketch the graph of the following vector functions:
   (a) $F(t) = (3-t)i + (2t)j + (3t-4)k$.
   (b) $F(t) = (2\sin t)i - (2\cos t)j + 3tk$.
   (c) $F(t) = 2ti + t^2 j$.
   (d) $F(t) = (\sin t)i - (\cos t)j$.

(e) $\boldsymbol{F}(t)=t\boldsymbol{i}-4\boldsymbol{k}$.

(f) $\boldsymbol{F}(t)=(\cos t)\boldsymbol{i}+(\sin t)\boldsymbol{j}+t\boldsymbol{k}$.

(g) $\boldsymbol{F}(t)=(1-t)\boldsymbol{i}+t^2\boldsymbol{j}+t\boldsymbol{k}$.

2. Find the domain of the vector following the vector functions:

(a) $\boldsymbol{F}(t)=2t\boldsymbol{i}-3t\boldsymbol{j}+\frac{1}{t}\boldsymbol{k}$.

(b) $\boldsymbol{F}(t)=(\sin t)\boldsymbol{i}+(\cos t)\boldsymbol{j}+(\tan t)\boldsymbol{k}$.

(c) $h(t)\boldsymbol{F}(t)$, where $h(t)=\sin t$ and $\boldsymbol{F}(t)=\frac{1}{\cos t}\boldsymbol{i}+\frac{1}{\sin t}\boldsymbol{j}+\frac{1}{\tan t}\boldsymbol{k}$.

(d) $\boldsymbol{F}(t)-\boldsymbol{G}(t)$, where $\boldsymbol{F}(t)=(\ln t)\boldsymbol{i}+3t\boldsymbol{j}-t^2\boldsymbol{k}$ and $G(t)=\boldsymbol{i}+5t\boldsymbol{j}-t^2\boldsymbol{k}$.

3. Let $\boldsymbol{F}(t)=t^2\boldsymbol{i}+t\boldsymbol{j}-(\sin t)\boldsymbol{k}$ and $\boldsymbol{G}(t)=t\boldsymbol{i}+\frac{1}{t}\boldsymbol{j}+5\boldsymbol{k}$, find

(a) $(\boldsymbol{F}+\boldsymbol{G})(t)$ (b) $(e^t\boldsymbol{F})(t)$ (c) $(\boldsymbol{F}\times\boldsymbol{G})(t)$ (d) $(\boldsymbol{F}\cdot\boldsymbol{G})(t)$.

4. Let $\boldsymbol{F}(t)=2t\boldsymbol{i}-5\boldsymbol{j}+t^2\boldsymbol{k}$ and $\boldsymbol{H}(t)=(\sin t)\boldsymbol{i}+e^t\boldsymbol{j}$, find

(a) $\boldsymbol{F}(t)\cdot\boldsymbol{H}(t)$ (b) $\boldsymbol{F}(t)\times\boldsymbol{H}(t)$ (c) $2e^t\boldsymbol{F}(t)+10\boldsymbol{H}(t)$.

5. Find each of the limit as following:

(a) $\lim_{t\to 2}\left[\left(t^2-3\right)\boldsymbol{i}+e^t\boldsymbol{j}+(\sin \pi t)\boldsymbol{k}\right]$.

(b) $\lim_{t\to 0}\left\{\frac{te^t}{1-e^t}\boldsymbol{i}+\frac{e^{t-1}}{\cos t}\boldsymbol{j}\right\}$.

6. Show that $\lim_{t\to 1}\{\boldsymbol{F}(t)\times\boldsymbol{G}(t)\}=\left\{\lim_{t\to 1}\boldsymbol{F}(t)\right\}\times\left\{\lim_{t\to 1}\boldsymbol{G}(t)\right\}$, where

$\boldsymbol{F}(t)=t\boldsymbol{i}+(1-t)\boldsymbol{j}+t^2\boldsymbol{k}$ and $\boldsymbol{G}(t)=e^t\boldsymbol{i}-\left(3+e^t\right)\boldsymbol{k}$.

7. Find the vector function **F** whose graph is the curve of intersection of the following:

(a) Hemisphere $z=\sqrt{4-x^2-y^2}$ and the parabolic cylinder $y=x^2$.

(b) Parabola $y=x^2$ and the plane $z=2$.

(c) Hemisphere $z=\sqrt{9-x^2-y^2}$ and the parabolic cylinder $x=y^2$.

8. Determine all values of t for which the following vector functions are continuous:

(a) $\boldsymbol{F}(t)=(\sin t)\boldsymbol{i}+(1-t)^{-1}\boldsymbol{j}+(\ln t)\boldsymbol{k}$.

(b) $\boldsymbol{F}(t)=t\boldsymbol{i}+3\boldsymbol{j}-(1-t)\boldsymbol{k}$.

(c) $F(t) = t\boldsymbol{i} - \frac{1}{t}\boldsymbol{k}$.

(d) $F(t) = \left(e^t \sin t\right)\boldsymbol{i} + \left(e^t \cos t\right)\boldsymbol{j}$.

9. For what values of t is $F(t) = |t|\boldsymbol{i} + (\cos t)\boldsymbol{j} + (t-5)\boldsymbol{k}$ differentiable?
10. Find the derivative of the following vector functions:
    (a) $F(t) = e^t\boldsymbol{i} + \sin t\boldsymbol{j} + \left(t^3 + 5t\right)\boldsymbol{k}$.
    (b) $F(t) = t\boldsymbol{i} + t^2\boldsymbol{j} + \left(t + t^3\right)\boldsymbol{k}$.
    (c) $F(t) = \left(t + 2t^2\right)\boldsymbol{i} + \left(t^2 - t\right)\boldsymbol{j} + \left(t^2 + 3\right)\boldsymbol{k}$.
    (d) $F(t) = \cos t\boldsymbol{i} + \sin t\boldsymbol{j} + 3\cos t\boldsymbol{k}$.
11. Find the first and second derivatives of the following vector functions:
    (a) $F(t) = \langle e^{2t}, \left(1 - t^2\right), \cos 2t\rangle$.
    (b) $F(t) = \langle t^2, t^{-1}, e^{2t}\rangle$.
    (c) $F(t) = \langle \sin t, \cos t, t^2\rangle$.
12. If $F(t) = \langle 1, t, t^2\rangle$ and $G(t) = \langle t, e^t, 3\rangle$, then show that

    $(F \times G)'(t) = (F' \times G)(t) + (F \times G')(t)$.
13. If $F(t) = \langle 1, e^t, t^2\rangle$ and $G(t) = \langle 3t^2, e^{-t}, -2t\rangle$, then, find
    (a) $\frac{d}{dt}\left(2F(t) + t^3 G(t)\right)$.
    (b) $\frac{d}{dt}(F(t) \cdot G(t))$.
14. A particle's position at time t is determined by the following position vectors. Analyze the motion of the particle and in particular, find the velocity, speed, acceleration and direction of motion of the particle:
    (a) $R(t) = \cos t\boldsymbol{i} + \sin t\boldsymbol{j} + t^3\boldsymbol{k}$, at time $t = 2$.
    (b) $R(t) = t\boldsymbol{i} + t^2\boldsymbol{j} + 2t\boldsymbol{k}$, at time $t = 1$.
    (c) $R(t) = e^t\boldsymbol{i} + e^{-t}\boldsymbol{j} + e^{2t}\boldsymbol{k}$, at time $t = \ln 2$.
15. Find the tangent vector on the graph of the following vector functions:
    (a) $F(t) = e^{2t}\boldsymbol{i} + \left(t^2 - t\right)\boldsymbol{j} + (\ln t)\boldsymbol{k}$, at $t = 0.2$.
    (b) $F(t) = t^2\boldsymbol{i} + 2t\boldsymbol{j} + \left(t^2 + t^3\right)\boldsymbol{k}$, at $t = 0, 1, -1$.
    (c) $F(t) = t^2\boldsymbol{i} + (\cos t)\boldsymbol{j} + \left(t^2 \cos t\right)\boldsymbol{k}$, at $t = 0, \frac{\pi}{2}$.

16. Find the parametric equations for the tangent line to the graph of the vector function $F(t)=\langle t^{-3},t^{-2},t^{-1}\rangle$, at $t=-1$.
17. Find the value of the following integrals:

    (a) $\int_0^{\pi}(t\boldsymbol{i}+3\boldsymbol{j}-\sin t\boldsymbol{k})dt$.

    (b) $\int_0^{a}\left(\left(t\sqrt{1+t^2}\right)\boldsymbol{i}+\left(\dfrac{1}{1+t^2}\right)\boldsymbol{j}\right)dt$.

    (c) $\int_0^{2a}(\cos t\boldsymbol{i}+\sin t\boldsymbol{j}+\sin t\cos t\boldsymbol{k})dt$.
18. Find the position of a particle at any time t, if the velocity $V(t)$ of a particle moving in space and the position vector at time t = 0, $R(0)$ are the following:
    (a) $\boldsymbol{V}(t)=e^t\boldsymbol{i}+t^2\boldsymbol{j}+(\cos 2t)\boldsymbol{k}$, $\boldsymbol{R}(0)=2\boldsymbol{i}+\boldsymbol{j}-\boldsymbol{k}$.
    (b) $\boldsymbol{V}(t)=t^2\boldsymbol{i}-e^{2t}\boldsymbol{j}+\sqrt{t}\boldsymbol{k}$, $\boldsymbol{R}(0)=\boldsymbol{i}+4\boldsymbol{j}-\boldsymbol{k}$.
    (c) $\boldsymbol{V}(t)=2\sqrt{t}\boldsymbol{i}+\cos t\boldsymbol{j}$, $\boldsymbol{R}(0)=\boldsymbol{i}+\boldsymbol{j}$.
    (d) $\boldsymbol{V}(t)=e^t\boldsymbol{i}+t^2\boldsymbol{j}$, $\boldsymbol{R}(0)=\boldsymbol{i}-\boldsymbol{j}$.
19. Determine whether the graph of the vector function $\boldsymbol{F}(t)=\langle t^2+1,\cos t,e^t+e^{-t}\rangle$ is smooth for all t.
20. Find the position vector $\boldsymbol{R}(t)$ and the velocity vector $\boldsymbol{V}(t)$, if the acceleration $\boldsymbol{A}(t)$, initial position $\boldsymbol{R}(0)$ and the initial velocity $\boldsymbol{V}(0)$ are the following:
    (a) $\boldsymbol{A}(t)=t^2\boldsymbol{i}-2\sqrt{t}\boldsymbol{j}+e^{3t}\boldsymbol{k}$, $\boldsymbol{R}(0)=2\boldsymbol{i}+\boldsymbol{j}-\boldsymbol{k}$ and $\boldsymbol{V}(0)=\boldsymbol{i}-\boldsymbol{j}-2\boldsymbol{k}$.
    (b) $\boldsymbol{A}(t)=24t^2\boldsymbol{i}+4\boldsymbol{j}$, $\boldsymbol{R}(0)=\boldsymbol{i}+2\boldsymbol{j}$ and $\boldsymbol{V}(0)=0$.
21. A boy standing at the edge of a cliff throws a ball upward at a 30° angle with an initial speed of 64 ft/s. Suppose when the ball leaves the boy's hand, it is 48 ft above the ground at the base of the cliff. Then
    (a) What are the time of flight of the ball and its range?
    (b) What are the velocity of the ball and its speed at impact?
    (c) What is the highest point reached by the ball during its flight?
22. A projectile is fired from ground level at an angle of 40° with muzzle speed 110 ft/s. Find the time of flight and the range.
23. Find the time of flight T_f (to the nearest tenth second) and the range R_f (to the nearest unit) of a projectile fired (in vacuum) from the ground level at the given angle α with the initial speed v_0. Assume that g = 32 ft/s^2 or g = 9.8 m/s^2

(a) $\alpha = 35°$, $v_0 = 128$ ft/s.

(b) $\alpha = 48.5°$, $v_0 = 850$ m/s.

(c) $\alpha = 23.74°$, $v_0 = 23.3$ m/s.

(d) $\alpha = 14.11°$, $v_0 = 100$ ft/s.

24. An object moves along the given curve in the plane. Find its velocity and acceleration in terms of the unit polar vectors $\boldsymbol{u}_r$ and $\boldsymbol{u}_\theta$:

    (a) $x = 2t$, $y = t$.

    (b) $r = 5(1 + \cos\theta)$, $\theta = 2t + 1$.

    (c) $r = e^{-\theta}$, $\theta = 1 - t$.

25. The position vector of a moving body is $\boldsymbol{R}(t) = 2t\boldsymbol{i} - t^2\boldsymbol{j}$ for $t \geq 0$. Express R and the velocity vector $\boldsymbol{V}(t)$ in terms of $\boldsymbol{u}_r$ and $\boldsymbol{u}_\theta$.

26. A shell fired from ground level at an angle of 45° hits the ground 2000 m away. What is the muzzle speed of the shell?

27. At what angle should a projectile be fired from the ground level, if its muzzle speed is 167.1 ft/s and the desired range is 600 ft.

28. A baseball player hits a baseball at a 30° angle with a speed of 144 ft/s. If the ball is 4 ft above the ground level when it is hit. Then,

    (a) What is the maximum height reached by the ball?

    (b) How far will it travel from the home plate it lands?

    (c) If it just barely clears a 5 feet wall in the outfield, how far is the wall from the home plate?

29. Find the tangential and normal components of the acceleration of an object that moves with position vector

    (a) $\boldsymbol{R}(t) = \langle t^3, t^2, t\rangle$.

    (b) $\boldsymbol{R}(t) = \langle \cos t, \sin t, t\rangle$.

    (c) $\boldsymbol{R}(t) = \langle t, t^2\rangle$.

    (d) $\boldsymbol{R}(t) = \langle t, t^2, t\rangle$.

30. The position of an object at time t is (x, y), where $x = 1 + \cos 2t$, $y = \sin 2t$. Find the velocity, the acceleration and the normal and tangential components of acceleration of the object at time t.

31. Find the tangential and normal components of the acceleration of an object that moves along the parabolic path $y = 4x^2$; at the instant the speed is $\frac{ds}{dt} = 20$.

32. Find the curvature of the curves whose position vectors are as follows:

    (a) $\boldsymbol{r}(t) = t^2\boldsymbol{i} + t^3\boldsymbol{j}$.

    (b) $\boldsymbol{r}(t) = e^{3t}\boldsymbol{i} + e^{-t}\boldsymbol{j}$

    (c) $\boldsymbol{r}(t) = 4\cos t\boldsymbol{i} + 4\sin t\boldsymbol{j} + t\boldsymbol{k}$.

    (d) $x = \cosh t$, $y = \sinh t$, $z = t$.

33. Find the radius of curvature and curvature at the stated point.

    (a) $r(t) = 3\cos ti + 4\sin tj + tk$, at $t = \pi/2$.

    (b) $r(t) = e^t\cos ti + e^t\sin tj + e^tk$, at $t = 0$.

    (c) $x = t^2$, $y = t^3$, at $t = \dfrac{1}{2}$.

    (d) $x = e^{3t}$, $y = e^{-t}$ at $t = 0$.

    (e) $x = t$, $y = \dfrac{1}{t}$, at $t = 1$.

11 CHAPTER

Practical Work

Here, we will discuss the practical works that perform on a computer using Mathematica software. In this chapter, we discuss the modeling of the some problems using Mathematica like: plotting of graphs of function of the type; plotting the graphs of polynomial of degree 4 and 5, the derivative graph, the second derivative graph and comparing them; sketching parametric curves; and tracing of conics in Cartesian coordinates.

11.1 INTRODUCTION

Mathematica notebook is an interactive environment. A notebook is a collection of Mathematica statements, outputs and graphics. When you type a command (like 4+4) and execute Mathematica (**Shift+Enter**), then Mathematica responds with the answer on the next line (8). You then type another command and so on. Each command you type in Mathematica will appear on screen in boldface font and its output will appear in a plain font. You can also use Mathematica just like a calculator for basic Arithmetic (Table 11.1).

Table 11.1 Commands for Basic Arithmetic

Arithmetic	Command for basic arithmetic
Addition	$a+b+c$
Subtraction	$a-b$
Multiplication	$a*b$ or ab
Division	a/b
Power	$a\wedge b$
Numerical value	$N[a,m]$, up to m significant digits

From Table 11.1, we have seen that for multiplication either we use * or *space*.

Here, we have some example for arithmetic operations:

```
In[1]:=2+2
Out[1]=4
In[2]:=4*3
Out[2]= 12
In[3]:=6/7
```

Out[3]= $\frac{6}{7}$

```
In[4]:= N[6/7]
Out[4]= 0.857143
In[5]:= N[6/7, 10]
Out[5]= 0.8571428571
```

In Mathematica, arithmetic operations are grouped according to the standard mathematical conventions.

```
In[6]:=3 * (4 + 3) / 5
```

Out[6]= $\frac{21}{5}$

When we perform a long calculation, it is convenient to give names to your variables or introducing named variables. For example, if we set the variable x to 3. Whenever x appears in syntax, Mathematica replace it with the value 3.

```
In[7]:= x = 3
Out[7]= 3
In[8]:= x^3
Out[8]=27
```

We want to remove any assigned value to x, we use command *Clear*[x].

Mathematica includes a large collection of mathematical functions. Here are a few of the common ones:

Sqrt[x]	square root ($\sqrt{x}$)
Exp[x]	exponential (e^x)
Log[x]	natural logarithm ($\log_e x$)
Log[b, x]	logarithm to base b ($\log_b x$)
Sin[x], Cos[x], Tan[x]	trigonometric functions (with arguments in radians)
ArcSin[x], ArcCos[x], ArcTan[x]	inverse trigonometric functions

$n!$	factorial (product of integers 1, 2, …, n)
Abs[x]	absolute value
Round[x]	closest integer to x
Mod[n, m]	n modulo m (remainder on division of n by m)
RandomReal[]	pseudorandom number between 0 and 1
Max[x, y, …], Min[x, y, …]	maximum, minimum of x, y, …
FactorInteger[n]	prime factors of n (see "Integer and Number Theoretic Functions")

11.2 DEFINING A FUNCTION, DERIVATIVES AND PLOTTING A FUNCTION

Mathematica has several built in functions, however, we can define our own functions in Mathematica like this:

```
In [1] := f [x_] := x ^ 3
In [1] := f [3]
Out [1] = 27
In [2] := f [x_] := x² + 2x − 4
In [2] := f [1]
Out [2] = −1
In [3] := f [π]
Out [3] = − 4 + 2π + π²
```

11.2.1 Plotting a Function

Mathematica is known for its graphical capability. We use *Plot* command to get a two-dimensional graph of the functions:

$$\text{Plot}\left[f[x],\ \{x, x_{min}, x_{max}\}\right]$$

The *Plot* command takes two arguments separated by comma, first $f[x]$ is the function to be graphed and second is called an iterator. We begin with a simple example

```
In[4]:= f[x_]:=x^2-2 x+4;
        Plot[f[x],{x,-1,3}]

Out[4]=
```

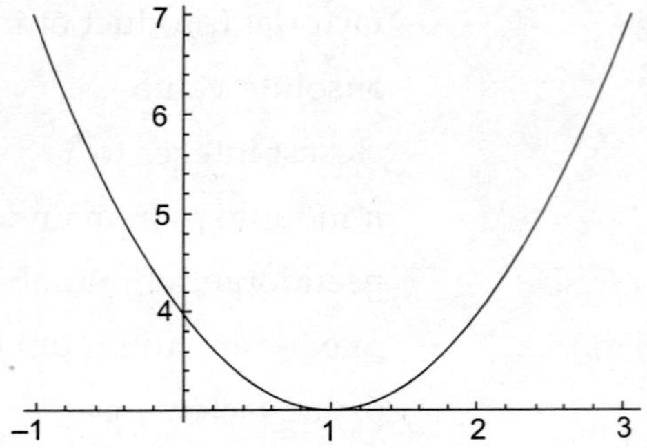


```
In [5] := Plot[Sin[x], {x, 0, 6 Pi}]
Out[5]=
```

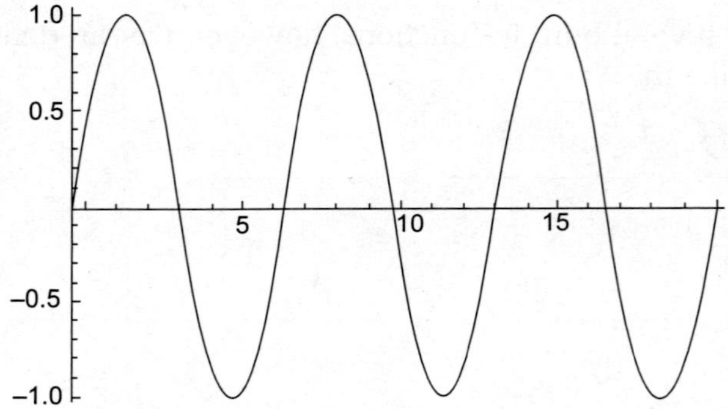

Plot several functions in single graph as

```
In [6] := Plot[{Sin[x], Sin[2x], Sin[3x]}, {x, 0, 2Pi}]
Out[6]=
```

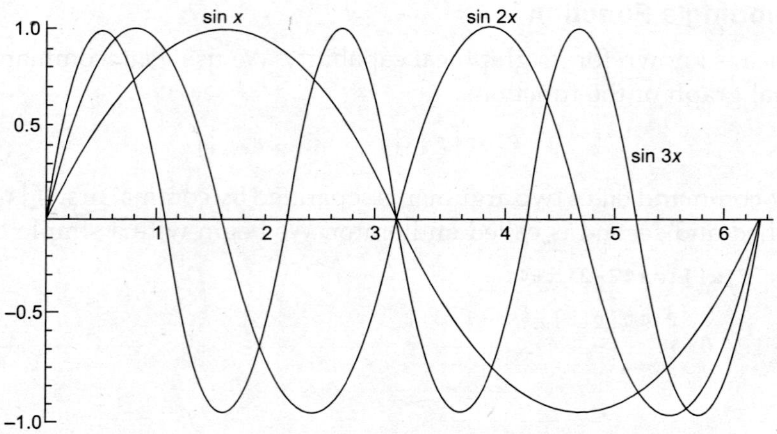

11.2.2 Plot Options

A plot of the graph often demands careful investigation. In this section, we will see how to twist the *Plot* command so that we get the most out of our graph. Plot[] has many options, we will give only few examples of plot options here.

<u>PlotRange:</u>

$$\texttt{Plot}\left[\texttt{function}, \{x, x_{min}, x_{max}\}, \texttt{PlotRange} \to \texttt{Full}\right]$$

The option **PlotRange** is set by default to an automated setting which will sometimes result in a graph with a truncated vertical scale. But suppose we want to see the function over the full domain. So, we can specify this by typing **PlotRange → Full** as an additional argument to the Plot command. It must be placed after the first two required arguments. For example, we want to show a graph between some specified values on the y axis and then we use PlotRange command as

```
In [7] := Plot[x ^ 2 + 2x - 4, {x, -2, 2}, PlotRange → {-10, 10}]
Out[7]=
```

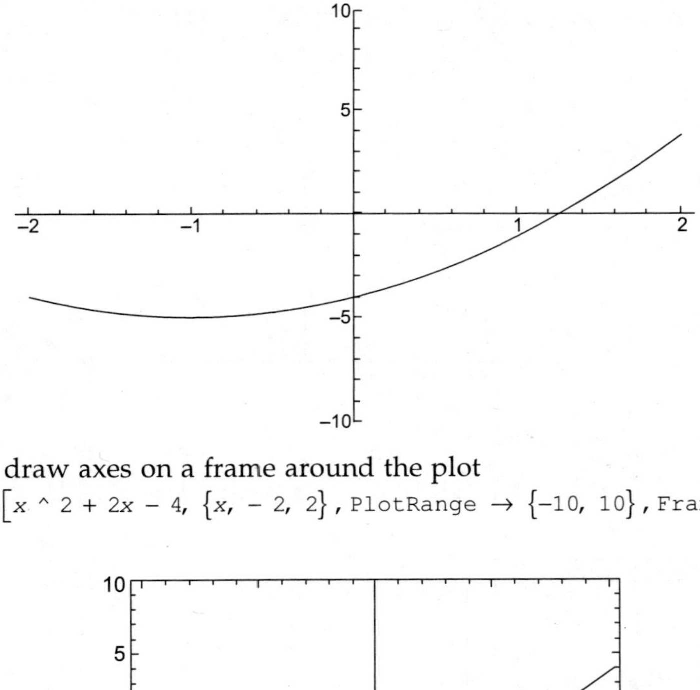

If we want to draw axes on a frame around the plot

```
In [8] := Plot[x ^ 2 + 2x - 4, {x, - 2, 2}, PlotRange → {-10, 10}, Frame → True]
Out[8]=
```

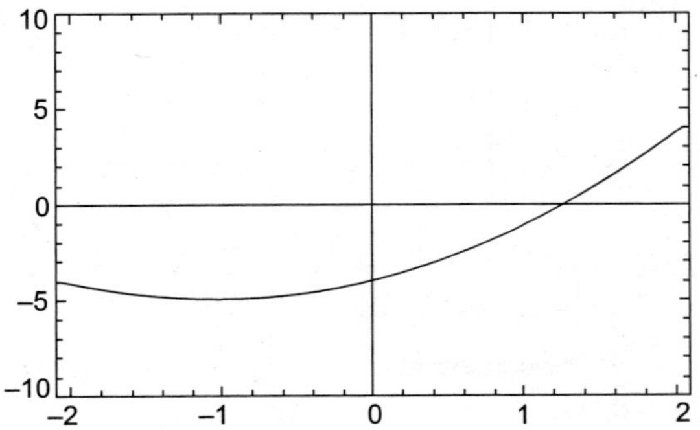

The option **Origin** will set the axes where we want them

```
In [9] := Plot[x ^ 6 - 5x, {x, - 5, 5}, PlotRange → {- 20, 20}, AxesOrigin → {1, 20}]
Out[9]=
```

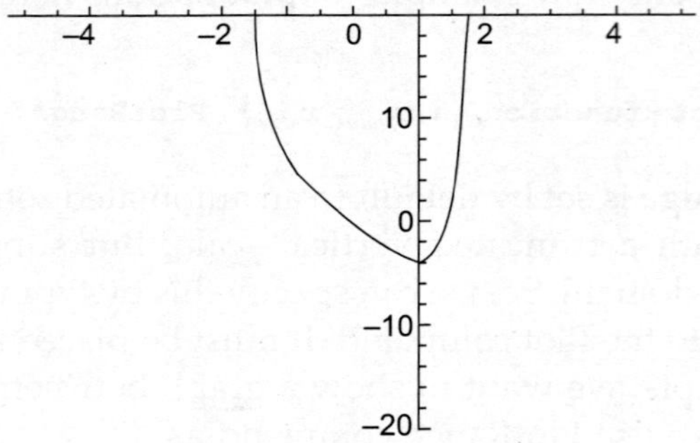

We can colour our graphs with the option **PlotStyle** and add the legends to the graphs by using option **PlotLegends**.

```
In [10] := Plot[x ^ 2, {x, - 2, 2}, PlotStyle → Red]
Out[10]=
```

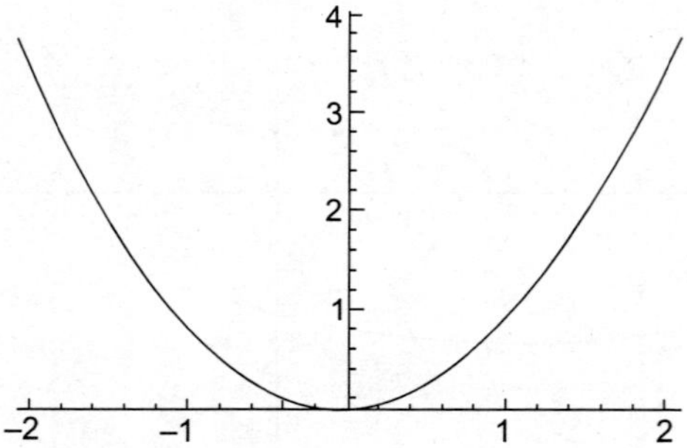

```
In [11] := Needs ["PlotLegends"]
In [12] := Plot [{Sin [x], Cos [x]}, {x, 0, 2Pi}, PlotLegend → {"x ^ 2", "cosine"}]
Out [12]=
```

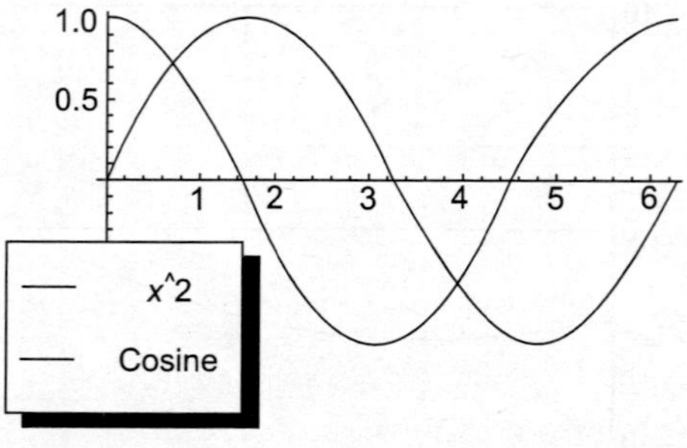

Labels can be added to the axes via the option **AxesLabel**

```
In [13] := Plot[x ^ 2, {x, - 2, 2}, AxesLabel → {"x", "x ^ 2"}]
Out[13]=
```

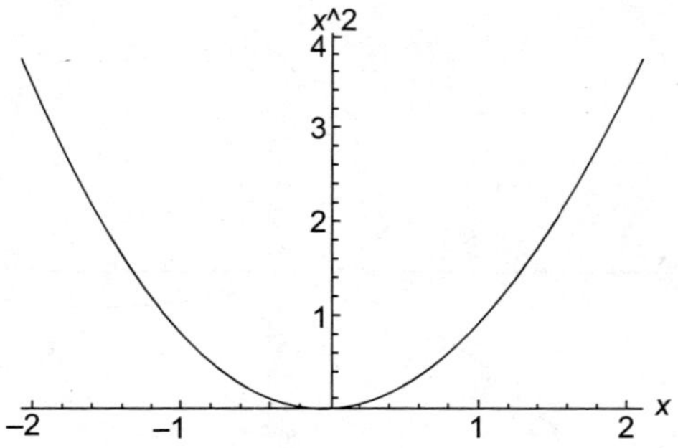

11.2.3 Derivatives and Plotting of its Graphs

The derivatives of the function can be calculated in Mathematica like:

The command for the nth order derivative of a function f typed as

$$D[f[x],\{x,n\}]$$

```
In [14] := f[x _] := x ^ 4 + 3x + 5;
           D[f [x], {x, 1}]
```

Out[14] = $3 + 4x^3$

```
In [15] := D[f [x], {x, 2}]
```

Out[15]= $12x^2$

If we want to calculate the value of 3rd order derivative at point $x = 2$, we can type command as

```
In [16] := D[f [x], {x, 3}] /. x → 2

Out[16]=48
```

We can calculate the derivatives of function and plot the graph of function and its derivative in Mathematica as

```
In[17]:= f[x_] := x^4 - 2 x^3 + x^2 - 2x + 4;
s1=D[f[x],x]
s2=D[f[x],{x,2}]
s3=D[f[x],{x,3}]
h1=Plot[s1,{x,-2,2}]
h2=Plot[s2,{x,-2,2}]
```

```
h3=Plot[s3,{x,-2,2}]
Show[h1, h2, h3]
```

Out[18]= $2 + 2x - 6x^2 + 4x^3$

Out[19]= $2 - 12x + 12x^2$

Out[20]= $-12 + 24x$

Out[21]=

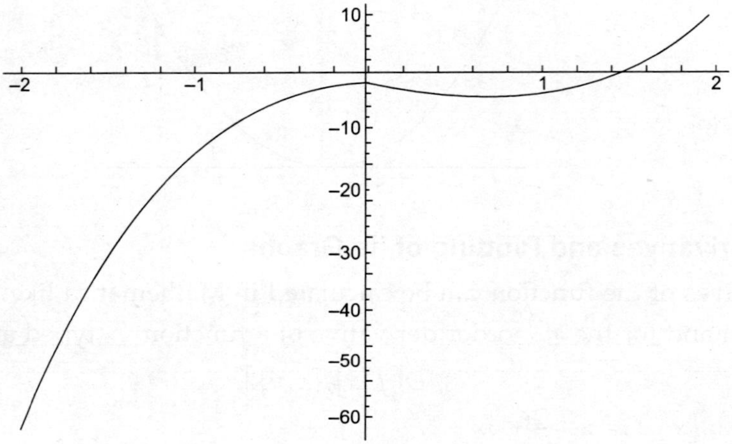

Out[22]=

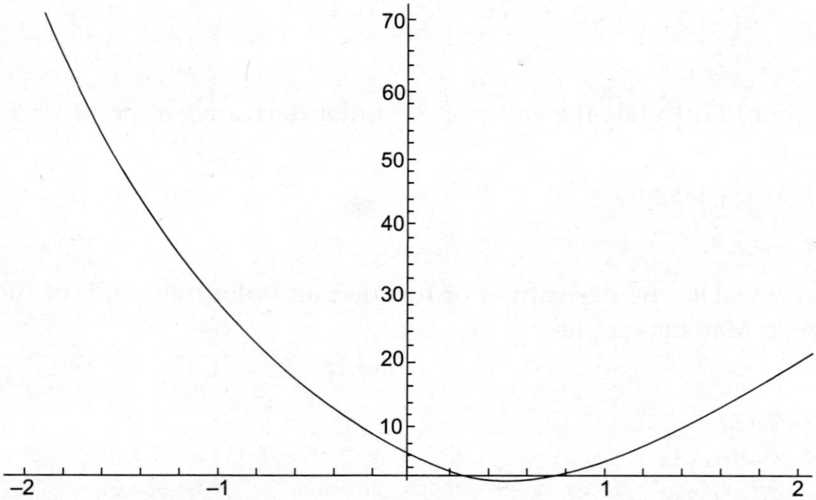

```
Out[23]=
```

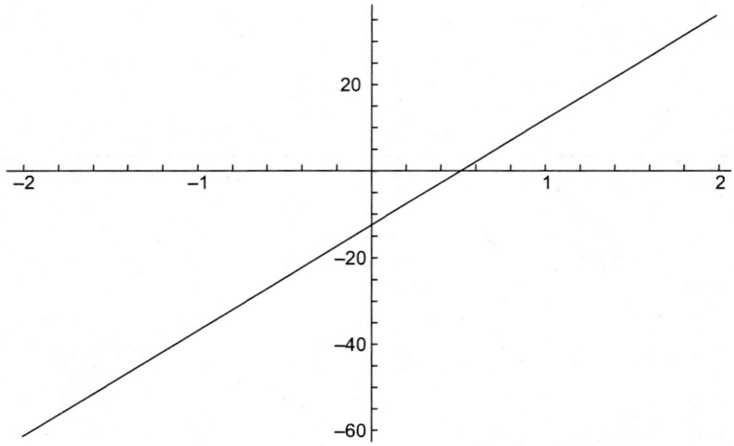


```
Out[24]=
```

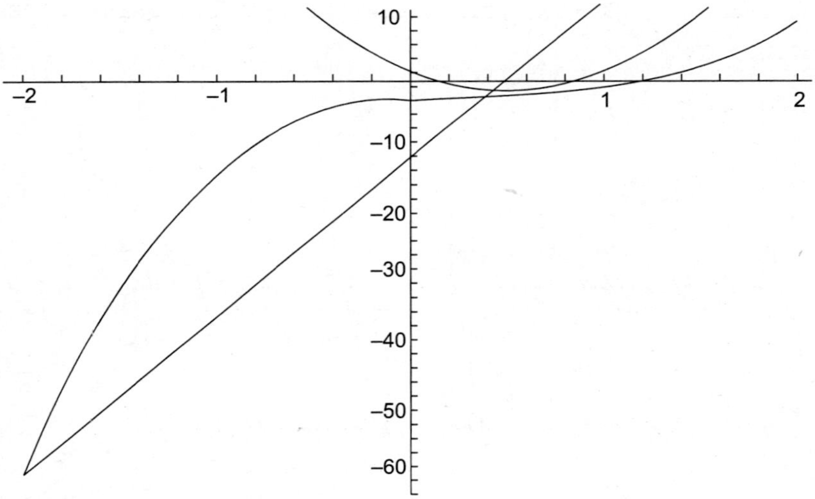

Show command shows several graphics combined. Some examples of plotting of graphs:

```
In[25]  := Needs PlotLegends;
In[26]  :=  s1=Plot[Log[2x+3],{x,-2,2},PlotStyle→Red,PlotLegends→{"log
            2x+3"}];
s2=Plot[Log[3x+4],{x,-2,2},PlotStyle→Blue,PlotLegends→{"log 3x+4"}];
s3=Plot[Log[5x+6],{x,-2,2},PlotStyle→Green,PlotLegends→{"log 5x+6"}];
Show[s1,s2,s3]
```

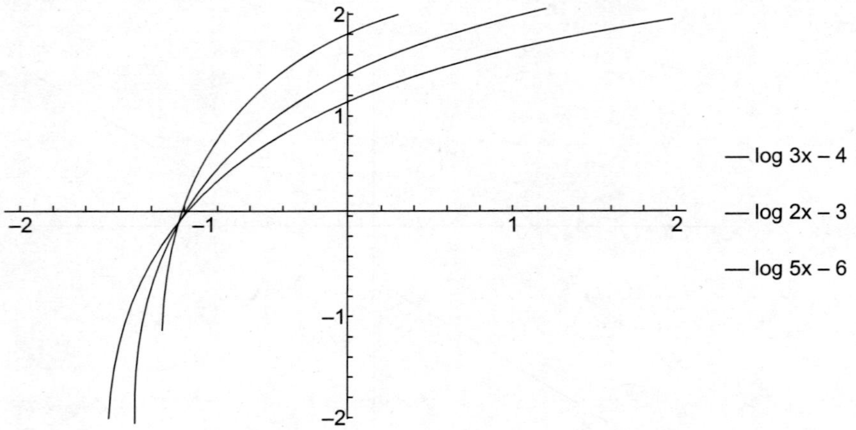


```
In[27]:=Needs "PlotLegends";
b1=Plot[Exp[2x+3],{x,-2,2},PlotStyle→Red,PlotLegends→ {"Exp[2x+3]"}];
b2=Plot[Exp[3x+4],{x,-2,2},PlotStyle→ Blue,PlotLegends→ {"Exp[3x+4]"}];
b3=Plot[Exp[5x+6],{x,-2,2},PlotStyle→Green,PlotLegends→ {"Exp[5x+6]"}];
Show[b1,b2,b3]
Out[27]=
```

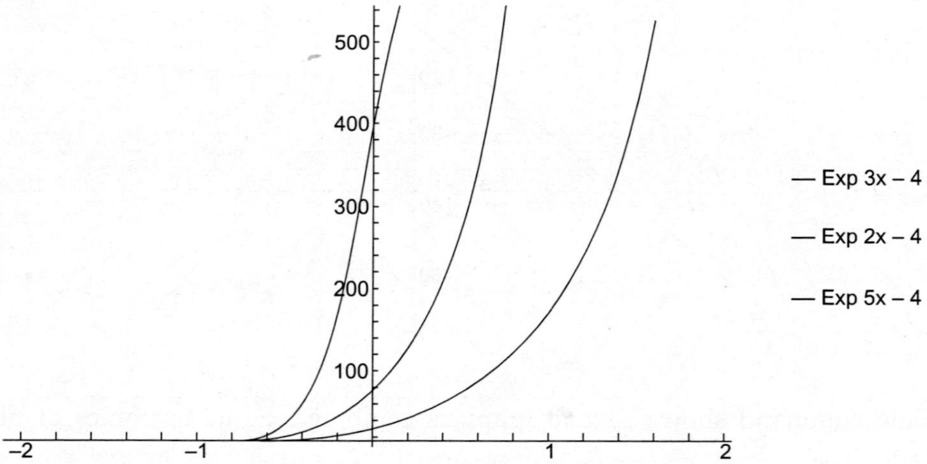


```
In[28]:=Needs "PlotLegends";
c1=Plot[Cosh[2x],{x,-4,4},PlotStyle→Red,PlotLegends→ {"Cosh[2x]"}]
c2=Plot[Sinh [3x],{x,-4,4},PlotStyle→ Blue,PlotLegends→ {"Sinh [3x]"}]
c3=Plot[Tanh [5x],{x,-4,4},PlotStyle→Green,PlotLegends→ {"Tanh [5x]"}]
Show[{c1,c2,c3}, AxesOrigin→{0,0}]
```

Out[28]=

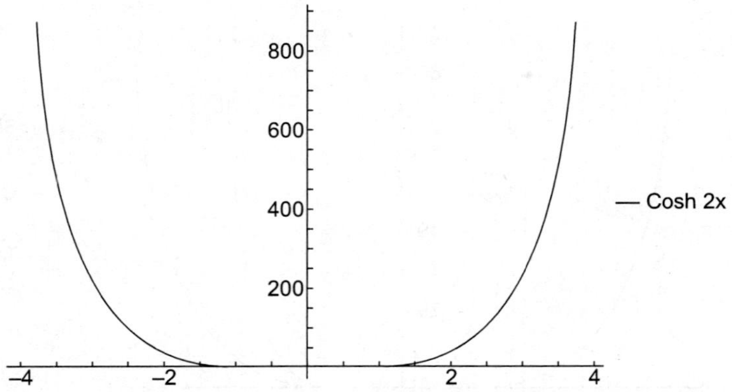

Out[29]=

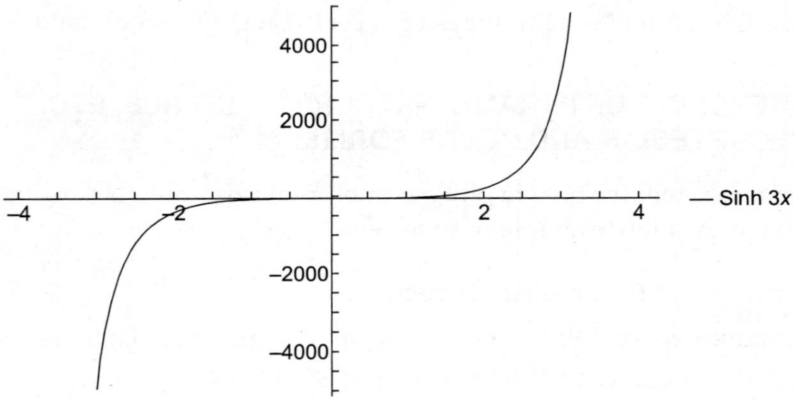

Out[30]=

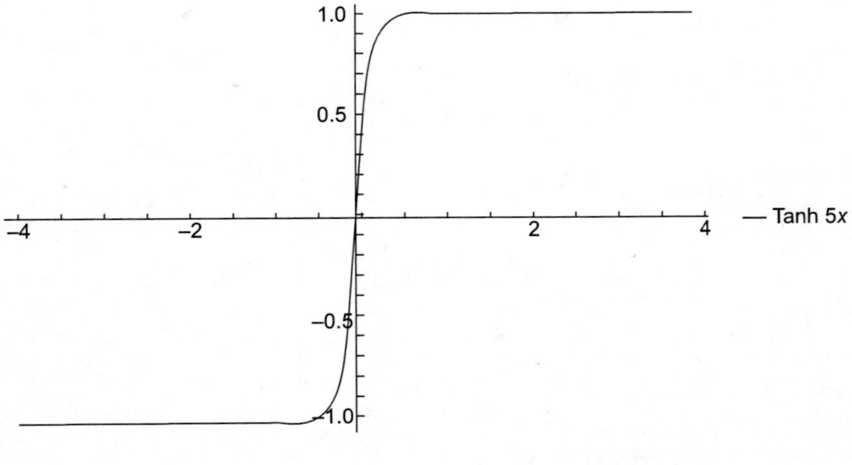

Out[31]=

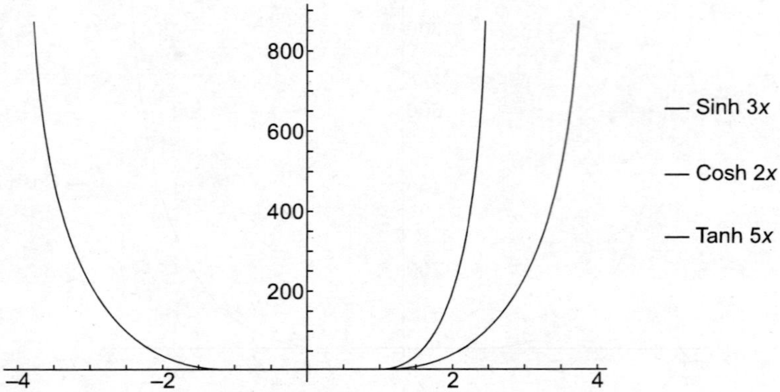

Similarly, we can plot several types of curves using *Plot* command.

11.3 SKETCHING OF PARAMETRIC CURVES, CONICS, ETC., IN CARTESIAN AND POLAR FORMS

In this section, we will see how to plot parametric curves, conic, etc., using Mathematica. We will give only a few example of each type.

11.3.1 Plotting of Parametric Curves

To plot parametric curves, we use *ParametricPlot* command in Mathematica. For example, plot the parametric curve $x = t^2 + t, y = 2t - 1, -6 \le t \le 6$. We can type command as follows:

```
In[32]:=  ParametricPlot[{t ^ 2 + t, 2t − 1}, {t, −6, 6}, PlotRange → {−6, 6}]
Out[32]=
```

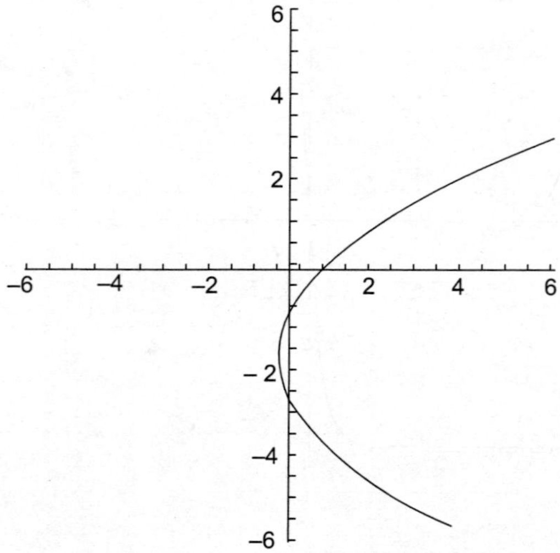

EXAMPLE 1 Plot the graph for the following parametric functions:

(a) $x = t, y = \sin t, -2\pi \le t \le 2\pi$.

(b) $x = 2(t - \sin t), y = 2(1 - \cos t), 0 \le t \le 4\pi$.

(c) $x = 1.5\cos t - \cos 30t, y = 1.5\ \sin t - \sin 30t, -2\pi \le t \le 2\pi$.

(d) $x = \sin(t + \cos 100t), y = \cos(t + \sin 100t), -2\pi \le t \le 2\pi$

Solution:

(a) In[33]:= ParametricPlot[{t, Sin[t]}, {t, −2π, 2π}, PlotRange → Full]

Out [33]=

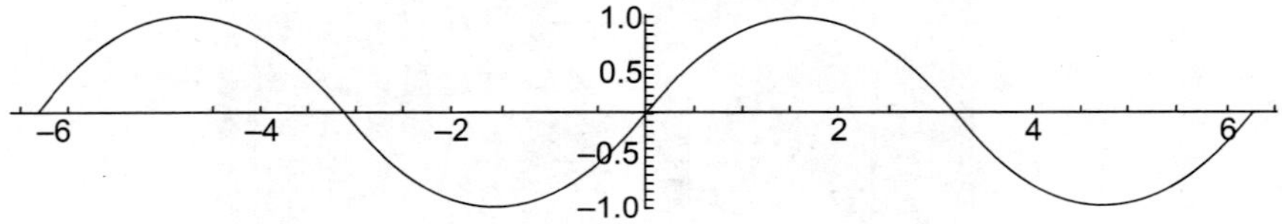

(b) In[34]:= ParametricPlot[{2t − 2Sin[t], 2 − 2Cos[t]}, {t, 0, 4π}]

Out[34]=

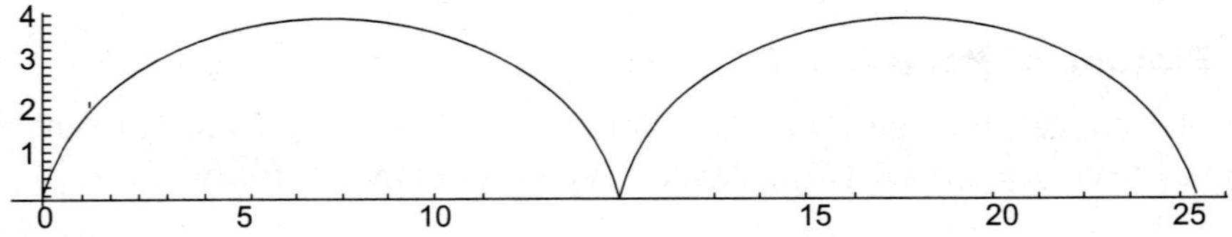

(c) In[35] := ParametricPlot[{1.5Cos[t] − Cos[30t], 1.5Sin[t] − Sin[30t]}, {t, −2Pi, 2Pi}]

Out[35]=

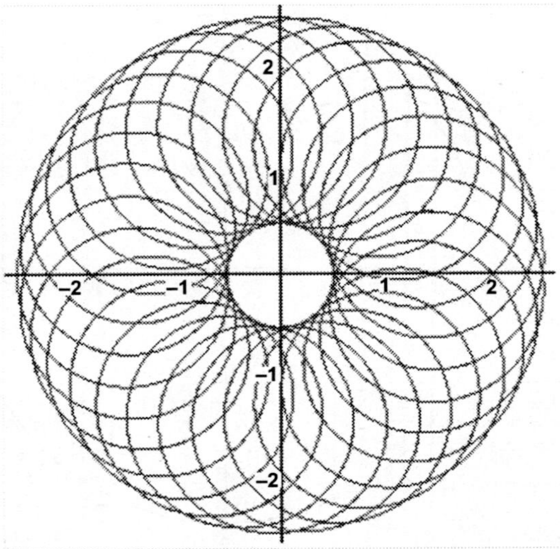

(d) In [36] := ParametricPlot [{Sin [t + Cos[100t]], Cos[t + Sin[100t]] }, {t, −2Pi, 2Pi}]

Out [36] =

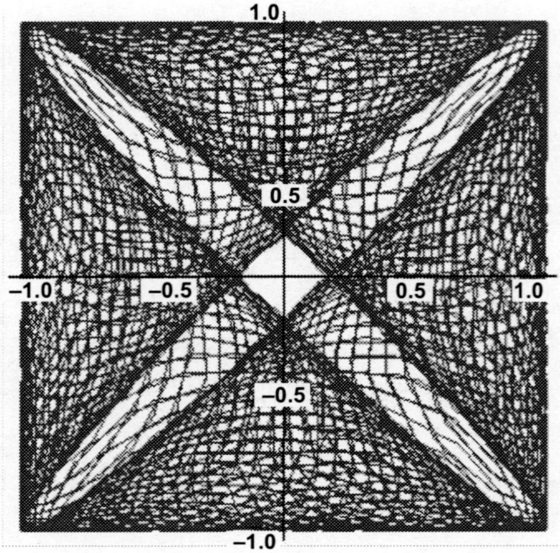

11.3.2 Plotting of Polar Curves

To plot polar curves, we use *PolarPlot* command in Mathematica. For example, plot the parametric curve $r = 2, 0 \leq \theta \leq 2\pi$. We can type command as follows:

In [37] := PolarPlot [2, {t, 0, 2π}]

Out [37] =

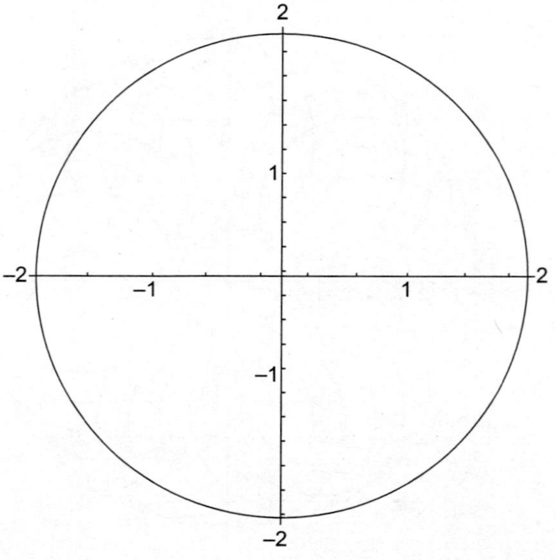

EXAMPLE 2 Plot the graph for the following polar functions:

(a) $r = \theta, 0 \le \theta \le 8\pi$.
(b) $r = \cos\theta, 0 \le \theta \le 2\pi$.
(c) $r = \cos 2\theta, 0 \le \theta \le 2\pi$.
(d) $r = 0.50 + \cos\theta, 0 \le \theta \le 2\pi$.
(e) $r = 1 + \sin\theta, 0 \le \theta \le 2\pi$.

Solution:

(a) `In[38]:= PolarPlot[θ, {θ, 0, 8π}]`

`Out[38]=`

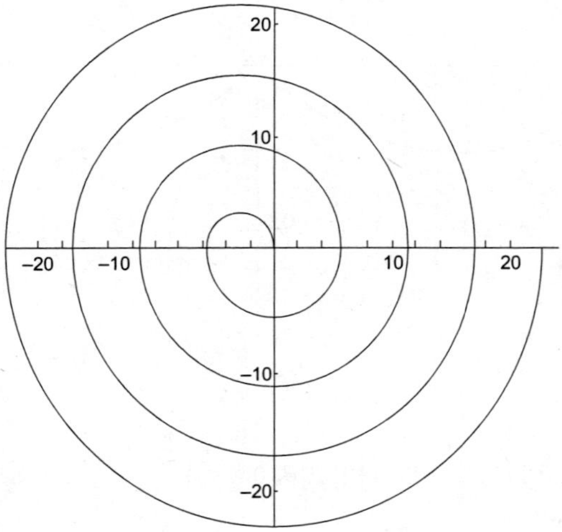

(b) `In [39] := PolarPlot [Cos [θ] , {θ, 0, 2π}]`

`Out[39]=`

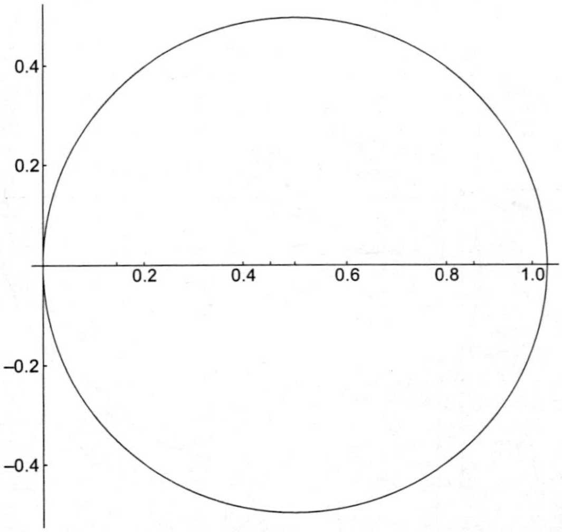

(c) In [40] := PolarPlot [Cos[2θ], {θ, 0, 2π}]

Out [40]=

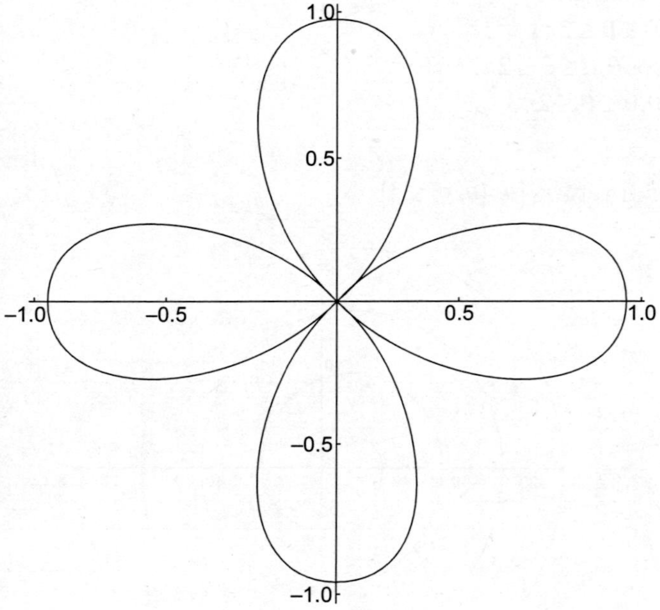

(d) In [41] := PolarPlot [0.5 + Cos [θ] , {θ, 0, 2π}]

Out [41]:=

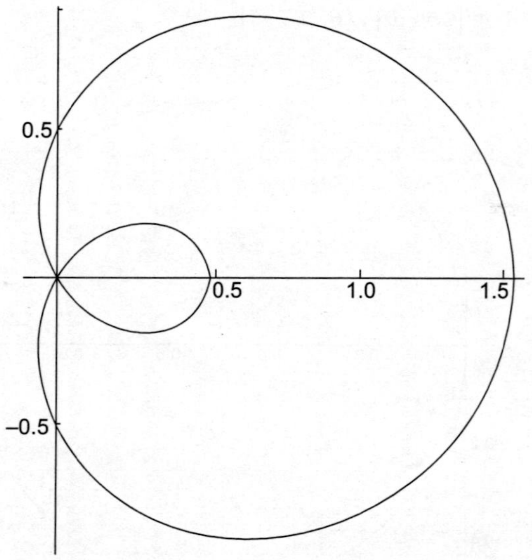

(e). In [42] := PolarPlot [1 + Sin [θ], {θ, 0, 2π}]

Out[42]=

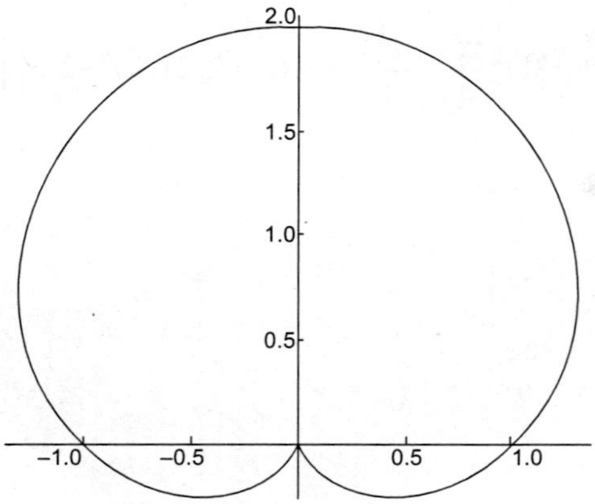

11.4 PLOTTING OF 3D CURVES

To plot or tracing of conics in Cartesian coordinates; sketching ellipsoid, hyperboloid of one and two sheets, elliptic cone, elliptic paraboloid, hyperbolic paraboloid using Cartesian co-ordinates, we use *Plot*3*D* command in Mathematica.

EXAMPLE 3 Sketch the surface $y^2 + z^2 = 1$

Solution:

```
In [43] := Plot3D
    [{-√(1 - y ^ 2), √(1 - y ^ 2)}, {x, -1, 1}, {y, -1, 1}, AxesLabel → {" x ", " y ", " z "}]
```

Out[43]=

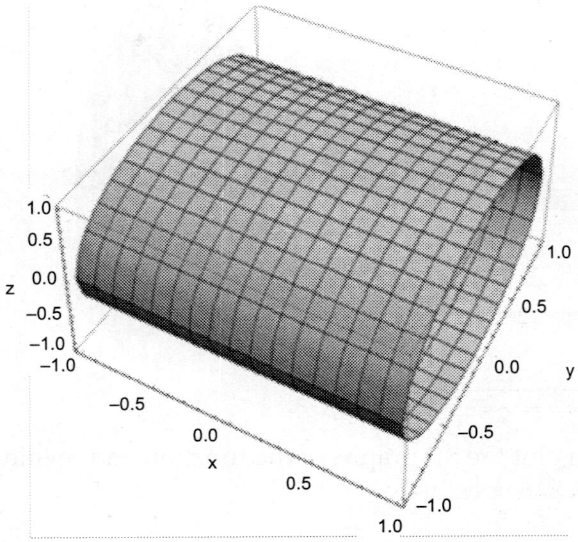

EXAMPLE 4 Plot the graph for the function $\frac{x^2}{4}+\frac{y^2}{8}=1$

Solution: $\text{In}[44]:=\ \text{Plot3D}\left[\frac{x^2}{4}+\frac{y^2}{8}==1,\{x,-2,2\},\{y,-2,2\}\right]$

Out [44]=

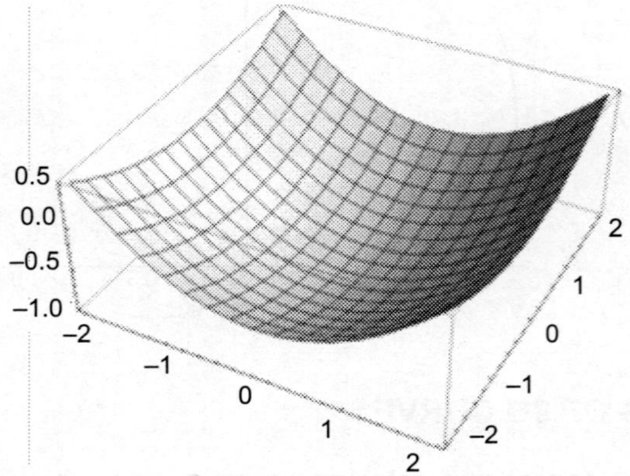

EXAMPLE 5 Plot the revolution of the curve $y=x^2$ about the x-axes in between $x=0$ and $x=2$.

Solution:

In [45] := RevolutionPlot3D[x^2, {x, 0, 2}, RevloutionAxis → {1, 0, 0}, AxesLabel → {"X", "Y", "Z"}, AxesOrigin → {0, 0, 0}, PlotStyle → Green]

Out[45]=

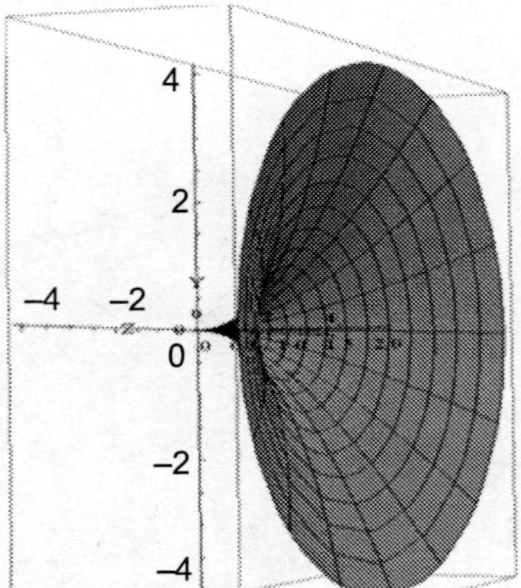

Similarly, we can plot the 3D graphs of the function and revolution of the curve about the x-axes and y-axes respectively.

Answers to the Exercises

Chapter 1

2. (a) $f+g=-2x-4, D=R; f-g=6x-4, D=R; fg=-4x(2x-4),$

$$D=R; \frac{f}{g}=\frac{2x-4}{-4x}, D=R-\{0\}.$$

(b) $f+g=5, D=R; f-g=2x+1, D=R; fg=(x+3)(2-x),$

$$D=R; \frac{f}{g}=\frac{x+3}{2-x}, D=R-\{2\}.$$

(c) $f+g=2+x^2-|x|, D=R; f-g=x^2+|x|, D=R;$

$$fg=(1+x^2)(1-|x|), D=R; \frac{f}{g}=\frac{1+x^2}{1-|x|}, D=R-\{-1,1\}.$$

3. (a) $fog=(2x-1)^2-1, gof=2(x^2-1)-1$.

(b) $fog=\sin(1-\sqrt{x}), gof=1-\sqrt{\sin x}$.

(c) $fog=\frac{1}{x}-1, gof=\frac{1}{x-1}$.

(d) $fog=\sqrt{2x^2+5}, gof=2x+4$.

4. (a) Increasing in $x\in(0,\infty)$; decreasing in $x\in(-\infty,0)$.
(b) Decreasing function.
(c) Increasing in $x\in(0,\infty)$; decreasing in $x\in(-\infty,0)$.
(d) Increasing in $x\in(\infty,-\infty)$.

7. (a) Injective function
(b) Bijective function
(c) Bijective function
(d) Bijective function
(e) Injective function
(f) Injective function

Chapter 2

2. 0.1975
3. 0.05
4. 0.02
5. 0.05
6. 0.02
7. 0.5
8. 0.05
9. 0.0033
10. 0.0025
11. 4
12. 12
13. 5
18. 2
20. 0
24. –22
25. –10
26. –703/2
27. –1000
28. –502
29. 502
30. 18
31. 10
32. 697/2
33. 1000
34. 0
35. $a=-\frac{3}{2}$, b is any real number except $b=0, c=\frac{1}{2}$
36. (d) discontinuous at $x=4$; (e) discontinuous at $x=-3,3$; (f) discontinuous at $x=0$
37. continuous at $x=0$
38. discontinuous at $x=c$; discontinuity of second kind
40. continuous at $x=0$
42. continuous at $x=0,1$
43. discontinuous at $x=0,1 \text{and} 2$
44. continuous at $x=1$ and discontinuous at $x=0$; discontinuity of first kind
48. Discontinuous at $x=0$; discontinuity of first kind

Chapter 3

2. (a) not differentiable (b) not differentiable (c) differentiable
3. $\frac{1}{3}$
4. $64-16t;-16$

6. (a) nx^{n-1}
   (b) $-\sin x$
   (c) $2x\cos x - x^2 \sin x$
   (d) $\cos\left(x^2+1\right)2x$

7. (a) $-4x^3 \sin x^4 \cos^4 x - 4\sin x \cos^3 x \cos x^4$
   (b) $(2+\log x)e^{x+x\log x}$
   (c) $\cos(\sin x)\cos x$
   (d) $2\left(x^2-x+1\right)(2x-1)$
   (e) $-\sin\left(a^3+x^3\right)3x^2$
   (f) $\sec^2(\tan(\tan x))\sec^2(\tan x)\sec^2 x$
   (g) $\exp(\exp(\exp x))\exp(\exp x)\exp x$
   (h) $\dfrac{1}{x\log x}$
   (i) $-\dfrac{12x^3}{\left(x^4+1\right)^4}$

10. (a) $\left[\cos x \cot x - \sin x \log \sin x\right](\cos x)^{\sin x}$
    (b) $\dfrac{2y^2}{x(1-2y\log x)}$, where $y=(x^2)^{(x^2)^{(x^2)^{(x^2)^{\cdots\cdots\infty}}}}$
    (c) $\dfrac{y^2 \cot x}{1-y\log \sin x}$, where $y=(\sin x)^{(\sin x)^{(\sin x)^{(\sin x)^{\cdots\cdots\infty}}}}$
    (d) $x^{\sin^{-1}x}\left[\dfrac{\sin^{-1}x}{x}+\dfrac{\log x}{\sqrt{1-x^2}}\right]$

11. (a) $-\cot t$
    (b) $3a^2t^2$
    (c) $\dfrac{\pi}{2}\dfrac{\cos\frac{\pi}{2}t-\sin\frac{\pi}{2}t}{2t+1}$

12. $-\sqrt{\dfrac{y}{x}}$

16. (a) $x=\dfrac{1}{2}$
    (b) $x=\dfrac{5}{2}$
    (c) $x=\dfrac{3}{2}$

17. (a) $c=1$
    (b) $c\approx 3.155934$
    (c) $c\approx 0.73$
    (d) $c\approx 0.54$
    (e) $c\approx 1.082$

20. (a) $x=-1,1$ (b) $x=0,2$

(c) $x=0,\frac{4}{5}$

21. (a) maximum value is 0 and minimum value is -4

(c) maximum value is 0 and minimum value is -2

(e) maximum value is 140 and minimum value is 60

22. (a) concave up for $x<0$; concave down for $x>0$; $x=0$

(c) concave up for $x>-1$; concave down for $x<-1$; $x=-1$

(d) concave up for $x>0$; concave down for $x<0$; no

(e) concave up for $\left(-\sqrt{3},0\right)\cup\left(\sqrt{3},-\infty\right)$; concave down for $\left(-\infty,-\sqrt{3}\right)\cup(0,\sqrt{3})$;

$\left(\sqrt{3},\frac{\sqrt{3}}{4}\right),\left(-\sqrt{3},\frac{\sqrt{3}}{4}\right)$

23. (a)

(b) maximum value at $x=\frac{\pi}{3}$

(d) relative maximum at $x=\frac{1}{2}$ and relative minimum at $x=1$

24. (a)

(b) maximum value at $x=2-\frac{1}{\sqrt{3}}$; minimum value at $x=2+\frac{1}{\sqrt{3}}$

(c) maximum value at $x=\tan^{-1}\left(\frac{3}{4}\right)$

(e) maximum value at $x=1$, minimum value at $x=3$; neither maxima nor minima at $x=0$

(f) minimum at $x = 2$, neither maxima nor minima at $x = -1$

(g) minimum at $x = b$, neither maxima nor minima at $x = a$

28. 0

30. (a) $\frac{1}{2}$ (b) $\frac{112}{3}$

(c) 0 (d) not defined

(e) $\frac{a}{b}$ (f) 0

(g) 0 (h) $-\frac{2}{3}$

(i) $-\frac{1}{2}$

(j) $e^{\frac{1}{3}}$

(k) e

(l) 0

(m) 0

31. price is equal to 29
32. level of production is 40
33. level of production is 26
34. level of production is 20
35. 208 years from now
38. Demand function is $p(x) = 450 - \frac{1}{2}x$; revenue function is $R(x) = 450 - \frac{1}{2}x^2$; 125

Chapter 4

1. (a) $a^2 e^{ax}\left(a^3x^3 + 15a^2x^2 + 60ax + 60\right)$

   (b) $x\left(x^2 - 60\right)\cos x + 15\left(x^2 - 4\right)\sin x$

   (c) $-\frac{6}{x^2}$

   (d) $e^x\left[\log x + \frac{5}{x} - \frac{10}{x^2} + \frac{20}{x^3} - \frac{30}{x^4} + \frac{24}{25}\right]$

   (e) $2^5 750(2x+3)^5\left[42x^2 + 35x(2x+3) + 5(2x+3)^2\right]$

   (f) $4^4 e^{4x}\left[4(3x+5)^2 + 30(3x+5) + 45\right]$

   (g) $e^x\left[41\cos 2x + 38\sin 2x\right]$

   (h) $121 \cos x \sin 2x + 122 \sin x \cos 2x$

2. (a) $\frac{(-1)^n n!}{x^{n+1}}\left[-\log x + 1 + \frac{1}{2} + \frac{1}{3} + \ldots\ldots + \frac{1}{n}\right]$

   (b) $\frac{(n-1)!}{x}\sum_{r=1}^{n}(-1)^{n+1}\frac{n!}{(n-r)!}$

   (c) $e^x\left[\log x + \frac{n}{x} - \frac{n(n-1)}{x^2} + \ldots\ldots\ldots + \frac{(-1)^n (n-1)!}{x^n}\right]$

(d) $2^{n-1}\left\{\left(2x^4 - {}^nc_2\, 6x^2 + {}^nc_4\, 3\right)\sin\left(\frac{n\pi}{2}+2x+3\right)\right\}$

$$-2^{n-1}\left\{x({}^nc_1\, 4x^2 - {}^nc_3\, 3)\cos\left(\frac{n\pi}{2}+2x+3\right)\right\}.$$

Chapter 5

3. (a) $x = t,\ y = 9t^2$ (b) $x = 5\cos^4\theta,\ y = 5\cos^3\theta\sin\theta$
4. $x = \sqrt{3}\cos t,\ y = \sqrt{3}\sin t,\ 0 \le t \le 2\pi$
5. $x = 3\cos t,\ y = 2\sin t,\ 0 \le t \le 2\pi$
6. $x = at^2 - \frac{9}{4},\ y = 2at$
7. $x = \frac{4}{\cos t},\ y = 3\tan t.$
8. (a) $y = 3 - 2x,\ -1 \le x \le 0$ (b) $y = x^{\frac{2}{3}},\ x \ge 0$

   (c) $(x-1)^2 + (y+2)^2 = 1,\ 0 \le x \le 2$ (d) $(x-1)^2 + (y+2)^2 = 1$

   (e) $\frac{y^2}{9} - \frac{x^2}{16} = 1,\ -\infty < x < \infty$
9. $x = 3 - t,\ y = 1 + t,\ z = 4 - 2t$

   (1, 3, 0) at *xy*-plane, (4, 0, 6) at *xz*-plane and (0, 4, −2) at *yz*-plane.
10. $\frac{x+1}{5} = \frac{y-3}{-1} = \frac{z-7}{-8}$,

    $\left(\frac{27}{8}, \frac{17}{8}, 0\right)$ on *xy*-plane, (14, 0, −17) on *xz*-plane and $\left(0, \frac{14}{5}, \frac{27}{5}\right)$ on *yz*-plane.
11. (a) parallel

    (b) intersecting at $\left(0, -\frac{3}{2}, 0\right)$

    (c) skew
12. $x = -1 - 23t,\ y = 3 + 2t,\ z = 1 - 19t$
13. (a) $x = -3,\ y = 3\sqrt{3}$ (b) $x = 2\sqrt{3},\ y = 2$

    (c) $x = -4,\ y = 4\sqrt{3}$ (d) $x = 3\sqrt{3},\ y = 3$

(e) $x = 0, y = 0$

(f) $x = -\dfrac{7\sqrt{3}}{2}$, $y = \dfrac{7}{2}$

(g) $x = -5, y = 0$

15. (a) $r = \pm 16,\ \theta = \dfrac{4\pi}{3} \pm 2n\pi$

(b) $r = \pm 7,\ \theta = \pi$

(c) $r = 4,\ \theta = \dfrac{11}{6}$

(d) $r = 2,\ r = \dfrac{3\pi}{2}$

(e) $r = 8\sqrt{2}$, $\theta = \dfrac{5\pi}{4}$

(f) $r = 6,\ \theta = \dfrac{2\pi}{3}$

17. (a) $r = 2 \sec\theta$

(b) $r = 4$

(c) $r = -6 \sin\theta$

(d) $r^2 \cos\theta \sin\theta = 4/9$

21. (a) $x^2 = \dfrac{25}{2} y$

(b) $y^2 = 12x$

(c) $y^2 = -28x$

(d) $x^2 = -12y$

(e) $y^2 = -5\left(x - \dfrac{19}{5}\right)$

22. (a) $\dfrac{x^2}{12} + \dfrac{y^2}{16} = 1$

(b) $\dfrac{x^2}{9} + \dfrac{y^2}{4} = 1$

(c) $\dfrac{x^2}{64} + \dfrac{y^2}{100} = 1$

(d) $\dfrac{x^2}{3} + \dfrac{y^2}{2} = 1$

(e) $\dfrac{x^2}{16} + \dfrac{y^2}{7} = 1$.

23. (a) $\dfrac{y^2}{64} - \dfrac{x^2}{36} = 1$

(b) $x^2 - \dfrac{y^2}{4} = 1$

(c) $\dfrac{y^2}{20} - \dfrac{x^2}{5} = 1$

(d) $\dfrac{(y-2)^2}{3} - \dfrac{4}{3}\left(x - \dfrac{1}{2}\right)^2 = 1$

24. $\dfrac{x'^2}{12} + \dfrac{y'^2}{4} = 1$ (an ellipse)

25. $x' = -1 + 3\sqrt{3}$, $y' = 3 + \sqrt{3}$

26. (a) $\dfrac{x'^2}{2} - \dfrac{y'^2}{2} = 1$ (a hyperbola)

(b) $\dfrac{(x'-1)^2}{9} + y'^2 = 1$ (an ellipse)

(c) $\frac{x'^2}{3}-\frac{y'^2}{2}=1$ (a hyperbola)

(d) $y' = x'^2$ (a parabola)

(e) $\frac{(x'+1)^2}{4}+y'^2=1$ (an ellipse)

27. (a) an ellipse

(b) an ellipse

(c) hyperbola

(d) hyperbola.

Chapter 6

1. (b) $16\sqrt{2}$
2. $\frac{\sqrt{2}}{3}a$

10. a
11. $a^2/3r$

15. (a) $p = 4,\ \rho = 85\sqrt{17}/2$ and $p = -1,\ \rho = 5\sqrt{2}$
16. $a\sqrt{2}$

18. $X = \frac{a^2+b^2}{a^4}x^3,\ Y = -\frac{a^2+b^2}{b^4}y^3$ and $(ax)^{\frac{2}{3}}-(by)^{\frac{2}{3}}=\left(a^2+b^2\right)^{\frac{2}{3}}$
19. $\left(x-\frac{3}{4}a\right)^2+\left(y-\frac{3}{4a}\right)^2=\frac{a^2}{2}$

Chapter 7

1. $6x-6y-7=0,\ 6x-2y-3=0,\ 3x+6y+5=0$
2. $x-y=0,\ 2x-y+1=0,\ 2x+y-1=0$
3. $x+y-1=0,\ 2x-2y+3+\sqrt{5}=0,\ 2x-2y+3-\sqrt{5}=0$
4. $x=\pm1,\ y=\pm1,\ x-y\pm\sqrt{2}=0,\ x+y\pm\sqrt{2}=0$
5. $4x-4y+1=0,\ 2x+2y-3=0,\ 4x+12y+9=0$
6. $x=\pm a,\ x-y=\pm\sqrt{2}a$
7. $y=-\frac{x}{2}-1,\ y=-x\pm2\sqrt{2}$
8. $y=x,\ y=x+1,\ y=x+2,\ y=-x$
9. $y=x,\ y=-2x-1,\ y=-\frac{x}{2}-\frac{1}{2}$

10. $y=-x$, $y=-x+1$, $5y=5x-1$, $15y=-10x+13$
11. $x=0$, $y=0$, $x-y=0$, $x+y=0$
12. $x-y=0$, $x+y=0$, $2x+3y=0$
13. $x-y=0$, $x+y=0$, $x+2y-1=0$
18. $r(\sin\theta+\cos\theta)+a=0$
19. $r\sin\theta=2$
20. $r\sin\theta=a$
21. $r\sin(\theta-1)=a$.
22. $r\sin\theta=a$
23. $r\sin(\theta-1)=a$
24. $a\left(1+(-1)^n\right)=(-1)^n r\cos\theta$
25. $r\cos\theta=a$
26. $r\sin\left(\theta-\frac{m\pi}{n}\right)=\frac{b}{n}, m$ being any integer.
27. $r\sin\left(\theta-\frac{m\pi}{n}\right)=\frac{a}{n\cos m\pi}, m$ being any integer

Chapter 8

1. $x = 0$, $y = 0$
2. $y = 0$
3. $x = 0$, $y = x/2$
4. $ax \pm iby = 0$
5. $x = 0$
6. $y+2c=\pm\left(\sqrt{3}/2\right)(x+c)$
8. origin is a node
9. $(a, 0)$ is a node
10. origin is a node
11. $(a, 0)$, $(-a, 0)$ and $(0, -(a)$ all are nodes
12. $(1, -1)$ is a cusp
13. $(2, 0)$ is a node
14. $(a, 0)$ is node, if $a > b$; cusp, if $a = b$; conjugate point, if $a < b$

15. origin is a conjugate point
16. $(a, 0)$ is a node
17. $(1, -1)$ is a cusp
18. origin single cusp with first species
19. origin double cusp with first species
20. origin single cusp with second species
21. origin single cusp with first species
22. origin single cusp with first species
23. origin double cusp with second species
24. single cusp with first species at $(a, 0)$
25. single cusp with second species at $(0, -1)$
26. origin is point of osco-inflexion
27. origin is point of osco-inflexion

Chapter 9

2. $-\frac{4}{15}\sin^2 x\cos x - \frac{8}{15}\cos x; 8/15$

3. $\frac{1}{8}\left[\frac{5}{2}x - \frac{15}{8}\sin 2x + \frac{3}{8}\sin 4x - \frac{1}{26}\sin 6x\right]; \frac{5\pi}{32}$

5. $8/315$

6. $\frac{32}{3}; \frac{23}{3}$

7. (a) $\frac{49}{192}$ (b) $\frac{1}{2}$

   (c) $\sqrt{2}$ (d) 24

8. $1+\frac{\sqrt{3}}{2}-\frac{5\pi^2}{24}$

9. $\frac{4}{3}$

10. $\frac{64}{3}$

13. $\frac{a^2}{6}$

15. (a) $\dfrac{32\pi}{6}$ (b) $\left(1-\dfrac{\sqrt{2}}{2}\right)\pi$

(c) $\dfrac{256\pi}{3}$ (d) $\dfrac{2048\pi}{15}$

16. (a) $\dfrac{3\pi}{5}$ (b) 8π

(c) 2π (d) $\dfrac{72\pi}{5}$

17. $\dfrac{15\pi}{2}$

18. $\dfrac{\pi}{2}$

20. $\dfrac{69\pi}{10}$

21. $\dfrac{32\pi}{5}$

22. $\pi a^3(8\log 2-3)$

24. $\dfrac{4\pi ab^2}{3}$

25. $\dfrac{\pi}{6}$

26. (a) $\dfrac{2\pi}{5}$ (b) $\dfrac{844\pi}{5}$

(c) $\dfrac{20\pi}{3}$ (d) $\dfrac{8\pi}{3}$

27. (a) $\dfrac{\pi}{2}$ (b) $\dfrac{\pi}{5}$

(c) $\dfrac{76\pi}{3}$

28. ≈ 3.8202

29. (a) $\dfrac{85\sqrt{85}-8}{243}$ (b) $\dfrac{80\sqrt{10}-13\sqrt{13}}{27}$

(c) $\frac{595}{144}$ (d) $\frac{17}{6}$

30. (a) $\frac{\left(2\sqrt{2}-1\right)}{3}$ (b) π

(c) $\frac{\pi^2}{2}$ (d) $\frac{\left(8\sqrt{10}-13\sqrt{13}\right)}{27}$

31. $\sqrt{2}\left(e^{\frac{\pi}{2}}-1\right)$

37. (a) $\frac{35\pi}{\sqrt{2}}$ (b) 8π

(c) $\frac{16\pi}{9}$ (d) $2\pi\left(\sqrt{2}+\log\left(\sqrt{2}+1\right)\right)$

38. (a) $40\pi\sqrt{82}$ (b) 24π

(c) $\frac{16911\pi}{1024}$

39. $\frac{8\pi}{3\sqrt{3}}\left(4\pi+3\sqrt{3}\right)$

40. 3π

Chapter 10

2. (a) $t \neq 0$

   (b) $t \neq (2n+1)\pi/2$, n being an integer

   (c) $t \neq n\pi/2$, n being an integer

   (d) $t > 0$

3. (a) $\left(t^2+t\right)\boldsymbol{i}+\left(t+\frac{1}{t}\right)\boldsymbol{j}+(5-\sin t)\boldsymbol{k}$

   (b) $e^t t^2\boldsymbol{i}+e^t t\boldsymbol{j}-e^t \sin t\boldsymbol{k}$

   (c) $\left(5t+\frac{\sin t}{t}\right)\boldsymbol{i}-\left(5t^2+t\sin t\right)\boldsymbol{j}+\left(t-t^2\right)\boldsymbol{k}$

   (d) $t^3+1-5\sin t$

4. (a) $2t\sin t-5e^t$

   (b) $-t^2e^t\boldsymbol{i}+t^2\sin t\boldsymbol{j}+\left(2te^t+5\sin t\right)\boldsymbol{k}$

(c) $\left(4te^t + 10\sin t\right)\boldsymbol{i} + 2e^t t^2 \boldsymbol{k}$

5. (a) $\boldsymbol{i} + e^2 \boldsymbol{j}$

   (b) $-\boldsymbol{i} + e^{-1} \boldsymbol{j}$

7. (a) $t\boldsymbol{i} + t^2 \boldsymbol{j} + \sqrt{4 - t^2 - t^4}\,\boldsymbol{k}$

   (b) $t\boldsymbol{i} + t^2 \boldsymbol{j} + 2\boldsymbol{k}$

   (c) $t^2\boldsymbol{i} + t\boldsymbol{j} + \sqrt{9 - t^2 - t^4}\,\boldsymbol{k}$

8. (a) $t > 0,\ t \neq 1$

   (b) continuous for all t

   (c) continuous for all t, except $t \neq 0$.

   (d) continuous for all t

9. (a) differentiable for all $t \neq 0$

10. (a) $e^t \boldsymbol{i} + \cos t\,\boldsymbol{j} + \left(3t^2 + 5\right)\boldsymbol{k}$

    (b) $\boldsymbol{i} + 2t\boldsymbol{j} + \left(1 + 3t^2\right)\boldsymbol{k}$

    (c) $(1 + 4t)\boldsymbol{i} + (2t - 1)\boldsymbol{j} + 2t\boldsymbol{k}$

    (d) $-\sin t\boldsymbol{i} + \cos t\boldsymbol{j} - 3\sin t\boldsymbol{k}$

11. (a) $\boldsymbol{F}'(t) = 2e^{2t}\boldsymbol{i} - 2t\boldsymbol{j} - 2\sin 2t\boldsymbol{k}$ and $\boldsymbol{F}''(t) = 4e^{2t}\boldsymbol{i} - 2\boldsymbol{j} - 4\cos 2t\boldsymbol{k}$

    (b) $\boldsymbol{F}'(t) = 2t\boldsymbol{i} - t^{-2}\boldsymbol{j} - 2e^{2t}\boldsymbol{k}$ and $\boldsymbol{F}''(t) = 2\boldsymbol{i} + 2t^{-3}\boldsymbol{j} + 4e^{2t}\boldsymbol{k}$

    (c) $\boldsymbol{F}'(t) = \cos t\boldsymbol{i} - \sin t\boldsymbol{j} + 2t\boldsymbol{k}$ and $\boldsymbol{F}''(t) = -\sin t\boldsymbol{i} - \cos t\boldsymbol{j} - 2\boldsymbol{k}$

13. (a) $15t^4\boldsymbol{i} + \left(2e^t + e^{-t}t^2(3 - t)\right)\boldsymbol{j} - 4t\left(2t^2 - 1\right)\boldsymbol{k}$

    (b) $-6t^2 + 6t$

14. (a) speed = 12.04, $\boldsymbol{V}(2) = -\sin 2\boldsymbol{i} + \cos 2\boldsymbol{j} + 12\boldsymbol{k}$,

    $\boldsymbol{A}(2) = -\cos 2\boldsymbol{i} - \sin 2\boldsymbol{j} + 12\boldsymbol{k}$, direction of motion $= -0.08\boldsymbol{i} - 0.03\boldsymbol{j} + \boldsymbol{k}$

    (b) speed = 3, $\boldsymbol{V}(1) = \boldsymbol{i} + 2\boldsymbol{j} + 2\boldsymbol{k}$, $\boldsymbol{A}(1) = 2\boldsymbol{j}$, direction of motion $= \frac{1}{3}\boldsymbol{i} + \frac{2}{3}\boldsymbol{j} + \frac{2}{3}\boldsymbol{k}$

    (c) speed $= \frac{\sqrt{273}}{2}$, $\boldsymbol{V}(\ln 2) = 2\boldsymbol{i} - \frac{1}{2}\boldsymbol{j} + 8\boldsymbol{k}$, $\boldsymbol{A}(\ln 2) = 2\boldsymbol{i} + \frac{1}{2}\boldsymbol{j} + 16\boldsymbol{k}$,

    direction of motion $= \frac{1}{\sqrt{273}}(4\boldsymbol{i} - \boldsymbol{j} + 16\boldsymbol{k})$

15. (a) $\boldsymbol{F}'(0.2) = 2e^{0.4}\boldsymbol{i} - 0.6\boldsymbol{j} + 5\boldsymbol{k}$

(b) $\boldsymbol{F}'(0) = 2\boldsymbol{j}$; $F'(1) = 2\boldsymbol{i} + 2\boldsymbol{j} + 5\boldsymbol{k}$; $F'(-1) = -2\boldsymbol{i} + 2\boldsymbol{j} + \boldsymbol{k}$

(c) $\boldsymbol{F}'(0) = 0$; $\boldsymbol{F}'\left(\frac{\pi}{2}\right) = \pi\boldsymbol{i} - \boldsymbol{j} - \frac{\pi^2}{4}\boldsymbol{k}$

16. $x = -1 - 3t$, $y = 1 + 2t$, $z = -1 - t$

17. (a) $\frac{\pi^2}{2}\boldsymbol{i} + 3\pi\boldsymbol{j} - 2\boldsymbol{k}$

(b) $\frac{1}{3}\left(2\sqrt{2} - 1\right)\boldsymbol{i} + \frac{\pi}{4}\boldsymbol{j}$

(c) $\boldsymbol{i} + \boldsymbol{j} + \frac{1}{2}\boldsymbol{k}$

18. (a) $\boldsymbol{R}(t) = \left(e^t + 1\right)\boldsymbol{i} + \left(\frac{t^3}{3} + 1\right)\boldsymbol{j} + \left(\frac{1}{2}\sin 2t - 1\right)\boldsymbol{k}$

(b) $\boldsymbol{R}(t) = \left(\frac{t^3}{3} + 1\right)\boldsymbol{i} + \left(-\frac{1}{2}e^{2t} + \frac{9}{2}\right)\boldsymbol{j} + \left(\frac{2}{3}t^{\frac{3}{2}} - 1\right)\boldsymbol{k}$

(c) $\boldsymbol{R}(t) = \left(\frac{4}{3}t^{\frac{3}{2}} + 1\right)\boldsymbol{i} + (\sin t + 1)\boldsymbol{j}$

(d) $\boldsymbol{R}(t) = e^t\boldsymbol{i} + \left(\frac{1}{3}t^3 - 1\right)\boldsymbol{j}$

19. The graph is not smooth for all t, but it is on any interval not containing $t = 0$

20. (a) $\boldsymbol{V}(t) = \left(\frac{t^3}{3} + 1\right)\boldsymbol{i} - \frac{4}{3}\left(t^{\frac{3}{2}} + 1\right)\boldsymbol{j} + \frac{1}{3}\left(e^{3t} - 7\right)\boldsymbol{k}$

$$\boldsymbol{R}(t) = \left(\frac{t^4}{12} + t + 2\right)\boldsymbol{i} + \frac{4}{3}\left(\frac{2}{5}t^{\frac{5}{2}} + t - 1\right)\boldsymbol{j} + \frac{1}{3}\left(\frac{e^{3t}}{3} - 7t - \frac{10}{9}\right)\boldsymbol{k}$$

(b) $\boldsymbol{V}(t) = 8t^3\boldsymbol{i} + 4t\boldsymbol{j}$

$\boldsymbol{R}(t) = \left(2t^4 + 1\right)\boldsymbol{i} + \left(2t^2 + 2\right)\boldsymbol{j}$

21. (a) $T_f = 3$ seconds and $R_f = 166$ ft

(b) $\boldsymbol{V}(3) = 32\sqrt{3}\boldsymbol{i} - 64\boldsymbol{j}$ and speed = 85 ft/s

(c) The highest point reached by the ball has coordinates (55,64).

22. $T_f = 4.42\,\text{sec}$ and $R_f = 372.38\,\text{ft}$

23. (a) $T_f = 4.6\,\text{sec}$ and $R_f = 481\,\text{ft}$

(b) $T_f = 129.9\,\text{sec}$ and $R_f = 73{,}175\,\text{metre}$

(c) $T_f = 1.9\,\text{sec}$ and $R_f = 41\,\text{metre}$

(d) $T_f = 1.5\,\text{sec}$ and $R_f = 148\,\text{ft}$

24. (a) $\boldsymbol{V}(t) = \sqrt{5}u_r$; $\boldsymbol{A}(t) = \boldsymbol{0}$

(b) $\boldsymbol{V}(t) = -10\sin(2t+1)\boldsymbol{u}_r + 10(1+\cos(2t+1)\boldsymbol{u}_\theta$;

$\boldsymbol{A}(t) = (-40\cos(2t+1) - 20)\boldsymbol{u}_r - 40\sin(2t+1)\boldsymbol{u}_\theta$

(c) $\boldsymbol{V}(t) = e^{-(1-t)}\boldsymbol{u}_r - e^{-(1-t)}\boldsymbol{u}_\theta$;

$\boldsymbol{A}(t) = -2e^{-(1-t)}\boldsymbol{u}_\theta$

25. $\boldsymbol{R}(t) = t\sqrt{t^2+4}\boldsymbol{u}_r$ and $\boldsymbol{V}(t) = \dfrac{(2t^2+4)\boldsymbol{u}_r - 2t\boldsymbol{u}_\theta}{\sqrt{t^2+4}}$

26. 140 m/s

27. 21.7°

28. (a) The maximum height is 85 ft

(b) The ball will land at a distance of 568 ft

(c) The distance to the home plate is 560 ft

29. (a) $A_T = \dfrac{18t^3+4t}{\sqrt{9t^4+4t^2+1}}$ and $A_N = 2\sqrt{\dfrac{9t^4+9t^2+1}{9t^4+4t^2+1}}$

(b) $A_T = 0$ and $A_N = 1$

(c) $A_T = \dfrac{4t}{\sqrt{1+4t^2}}$ and $A_N = \dfrac{2}{\sqrt{1+4t^2}}$

(d) $A_T = \dfrac{4t}{\sqrt{2+4t^2}}$ and $A_N = \dfrac{2}{\sqrt{1+2t^2}}$

30. $\boldsymbol{V}(t) = -2\sin 2t\boldsymbol{i} + 2\cos 2t\boldsymbol{j}$; $\boldsymbol{A}(t) = -4\cos 2t\boldsymbol{i} - 4\sin 2t\boldsymbol{j}$; $A_T = 0$ and $A_N = 4$

31. $A_T\left(\dfrac{\sqrt{399}}{8}\right) \approx 7.98$ and $A_N\left(\dfrac{\sqrt{399}}{8}\right) \approx 0.4$

32. (a) $\dfrac{6}{t\left(4+9t^2\right)^{\frac{3}{2}}}$ (b) $\dfrac{12e^{2t}}{\left(9e^{6t}+e^{-2t}\right)^{\frac{3}{2}}}$

(c) $\dfrac{4}{17}$ (d) $\dfrac{1}{2\cosh^2 t}$

33. (a) $k=\dfrac{2}{5}$ and $\rho=\dfrac{5}{2}$ (b) $k=\dfrac{\sqrt{2}}{3}$ and $\rho=\dfrac{3\sqrt{2}}{2}$

(c) $\dfrac{96}{125}$ (d) $\dfrac{6}{5\sqrt{10}}$

(e) $\dfrac{1}{\sqrt{2}}$

Index